AF793318

Proceedings of the 2nd International Symposium on Trichoptera

Proceedings of the 2nd International Symposium on Trichoptera

University of Reading, England, 25-29 July 1977

Edited by M. Ian Crichton

Springer-Science+Business Media, B.V. 1978

Symposium photograph, outside the Palmer Building, University of Reading, 25 July 1977: 1. J.O. SOLEM, 2. NANCY E. WILLIAMS, 3. D.D. WILLIAMS, 4. O.S. FLINT, 5. N. CASPERS, 6. L.W.G. HIGLER, 7. R.A. JENKINS, 8. R.Y. OBERNDORFER, 9. H. ZINTL, 10. A. GÖTHBERG, 11. J. LHONORÉ, 12. K. KUMANSKI, 13. H. MALICKY, 14. J.C. MORSE, 15. N.H. ANDERSON, 16. H.H. ROSS, 17. J. GREENWOOD, 18. JUDITH M. SUTCLIFFE, 19. B. STATZNER, 20. MARY A. NORTON, 21. J.P. O'CONNOR, 22. W. WICHARD, 23. N.V. JONES, 24. L.S.W. TERRA, 25. TESSA GIDLÖF, 26. J.F. FLANNAGAN, 27. N.E. HICKIN, 28. A. NIELSEN, 29. DOROTHY B. FISHER, 30. O.I. KISS, 31. MARGIT KISS, 32. MATHILDE MARLIER, 33. G. MARLIER, 34. P.C. BARNARD, 35. A. NEBOISS, 36. A.P. NIMMO, 37. I.D. WALLACE, 38. R. JEAN CONRAN, 39. RUTH M. BADCOCK, 40. CLAUDINE MOREILLON, 41. R.W. BAUMANN, 42. YVETTE BOUVET, 43. C. DENIS, 44. H.B. BUHOLZER, 45. V.H. RESH, 46. MARA MARINKOVIĆ-GOSPODNETIĆ, 47. G.B. WIGGINS, 48. ROSEMARY J. MACKAY, 49. E. MARLIER, 50. M.I. CRICHTON, 51. ELAINE WATTS, 52. FERNANDA CIANFICCONI, 53. G.P. MORETTI, 54. CLARA BICCHIERAI, 55. H. TACHET, 56. G.N. PHILIPSON, 57. M.J. WINTERBOURN, 58. A.G. HILDREW, 59. G.M. GISLASON, 60. P.D. HILEY, 61. DENISE L. LHONORÉ, 62. M. BOURNAUD.

ISBN 978-90-481-8516-0 ISBN 978-94-017-2778-5 (eBook)
DOI 10.1007/978-94-017-2778-5

Originally published by Dr. W. Junk b.v. — Publishers — The Hague in 1978

Cover design: Sally McIntosh, from a drawing by Norman E. Hickin

Contents

Tuesday, 26 July

Morning session Chairman: O.S. FLINT

Afternoon session Chairman: G. MARLIER

Wednesday, 27 July

Morning session Chairman: H.H. ROSS

Thursday, 28 July

Morning session Chairman: J.M. EDINGTON

Afternoon session Chairman: G.N. PHILIPSON

Friday, 29 July

Morning session Chairman: G.B. WIGGINS

Demonstrations

The following demonstrations were on display in the Department of Zoology during the Symposium:

P.C. BARNARD: (1) Scent organs in Trichoptera. (2) Emergence of *Rhyacophila dorsalis* (Curt.)
C. DENIS: (1) Labial morphology of trichopteran larvae; morphology of the silk threads. (2) Organization of the Malpighian tubules in Glossosomatidae and Hydroptilidae.
N.E. HICKIN: The scraper board technique for biological illustration.
G.P. MORETTI, FERNANDA CIANFICCONI & CLARA BICCHIERAI: Scent organs in some species of male Trichoptera.
H. TACHET: Differences in the ornamentation of the larval head capsules of *Hydropsyche siltalai* DÖHLER and *H. angustipennis* (CURT.)

Preface

The 2nd International Symposium on Trichoptera was held at the University of Reading, England, 25-29 July 1977. It attracted 68 participants from 22 countries, which was a gratifying response to the circulation of about 250 workers on caddisflies. It was H. MALICKY who appreciated the need for a specialized meeting of this kind and organized the 1st International Symposium on Trichoptera, which was held at Lunz am See, Austria, 16-20 September 1974.

This volume of Proceedings includes 38 papers; all except one were presented and discussed in the sessions listed in the programme. The papers were given in a lecture theatre of the Palmer Building, and demonstrations were laid out in a laboratory of the Department of Zoology where members met for their morning and afternoon breaks. Members were accommodated in St Patrick's Hall, one of the University Halls of Residence. They were the guests of the University at an informal reception on 25 July. On the afternoon of 27 July an excursion was made to the River Lambourn at Bagnor near Newbury. This chalk stream has been the subject of an ecological study by a team from the Department of Zoology since 1970. The excursion was also an opportunity to see something of the local caddis fauna, and to do some collecting.

The final session on 29 July, under the chairmanship of G.B. WIGGINS, was followed by a discussion on future plans. There was a unanimous wish to hold further symposia, and the majority opinion was that three years was a suitable interval. North America was considered an appropriate region for a symposium, but it was clear that not enough workers from Europe would be able to find the necessary funds. The same problem ruled out acceptance of an offer from A. NEBOISS to hold the next symposium in Australia. G.P. MORETTI then invited members to meet in Perugia in 1980. This welcome offer was accepted and it was agreed that the symposium should be held at a time convenient to the host University between late July and the end of August. H. MALICKY and I expressed our readiness to give any help needed in organization.

H. MALICKY reported that he was happy to continue to produce the *Trichoptera Newsletter*, which was now receiving financial support from the Biologische Station Lunz of the Austrian Academy of Sciences. He proposed to include titles of new publications on Trichoptera as a regular feature. Members expressed their appreciation of the Newsletter.

In the absence of L.W.G. HIGLER, who had taken on the task of compiling the *Trichopterorum Catalogus* after the death of F.C.J. FISCHER, H. MALICKY reported that references for 1960-70 were nearly complete. They would be circulated to recipients of the Newsletter for comments and additions.

H.H. ROSS, on behalf of all members, thanked me as Convener for the organization of the symposium. In reply, I expressed my appreciation of all the support received from colleagues in the University, and especially from K. SIMKISS as Head of the Department of Zoology, and from DOROTHY FISHER for her invaluable help with the organization and running of the symposium. I was also grateful to H. MALICKY, L. BOTOSANEANU and G.B. WIGGINS for advice in the initial planning and in preparation of the programme. Finally, I reminded members that it was their active and friendly participation that had made the symposium a success.

Reading, October 1977

M. IAN CRICHTON

List of participants

ANDERSON, Dr Norman H., Department of Entomology, Oregon State University, Corvallis, Oregon 97331, U.S.A.
BADCOCK, Dr Ruth M., Department of Biology, The University, Keele, Staffordshire, ST5 5BG, England.
BARNARD, Dr Peter C., Department of Entomology, British Museum (Natural History), Cromwell Road, London SW7 5BD, England.
BAUMANN, Dr Richard W., Department of Zoology & Entomology, Brigham Young University, Provo, Utah 84602, U.S.A.
BICCHIERAI, Dr Clara, Istituto di Zoologia, Università degli Studi, Via Elce di Sotto, I-06100 Perugia, Italy.
*BOON, Mr Philip J., Department of Zoology, The University, Newcastle upon Tyne, NE1 7RU, England.
*BOTOSANEANU, Dr Lazare, Institut de Spéleologie, C.P. 2021, R-78101 Bucuresti 12, Romania.
BOURNAUD, Dr Michel, Département de Biologie Animale et Zoologie, Université Claude Bernard, Lyon 1, 43 Boulevard du 11 Novembre 1918, F-69621 Villeurbanne, France.
BOUVET, Dr Yvette, Département de Biologie Animale et Zoologie, Université Claude Bernard, Lyon 1, 43 Boulevard du 11 Novembre 1918, F-69621 Villeurbanne, France.
BUHOLZER, Mr Hubert B., Entomologisches Institut, Eidgenössische Technische Hochschule, ETH-Zentrum, CH-8092 Zürich, Switzerland.
CASPERS, Dr Norbert, Institut für landwirtschaftliche Zoologie und Bienenkunde, Rheinische Friedrich Wilhelms Universität, Melbweg 42, D-5300 Bonn, Germany.
*CHAPIN, Dr Jay W., Department of Entomology & Economic Zoology, Clemson University, Clemson, South Carolina 29631, U.S.A.
CIANFICCONI, Dr Fernanda, Instituto di Zoologia, Università degli Studi, Via Elce di Sotto, I-06100 Perugia, Italy.
CONRAN, Miss R. Jean, 14 Milton Court, Parkleys, Ham, Richmond, Surrey, TW10 5LY, England.
CRICHTON, Dr M. Ian, Department of Zoology, The University, Whiteknights, Reading, RG6 2AJ, England.
DENIS, Dr Christian, Laboratoire de Biologie Animale, Faculté des Sciences, Université de Rennes, B.P. 25A, F-35031 Rennes Cedex, France.
EDINGTON Dr John M. and Dr Ann M., Department of Zoology, University College, P.O. Box 78, Cardiff, CF1 1XL, Wales.
FISHER, Mrs Dorothy B., Department of Zoology, The University, Whiteknights, Reading RG6 2AJ, England.
FLANNAGAN, Mr John F., Freshwater Institute, 501 University Crescent, Winnipeg, Manitoba, R3T 2N6, Canada.
FLINT, Dr Oliver S. Jr., Department of Entomology, Smithsonian Institution, Washington, D.C. 20560, U.S.A.
GÍSLASON, Dr Gísli M., Institute of Biology, University of Iceland, Grensásvegur 12, Reykjavik, Iceland.
GÖTHBERG Mr Anders, Section of Ecological Zoology, Umeå Universitet, S-901-87-Umeå, Sweden.
GOWER, Mr Anthony M., Department of Environmental Sciences, Plymouth Polytechnic, Plymouth, PL4 8AA, England.

GREENWOOD, Mr John, Department of Zoology, The University, Newcastle upon Tyne, NE1 7RU, England.

HICKIN, Dr Norman E., Kateshill, Bewdley, Worcestershire, DY12 2DR, England.

HIGLER, Dr Lambertus W.G., Spotvogellaan 12, P.O. Box 184, Bilthoven, The Netherlands.

HILDREW, Dr Alan G., Department of Zoology, Queen Mary College, Mile End Road, London, E1 4NS, England.

HILEY, Dr Peter D., Yorkshire Water Authority, Olympia House, Gelderd Road, Leeds, LS12 6DD, England.

HOBDAY, Mr Colin A., Severn-Trent Water Authority, Directorate of Scientific Services, Regional Laboratory, Meadow Lane, Nottingham, NG2 3HN, England.

JENKINS, Mr R. Anthony, Department of Water Quality & Fisheries, Welsh National Water Development Authority, 19 Penyfai Lane, Furnace, Llanelli, Dyfed, SA15 4EL, Wales.

JONES, Dr Neville V., Department of Zoology, The University, Hull, Yorkshire, HU6 7RX, England.

KISS, Dr Ottó I. and Mrs Margit, Cifrakapu u. 120 I/5, H-3300 Eger, Hungary.

KUMANSKI, Dr Krassimir, Bulgarian Academy of Science, National Natural History Museum, Boulevard Russki 1, BG-1000 Sofia, Bulgaria.

LHONORÉ, Dr Denise L., Laboratoire d'Histophysiologie des Insectes, 12 rue Cuvier, F-75005 Paris, France.

MACKAY, Dr Rosemary J., Department of Zoology, University of Toronto, Toronto, Ontario, M5S 1A1, Canada.

MALICKY, Dr Hans, Biologische Station Lunz, A-3293 Lunz am See, Niederösterreich, Austria.

MARINKOVIĆ-GOSPODNETIĆ, Dr Mara, Prirodno-Matematički Fakultet, YU-71000 Sarajevo, Yugoslavia.

MARLIER, Dr Georges, Institut Royal des Sciences Naturelles de Belgique, Rue Vautier 31, B-1040 Bruxelles, Belgium.

MOREILLON, Miss Claudine, Musée Zoologique, Place Riponne, CH-1005, Lausanne, Switzerland.

MORETTI, Dr Giampaolo P., Istituto die Zoologia, Università degli Studi, Via Elce di Sotto, I-06100 Perugia, Italy.

MORSE, Dr John C., Department of Entomology & Economic Zoology, Clemson University, Clemson, South Carolina 29631, U.S.A.

NEBOISS, Dr Arturs, Department of Entomology, National Museum of Victoria Annexe, 71 Victoria Crescent, Abbotsford, Victoria 3067, Australia.

NIELSEN, Dr Anker, Zoological Museum, Universitetsparken 15, DK-2100 København, Denmark.

NIMMO, Dr Andrew P., Department of Entomology, University of Alberta, Edmonton, Alberta T6G 2E3, Canada.

NORTON, Miss Mary A., Natural History Division. National Museum of Ireland, Kildare Street, Dublin 2, Eire.

OBERNDORFER, Mr Reed Y. Department of Zoology & Entomology, Brigham Young University, Provo, Utah 84602, U.S.A.

O'CONNOR, Dr James P., Natural History Division, National Museum of Ireland, Kildare Street, Dublin 2, Eire.

PHILIPSON, Dr G. Norman, Department of Zoology, The University, Newcastle upon Tyne, NE1 7RU, England.

RESH, Dr Vincent, H. Division of Entomology & Parasitology, University of California, Berkeley, California 94720, U.S.A.

ROSS, Dr Herbert H., Department of Entomology, University of Georgia, Athens, Georgia 30602, U.S.A.
SOLEM, Dr John O., Det Kongelige Norske Videnskabers Selskab Museet, Universitetet i Trondheim, N-7000 Trondheim, Norway.
STATZNER, Mr Bernhard, Zoologisches Institut der Universität, Hegewischstrasse 3, D-2300 Kiel, Germany.
SUTCLIFFE, Miss Judith M., Department of Zoology, The University, Newcastle upon Tyne, NE1 7RU, England.
TACHET, Dr Henri, Département de Biologie Animale et Zoologie, Université Claude Bernard, Lyon 1, 43 Boulevard du 11 Novembre 1918, F-69621 Villeurbanne, France.
TERRA, Mr Luiz S.W., Estação Aquícola, Vila do Conde, Portugal.
TOWNSEND, Dr Colin R., School of Biological Sciences, University of East Anglia, Norwich, NR4 7TJ, England.
*VAILLANT, Dr François, Allée de Pont Croissant, F-38330 Montbonnot par Saint-Ismier, France.
VERMEHREN, Mr Hans-Jürgen, Claudiusstrasse 13, D-2300 Kiel 17, Germany.
WALLACE, Dr Ian D., Department of Invertebrate Zoology, Merseyside County Museums, William Brown Street, Liverpool, L3 8EN, England.
WATTS, Dr Elaine, Imperial College Field Station, Ashurst Lodge, Ascot, Berkshire, SL5 7DE, England.
WICHARD, Dr Wilfried, Institut für Cytologie-Mikromorphologie, Ulrich Haberland-strasse 61a, D-5300 Bonn, Germany.
WIGGINS, Dr Glenn B., Department of Entomology, Royal Ontario Museum, 100 Queen's Park, Toronto, Ontario M5S 2C6, Canada.
WILLIAMS, Dr D. Dudley and Mrs Nancy E., Division of Life Sciences, Scarborough College, University of Toronto, West Hill, Toronto, Ontario M1C 1A4 Canada.
WINTERBOURN, Dr Michael J., Department of Zoology, University of Canterbury, Christchurch 1, New Zealand.
ZINTL, Mr Heribert, Grossherzogin-Maria-Anna-Weg 16a, D-8172 Lenggries, Germany.

* not present in Reading

Proc. of the 2nd Int. Symp. on Trichoptera, 1977, Junk, The Hague

The present disposition of components of the Sericostomatidae *s. lat.* (Trichoptera)[1]

H.H. ROSS

Abstract

By 1940, the family Sericostomatidae had become an extremely diverse group, containing 8 subfamilies or tribes and many disparate genera. Since then, primarily on the basis of larval studies, many taxa have been removed from the family either as full new families or as entities within other families. Knowledge of the larvae throughout the superfamily Limnephiloidea has demonstrated that these transfers were not a case of simple hierarchal inflation, but were necessitated because the Sericostomatidae in the broad sense was polyphyletic in the extreme.

Two of the basic functions of a classification are (1) to provide names for the entities of the universe and (2) to arrange these in some meaningful fashion. In biology this second function is expressed as a hierarchal system of categories, centering on the species and proceeding upward through various supraspecific categories and downward through infraspecific ones. Concerning the insects, the categories family and subfamily have been especially important in grouping taxa within almost every order.

By and large, the older family and subfamily groupings have indeed been of 'meaningful fashion', in the sense that in a remarkable number of instances they have proven to be monophyletic units (or nearly so) expressing probable monophyletic groupings. The fact that many of these groups have been subject to hierarchal escalation does not change the situation. Here and there, however, certain of these groups have proven to be artificial in the extreme, their components being disparate elements bearing little relationship with each other.

One of the classic examples of this latter situation certainly is the trichopteran family Sericostomatidae. Since it was established by STEPHENS in 1836, it seems that genera of Trichoptera Integripalpia that fitted no other families were placed in it. Originally it was diagnosed essentially as having no ocelli and maxillary palps with a reduced number of segments, often held in front of the face like a mask. But these criteria were not always applied, for example, the included genus *Thremma*

[1] This study was supported by a research grant from the National Science Foundation, U.S.A.

which had definite ocelli. The chief criterion for inclusion in the Sericostomatidae appears to have been that the genus was peculiar in appearance, often in antennal or wing characters. McLACHLAN (1876) was well aware of this, calling the family the 'curiosity shop' of the Trichoptera. I would like to review the history of the more distinctive entities of the family, treating first the groups or genera placed in the family, then the later placement of these groups. Space does not allow a consideration of all the elements in the family nor is such necessary at this time; rather, I will treat only those elements that were formerly placed in the Sericostomatidae and subsequently removed to other families.

Inclusions in the Sericostomatidae

The first comprehensive treatment of the family was that of McLACHLAN (1876), who recognized four sections categorized only by number. These sections were later given the status of subfamilies: Sericostomatinae (STEPHENS, 1836), Lepidostomatinae (ULMER, 1903), Brachycentrinae (ULMER, 1903), and Goerinae (KLAPALEK, 1904). Other segregates were added later: Helicopsychinae (ULMER, 1906), Uenoinae (IWATA, 1927), Thremmatinae (MARTYNOV, 1935), and Oeconesini (TILLYARD, 1921).

In addition, a large number of genera were added to these sericostomatid entities. A few of these, placed simply in the family Sericostomatidae as a whole, included the genera *Plectrotarsus* KOLENATI (1848), *Pseudogoera* CARPENTER (1933), *Tasimia* MOSELY (1936), and *Antipodoecia* MOSELY (1934).

Earlier in this century, many of these taxa were known only from the adult stages. Of those known from both immature and adult stages, the significance of the added content of character states was lost because of the lack of a sufficient basis for comparative studies of larval characters throughout the diverse elements of not only the Sericostomatidae but of other groups of Trichoptera. Through the pioneering works especially of ULMER, THIENEMANN, SILFVENIUS, and BETTEN, an excellent base was established for comparing the immature stages throughout the order Trichoptera. The ensuing comparisons as applied to the classification of the order supported a large number of earlier conclusions, but added new evidence that the Sericostomatidae were indeed the 'curiosity shop' of the Trichoptera.

The dismemberment of the Sericostomatidae

In essence, the new larval evidence indicated that each disparate sericostomatid entity such as the Goerinae or Brachycentrinae was each a closely-knit phylogenetic unit, but that these units did not together constitute a phylogenetic unit. Instead, the evidence indicated that many of these entities were much more closely related to non-sericostomatid entities than to any with which they had hitherto been placed. A brief documentation of presently-known examples follows.

Rather than giving only a listing of these, they are delineated in relation to a family tree of the Limnephiloidea, to which all the 'sericostomatids' belong. The base tree is primarily that of ROSS (1967) but with changes and additions suggested by various authors. The various 'sericostomatid' entities are inserted into this tree according to the best information now available to me (Fig. 1) and are indicated by arrows. This tree has three major features that seem to be well established as evolutionary developments: (1) an array of three families that appear to be the most primitive of the Limnephiloidea, and beyond these (2) a large limnephilid branch, and (3) an equally large leptocerid branch. One or two families seem to represent primitive forms that arose from or near the division that led to the two major branches.

At present, 13 entities have been removed from the Sericostomatidae *s. lat.* and placed in other families, as follows:

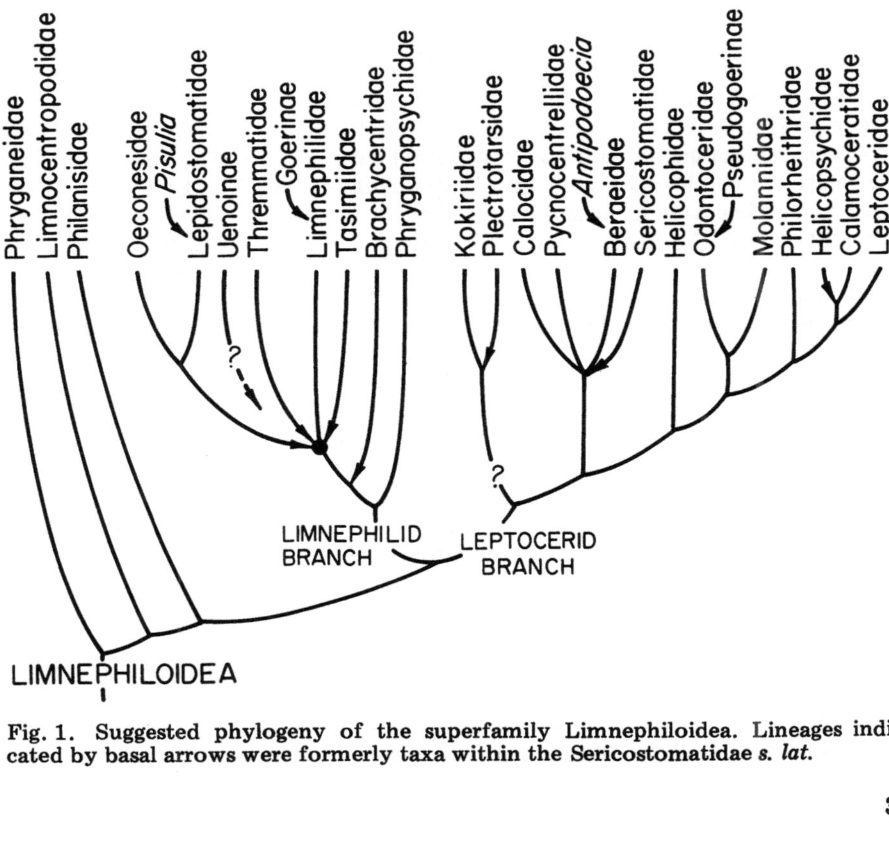

Fig. 1. Suggested phylogeny of the superfamily Limnephiloidea. Lineages indicated by basal arrows were formerly taxa within the Sericostomatidae *s. lat.*

ROSS (1944) introduced the family terms Brachycentridae, Goeridae, Helicopsychidae, and Lepidostomatidae, and restricted the Sericostomatidae to the former Sericostomatinae of auct., and (1967) proposed Pisuliidae for *Pisulia* MARLIER and Antipodoeciidae for *Antipodoecia*

SCHMID (1952) introduced the name Thremmidae, emended to Thremmatidae by Fischer (1970); MOSELY & KIMMINS (1953) removed *Plectrotarsus* to a new family Plectrotarsidae; RIEK (1968) transferred *Tasimia* to a new family Tasimiidae; and (1970) transferred *Antipodoecia* to the Beraeidae; WALLACE & ROSS (1971) transferred *Pseudogoera* CARPENTER to the Odontoceridae as the subfamily Pseudogoerinae; WIGGINS (1973) placed the Goeridae as the subfamily Goerinae of the Limnephilidae; NEBOISS (1975) raised Tillyard's Oeconesini to the family Oeconesidae.

Seven of these sericostomatid taxa have been placed in the limnephilid branch, three (plus Sericostomatidae s. st.) in the leptocerid branch, and the placement of the last two is open to question.

Of those placed in the limnephilid branch, the Brachycentridae are certainly the most primitive, probably representing the first lineage with 3-segmented adult maxillary palps but retaining the primitive position of the larval antenna near the base of the mandible. Beyond this is a cluster of five families having the larval antenna midway between the mandibular base and the eye. The ancestor of this group gave rise to 3 families exhibiting this condition, the Limnephilidae and two previous sericostomatid entities, the Tasimiidae and the Thremmatidae. The Uenoinae are considered a subfamily of this family (FISCHER, 1970); but I have seen no specimens of them and have placed them only tentatively in the family tree. ROSS (1967) placed the goerids as a separate family also arising from this ancestor, but WIGGINS (1973, 1974) adduced larval evidence indicating that they are a branch of the Limnephilidae.

From this same pre-limnephid ancestor arose another lineage in which the larval antenna became adjacent to the eye. This lineage divided into two families, the Lepidostomatidae and the Oeconesidae. The genus *Pisulia* is difficult to place because its larva is unknown, but Dr. SCOTT has suggested (in litt.) that the adult structures are suggestive of its being a member of the Lepidostomatidae rather than belonging where I placed it in 1967, a view with which I now agree.

The four 'old sericostomatid' entities placed in the leptocerid branch are scattered through it (Fig. 1), *Antipodoecia* in the Beraeidae, Sericostomatidae *s. st.* in the same primitive cluster, *Pseudogoera* as a subfamily of the Odontoceridae, and Helicopsychidae in the specialized end cluster.

The most puzzling problem thus far encountered with 'old sericostomatid' taxa concerns the family Plectrotarsidae. It shares with the Kokiriidae a most unusual adult feature, an extensile maxillary-labial structure with associated specialized structures in the head. These shared derived structures indicate that the two are sister families of the same fork. The larvae of *Kokiria* (those of *Plectrotarsus* are unknown) are also extremely highly specialized, presumably for predation (Mc-

FARLANE, 1964). The lines of abdominal spicules above the lateral fringe are lacking, hence this character gives no help in placing them phylogenetically. Other evidence suggests two choices. In the *Kokiria* larva, setal area 3 on segments II and III form membranous raised warts bearing many long hairs, and the posterior part of the lateral abdominal fringe forms a gill-like appendage at each posterior corner. To my knowledge these two character states occur only in the Phryganeidae. On the other hand, the larval head has long antennae, and many features of the larval thorax are similar to the conditions found in the Molannidae. The cases of *Kokiria* and *Molanna* are also similar, but similar cases are made by other groups also. Tentatively, I am placing the Kokiriidae and Plectrotarsidae as a primitive lineage of the leptocerid branch.

This account brings us up to the present concerning the disposition of members of the Sericostomatidae in its older broad sense. There are of course many genera that need more study to resolve their place in the trichopteran phylogeny. Many of these will not be well placed until their immature stages are known. But I think we have enough evidence to assert that the old Sericostomatidae was a highly polyphyletic unit. On this basis it is seen that in this group the elevation of subfamily taxa to the category of families has not been a simple example of hierarchal inflation. Rather, this extensive realignment of taxa was necessary to express a more probable family tree for the entire superfamily.

References

CARPENTER, F.M. 1933. Trichoptera from the mountains of North Carolina and Tennessee. Psyche 40: 32-47.

FISCHER, F.C.J. 1970. Trichopterorum Catalogus 9: 1-316. Amsterdam: New. Ent. Veren.

IWATA, M. 1927. Trichopterous larvae from Japan. Annot. Zool. Jap. 11: 203-233.

KLAPALEK, F. 1904. Die Morphologie der Genitalsegmente und Anhänge bei Trichoptera. Bul. Int. Acad. Boheme 8: 161-197.

KOLENATI, F.A. 1848. Gen. et Sp. Trich. 1: 94.

MARTYNOV, A.V. 1935. Trichoptera of the Amur Region. Part I. Trav. Inst. Zool. Leningrad 2: 205-395.

McFARLANE, A.G. 1964. A new endemic subfamily, and other additions and emendations to the Trichoptera of New Zealand Rec. Canterbury Mus. 8: 55-79.

McLACHLAN, R. 1874-80. A monographic revision and synopsis of the Trichoptera of the European fauna. (Reprint) Hampton: Classey.

MOSELY, M.E. 1934. A new Australian caddis-fly (Trichoptera). Entomologist 67: 178-180.

——. 1936. Tasmanian Trichoptera or caddis-flies. Proc. Zool. Soc. Lond. 1936. 395-424.

——. & KIMMINS, D.E. 1953. The Trichoptera (caddis-flies) of Australia and New Zealand. Br. Mus. (nat. Hist.), London.

NEBOISS, A. 1975. The family Oeconesidae (Trichoptera) from New Zealand and Tasmania. Austr. Ent. Mag. 2: 79-84.

RIEK, E.F. 1968. A new family of caddis-flies from Australia (Trichoptera: Tasimiidae). J. Austr. Ent. Soc. 7: 109-114.

— 1970. Trichoptera (Chapter 35). In: *The insects of Australia.* Melbourne Univ. Press,
ROSS, H.H. 1944. The caddis flies, or Trichoptera, of Illinois. Ill. St. nat. Hist. Surv. Bull. 23: 1-326.
— 1967. The evolution and past dispersal of the Trichoptera. of Rev. Ent. 12: 169-206.
SCHMID, F. 1952. Contribution à l'étude des Trichoptères d'Espagne. Pirineos 8: 627-695.
STEPHENS, J.F. 1836-1837. Illustrations of British Entomology 6: 146-234.
TILLYARD, R.J. 1921. Studies of New Zealand Trichoptera, or caddis-flies. No. 1. Description of a new genus and species belonging to the family Sericostomatidae. Trans. N. Z. Inst. Wellington 53: 346-350.
ULMER, G. 1903. Ueber die Metamorphose der Trichopteren. Abh. Naturw. Ver. Hamburg 18: 1-154.
—. 1906. Neuer Beitrag zur Kenntnis Aussereuropäischer Trichopteren. Notes Leyden Mus. 28: 1-116.
WALLACE, J.B., & ROSS, H.H. 1971. Pseudogoerinae: a new subfamily of Odontoceridae (Trichoptera). Ann. Ent. Soc. Am. 64: 890-894.
WIGGINS, G.B. 1973. New systematic data for the North American caddisfly genera *Lepania, Goeracea* and *Goerita* (Trichoptera; Limnephilidae). Life Sci. Contr., R. Ont. Mus. 91: 1-33.
—. 1974. Contributions to the systematics of the caddis-fly family Limnephilidae (Trichoptera). III; the genus *Goereilla.* Proc. 1st. int. Symp. Trich.: 7-19.

Discussion

MARLIER: *Pisulia* is a Lepidostomatidae; the larva is very close to that of *Dyschimus* Barnard.

(Since the Symposium, Dr. Marlier has published this information in Trichoptera Newsletter No. 4, Jan., 1978, in an article entitled 'The larva of the genus *Pisulia* and its affinities.' H.H.R.)

Proc. of the 2nd Int. Symp. on Trichoptera, 1977, Junk, The Hague

The *Sericostoma* Latr. genus in Italy

G.P. MORETTI and FERNANDA CIANFICCONI

As all Trichopterologists know, the problem of systematics of the *Sericostoma* LATR. genus is still far from clear. The lack of clear taxonomic characters in the genital armature and, conversely, the extreme variability in the form of the appendages of segment X in the ♂ of certain species left, and still leaves, specialists puzzled in safely attributing specimens to this or that species. Already McLACHLAN in 1884 (McLACHLAN, 1874-1880) furnished numerous drawings of the appendages of segment X of the ♂ which, in his opinion, were able to distinguish, just to give an example, two very similar species such as *S. personatum* KIRBY & SPENC. and *S. pedemontanum* McL.

In another publication (1898) the same author (McLACHLAN, 1898) created the *S. subaequale* species by separating it, with the same criteria, from *S. pedemontanum*.

Other authors continued in the same vein and BOTOSANEANU, in trying to put some systematic and zoogeographical order into the genus, in 'Limnofauna Europaea' (1967) (BOTOSANEANU, 1967) was able to show this unsolved problem in the correct light. He associated various uncertain taxa under a provisional specific common name taken from preceding literature (*S. flavicorne* SCHNEID) and abolished *S. pedemontanum* which he considered to be synonymous with *personatum*. In 1973, BOTOSANEANU and SCHMID (1973), while inspecting the Trichoptera at the Museum d'Histoire Naturelle de Genève where this genus is fairly well represented, found that a total reclassification of *Sericostoma* was necessary. They proposed a certain new setting up of various species, thus clarifying at least one aspect of this complex subject. In this way *S. baeticum* ED. PICT, was once more brought to light; *S. personatum* and *S. pedemontanum* were provisionally labelled with a double denomination, even though the authors were convinced that they belonged to the same species; *S. pyrenaicum* ED. PICT., *S. selysii* ED. PICT. and *S. timidum* HAG. returned to an elevated position while in 'Limnofauna Europaea' the first and the third were put under the old name of *S. flavicorne* SCHNEID. Of the above quoted species: *S. baeticum*, *S. pyrenaicum*, *S. selysii* and *S. flavicorne* are not part of the Italian fauna.

We set about studying the situation in the geonemic territory of Italy, inspecting numerous specimens of *Sericostoma* coming partly from museums and old private collections but mostly from material collected by ourselves and our students from the different regions of our peninsula and islands. This work, although still incomplete, makes it possible for us to furnish a preliminary list of the species known in Italy.

Sericostoma is a genus that is extremely uniform and monotonous in its external morphological characters. As everyone knows, these Trichoptera are of average size with pubescent bronzish-brown coloured wings, excepting infrequent examples of chromatic variations. The body is dark brown, the thorax black, the legs are pale, the antennae a little longer than the anterior wing, thick and exceptionally annulated. However, the possibilities of classification based only on the above chromatic and morphological characters are limited. In rare cases the protrusion of the androconial mask formed by the maxillary palpi of the ♂ allows for recognition at a specific level depending on the degree of protrusion.

The genital armature of the genus is represented by: a. segment X, dorsally conical; b. the lateral spiniform processes of segment X; c. the small superior appendages, oval and dark; d. the dorsal branches of the inferior appendages, large, tawny, narrow at the base and widely expanded in lobe form and cut off at the apex. They are constantly furnished with an indentation along the inferior margin as well as a sinuous edging hollowed out of the internal face, in a superior preapical position; e. the ventral branches of the inferior appendages with their lyre-shaped character; f. the phallus; g. finally, the ventral process of segment IX.

When carefully observing these structures, it can be seen that the most marked differences concern the spiniform processes of segment X (b). These show taxonomic characters that merit a systematic classification because of their form in different species. In fact the present classification of the species of the *Sericostoma* genus is mainly, if not completely, based by all Trichopterologists on the form of the spines of segment X.

However, with the increasing number of specimens examined, these spines were found to be so varied amongst the same species coming from same or different localities that Trichopterologists, from McLACHLAN onwards, ended up by losing track, creating new species or merging them together under one species. New outlines were given, rich in contours for this or that species, to such an extent that some *taxa* cannot be safely classified and neither can they be considered good species.

It is therefore quite true that, as BOTOSANEANU & SCHMID (1973) said: 'La systématique des *Sericostoma* d'Europe est à refaire et que cela ne serait possible que dans le cadre d'une revision complète.'

While waiting for some Trichopterologist to tackle the entire problem, we propose bringing Italian taxa up to date, also showing some accessory characters such as the presence of light and dark annulations on the antennae; the shape, number and disposition of the proximal and lateral teeth of the appendages of segment X; the shape of the ventral branch of the inferior appendages and the prominence of urosternite IX.

For the sake of simplicity, we shall show the species found in Italy in alphabetical order, it not being possible to show them in phyletic lines until the entire *Sericostoma* genus has been generally revised, as previously mentioned. Neither would the geonemic criteria be clear, since the same species can be found in different and distant localities of the peninsula and islands.

We shall therefore begin reviewing the species found in Italy[1].

1. *Sericostoma cianficconii* MORET.

This belongs to the *siculum* McL. group, the ♂ having the posterior apex of segment X lanceolate, simple and not divided, with the dark points facing downwards and outwards (Fig. 1: b, c, d). The proximal teeth are short, small, triangular, cut short and either moulded together or separate (Fig. 1: a). The ventral branch of the inferior appendages, shaped like a lyre, is shortened and widened (Fig. 12).

Wing expanse 20-27 mm (♂), 22-26 mm (♀); length of body 6-7 mm (♂), 6-8 mm (♀). Therefore, *S. cianficconii* is smaller and lighter than *S. siculum*.

This species was previously assigned, with reserve, by one of us (MORETTI) to *S. siculum* (MORETTI, 1949, 1950, 1952; MORETTI & MICHELETTI 1952; MORETTI, CIANFICCONI, GIANOTTI, PIRISINU & VIGANÒ 1970; SECONDARI, 1950; ZANGHERI, 1966), but the differences that emerged from a subsequent direct comparison with specimens belonging to *S. siculum* made it necessary to separate it from this species and to create a new one. It has been found in Emilia, Apuanian Mountains, the Marches and in Umbria (Fig. 16: 1).

The larvae live in both clear and fast-running spring-water in the hills, both in streams that flow through shaded woods and in muddy, low altitude rivers. This is therefore an eurybiont species.

2. *Sericostoma clypeatum* HAG. (ANGELIER, 1959; BERLAND & MOSELY, 1963; BRAUER, 1876; ESBEN-PETERSEN, 1912; ESBEN-PETERSEN, 1913; HAGEN, 1864; KLAPALEK, 1917; MARTIN, 1893; MORTON, 1934; MOSELY, 1930, 1932; ROSTOCK, 1888; ULMER, 1905, 1907).

This is a good species characterized by its spiniform appendage to segment X, thin, scythe-shaped, turned downwards and preceded by a basal spine on the lower border (Fig. 2).

Wing expanse 23-25 mm. It is an insular endemic of Corsica (Fig. 16: 2).

3. *Sericostoma galeatum* RAMB. (BRAUER, 1876; KIMMINS, 1957; MALICKY, 1971; MORETTI, 1940; RAMBUR, 1842; SCHMID, 1947).

Characterized by the extremity of the appendages of segment X, which terminates in the shape of a fish-hook (Fig. 3: b, c, d); there is a preapical dorsal tooth preceded by a crest and a ventral tooth larger than the preceding one (Fig. 3: a, c, d).

Wing expanse 20-21 mm, length of body 7-8 mm.

It has been found in Piedmont, Trentino-Alto Adige, Tuscany, Latium, Basilicata and Sardinia (Fig. 16: 3). (see Note after Discussion, p. 30)

[1] The *S. cianficconii* and *S. italicum* species have already been described in a special paper, now being printed.

4. *Sericostoma italicum* MORET.

This is a beautiful insect with large wings (expanse 25-29 mm; length of body 8-10 mm), covered with bronze pubescence, with a hairy black body and pale legs (Fig. 4).

From the side view it shows the superior branches expanded from the inferior appendages more or less markedly serrated at the lower edge (Fig. 5: b). The appendages of segment X are very large, supple, always have a strong medial inferior tooth turned downwards and terminating in a large point turned upwards and outwards, forming an open angle with the remaining profile of the appendage

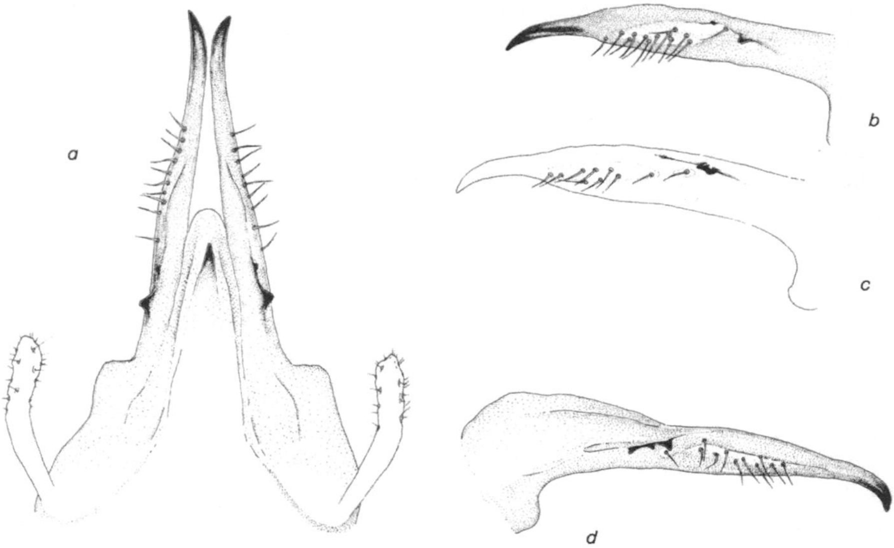

Fig. 1. *Sericostoma cianficconii* MORET.: ♂ — a = Segment X with lateral spiniform processes, superior appendages, from above; b, c, d = lateral spiniform processes, side view.

Fig. 2. *Sericostoma clypeatum* Hag.: ♂ — Lateral spiniform process of the segment X, side view (after McLACHLAN).

(Fig. 6: b, c)[2]. The appendages of segment X, seen from above, appear to be convergent or crossed and the ventral teeth are strong, robust, long, turned outwards and a little backwards. (Figs. 5, 6: a). There are rare cases of anomaly in the shape of the spines of segment X (Fig. 5: c, e).

It is found in a well defined and continuous area, namely Tuscany, and the Apuanian Mountains, Umbria, Latium, Abruzzo, Molise and Campania (Fig. 16: 4).

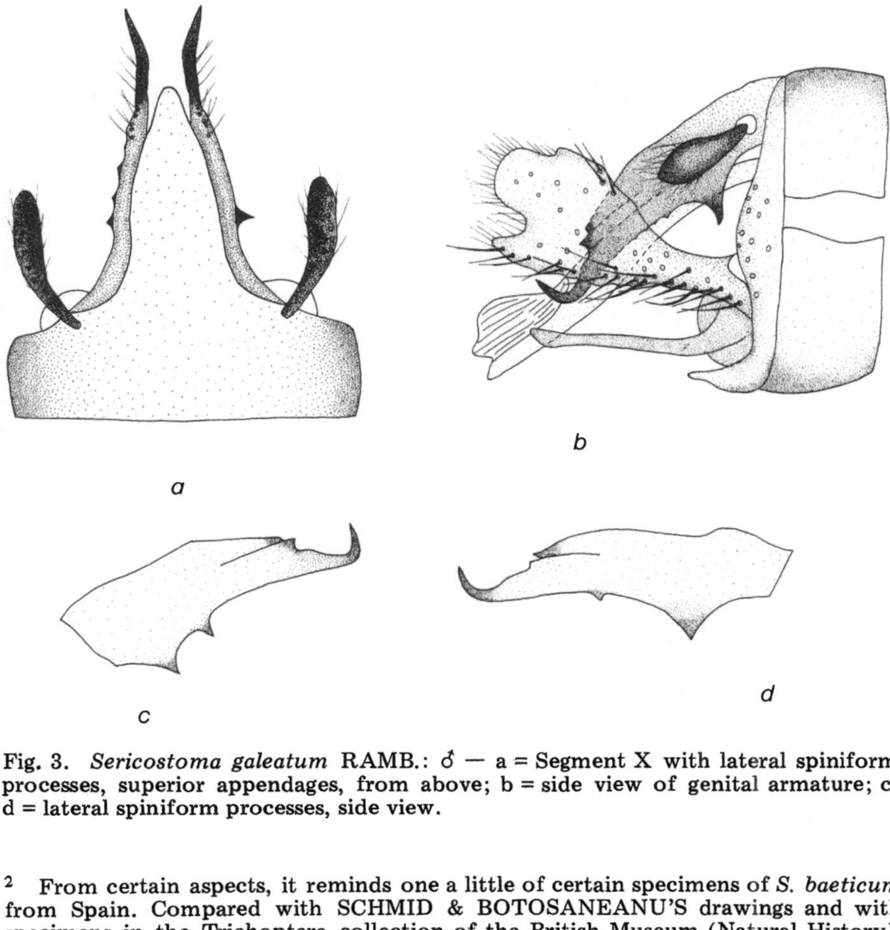

Fig. 3. *Sericostoma galeatum* RAMB.: ♂ — a = Segment X with lateral spiniform processes, superior appendages, from above; b = side view of genital armature; c, d = lateral spiniform processes, side view.

2 From certain aspects, it reminds one a little of certain specimens of *S. baeticum* from Spain. Compared with SCHMID & BOTOSANEANU'S drawings and with specimens in the Trichoptera collection of the British Museum (Natural History), by kind permission of Dr. BARNARD, the morphological differences seem to be unquestionable.

An Apennine Central-Italy term can be defined, going from low and average levels to over 1000 m above sea level.

5. *Sericostoma maclachlanianum* COSTA (COSTA, 1884, 1885; MORETTI, 1940; ULMER, 1905, 1907).

This is clearly recognizable by the marked widening in the marginal branches of the appendages of segment X (Fig. 7: a, b) which are neither pointed at the apex nor in relation to the superior branch which is blunt, not pointed and minutely serrated (Fig. 7: c, d, e, f). The inferior point forms a right angle with the superior one and is minutely indented, as is all the profile of the piece.

Wing expanse 23 mm; length of body 9 mm.

It is found on the island of Sardinia (Fig. 17: 1)

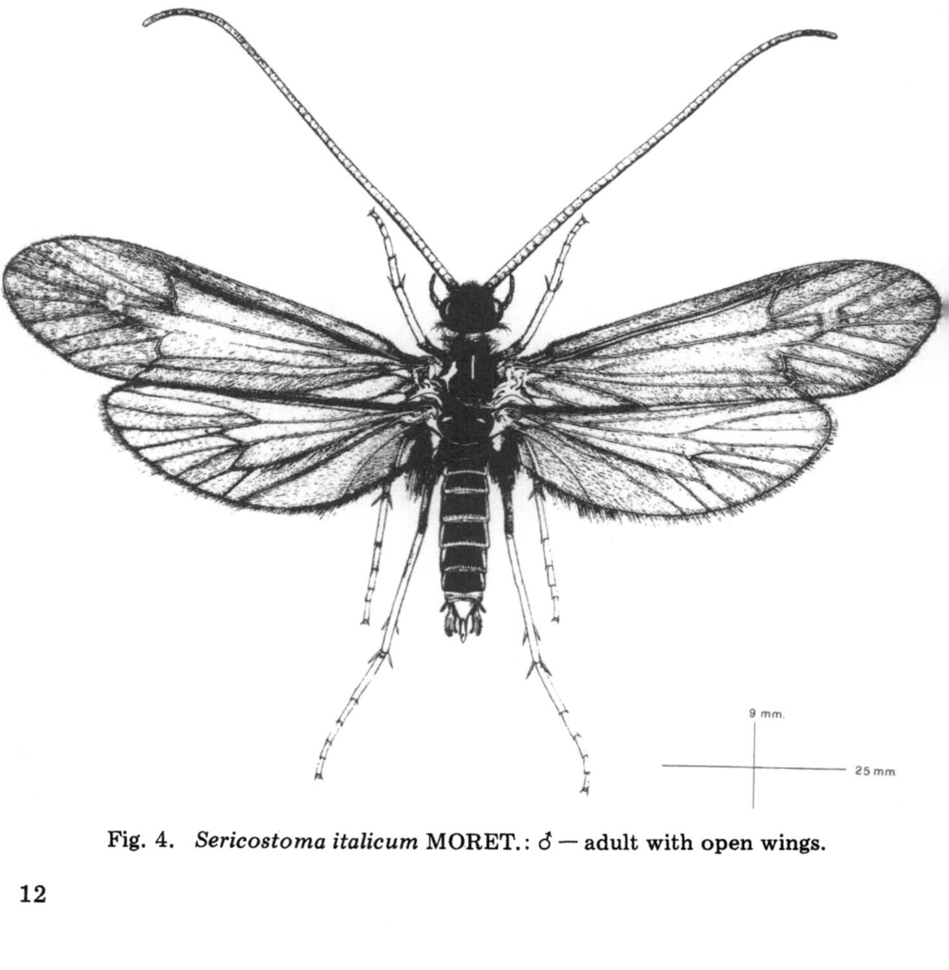

Fig. 4. *Sericostoma italicum* MORET.: ♂ — adult with open wings.

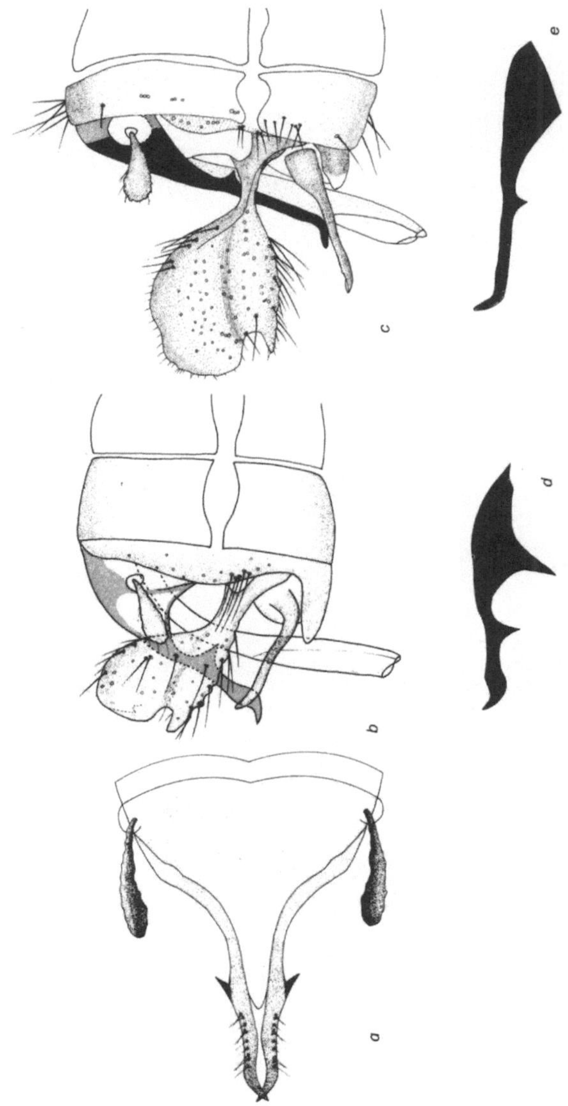

Fig. 5. *Sericostoma italicum* MORET.: ♂ — a = Segment X with lateral spiniform processes, superior appendages, from above; b = side view of genital armature; c = idem b in anomalous specimen; d = lateral spiniform process of segment X, side view; e = idem d in anomalous specimen.

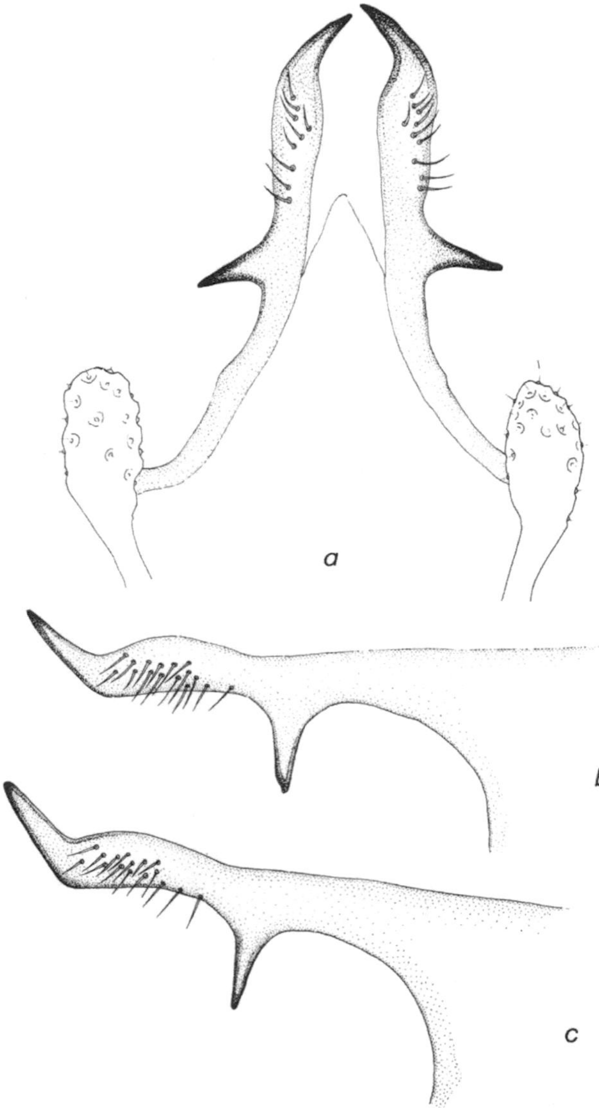

Fig. 6. *Sericostoma italicum* MORET.: ♂ — a = Segment X with lateral spiniform processes, superior appendages, from above; b, c = lateral spiniform process, side view.

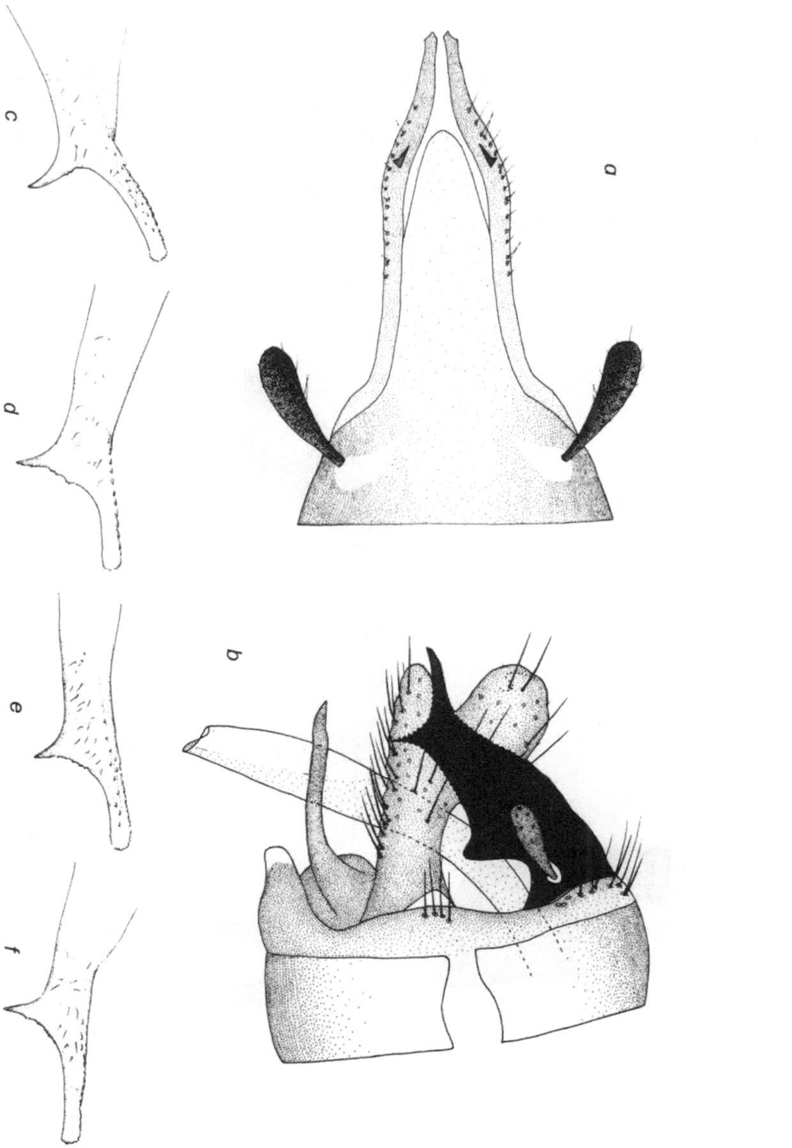

Fig. 7. *Sericostoma maclachlanianum* McL.: ♂ — a = Segment X with lateral spiniform processes, superior appendages, from above; b = side view of genital armature; c, d, e, f = various aspects of lateral spiniform process.

6-7. *Sericostoma pedemontanum* McL.; *S. personatum* KIRBY & SPENC.

The question of validity of these two species is, as has already been said, debatable. In our opinion, the *S. pedemontanum* form (ESBEN-PETERSEN, 1934; FELBER, 1908; McLACHLAN, 1874-1880; MORETTI, 1937, 1939, 1952; MORETTI & MICHELETTI, 1952; MORETTI, 1955; MORETTI & VIGANO', 1959, 1961; MORETTI, CIANFICCONI, GIANOTTI, PIRISINU & VIGANO', 1970; NAVAS, 1934; NOCENTINI, 1963; RIS, 1889; VIGANO', 1958; ZANGHERI, 1966) clearly prevails in Italy, as shown by the only slight protrusion of the facial mask and above all by the profiles of the appendages of segment X which faithfully repeat models 3, 4, 7, 11 and 12 drawn by McLACHLAN in 1884 (1874-1880) rather

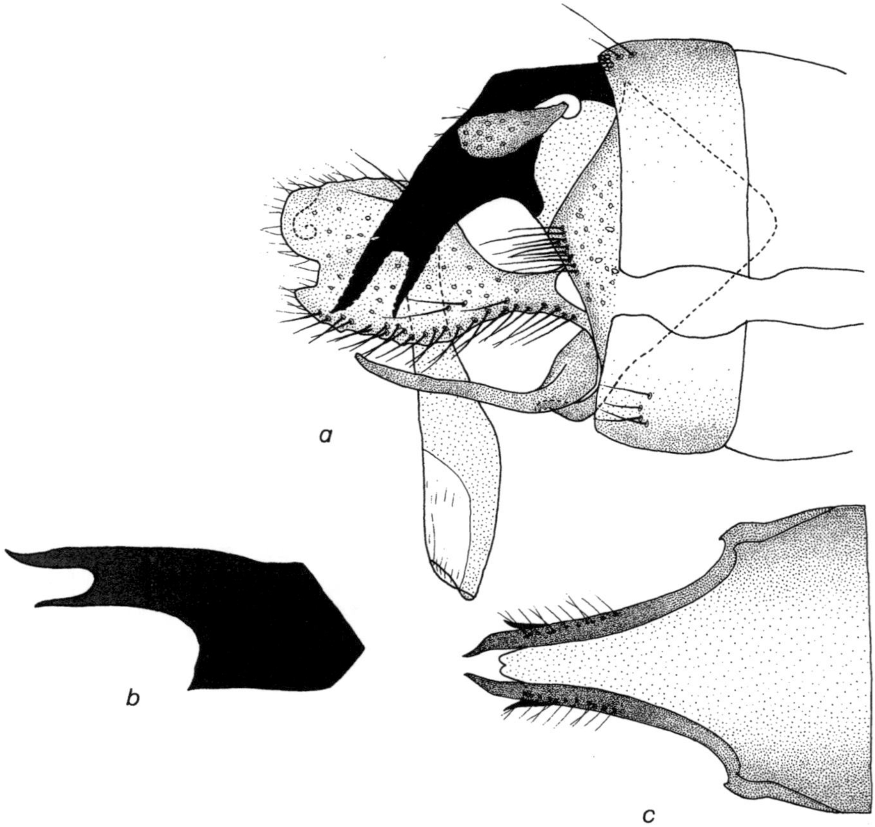

Fig. 8. *Sericostoma pedemontanum* McL.: ♂ — a = side view of genital armature; b = lateral spiniform process of segment X, side view; c = segment X with lateral spiniform processes, from above.

than the figures shown by BOTOSANEANU & SCHMID (1973) in which the two branches are more or less indicated by subequal length.

The most frequent models are shown (Figs. 8, 9). Fig. 9 shows the thin and varied indentations of the appendages of segment X (a, b, c) and an anomalous form found on the Isle of Elba (d).

Wing expanse 25-31 mm, length of body 9-10 mm.

S. pedemontanum is mainly found along the pre-alpine area from Piedmont to Veneto and in the Apennines from Romagna-Tuscany and Umbria-the Marches (Fig. 17: 2) where it is then substituted by *S. italicum* and further south by *S. siculum.*

Sericostoma personatum KIRBY & SPENC. (COSTA, 1869, 1871; GRIFFINI, 1897; HAGEN, 1859, 1860; KOLENATI, 1859; McLACHLAN, 1869, 1870, 1875; MEYER-DÜR, 1875; ROSTOCK, 1874).

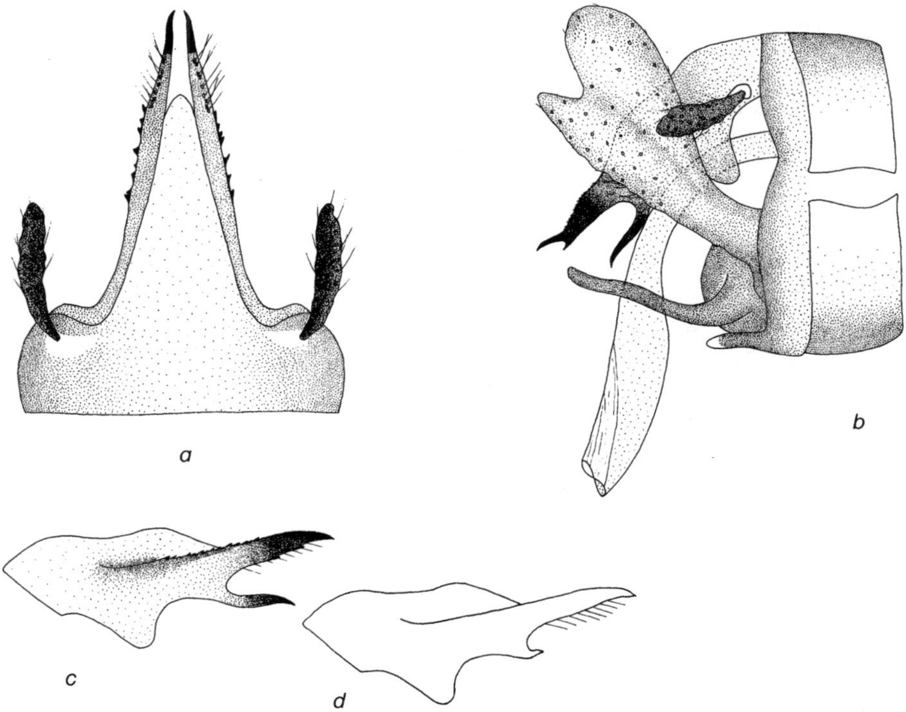

Fig. 9. *Sericostoma pedemontanum* McL.: ♂ — a = Segment X with lateral spiniform processes, superior appendages, from above; b = side view of genital armature; c = lateral spiniform process, side view; d = idem c in an Isle of Elba specimen.

We are not in a position to be able to indicate its distinctive characters with accuracy, since we have been unable to see any specimens clearly attributable to this species.

Therefore we shall limit ourselves to a geographical map which shows the localities where other authors have found this species, namely the Piedmont and Ligurian Alps and Calabria (Fig. 17: 3).

8. *Sericostoma romanicum* NAV. (NAVAS, 1930; ZANGHERI, 1966).

We were unable to see the specimen which is in the Navas collection. We show the author's drawing (Fig. 10) but with reserves on the validity of this species which could be assigned both to *S. turbatum* and to certain models of *S. pedemontanum.*

Wing expanse 12,3 mm.

The region where it is found is shown on the map (Fig. 17: 4)

9. *Sericostoma siculum* McL. (HAGEN, 1860; McLACHLAN, 1874-1880; MALICKY, 1971; ULMER, 1907).

This shows apex processes of segment X lanceolate, not furcate, pointed and curved at the apex, slightly apart and from this aspect is similar to *S. cianficconii*, but is nevertheless larger. (Wing expanse 22-28 mm, length of body 8-12 mm).

It is very dark with very pubescent wings.

Specimens coming from Sicily have only one proximal tooth at the base of the process of segment X (Fig. 11: d, e); those found in Calabria and Basilicata, how-

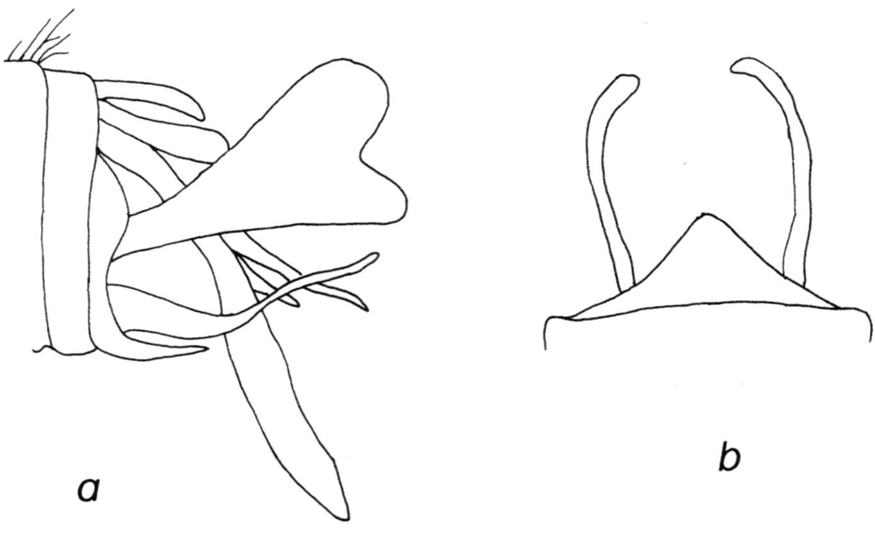

Fig. 10. *Sericostoma romanicum* NAV.: ♂ — a = Side view of genital armature; b = ventral branches of inferior appendage (after NAVAS).

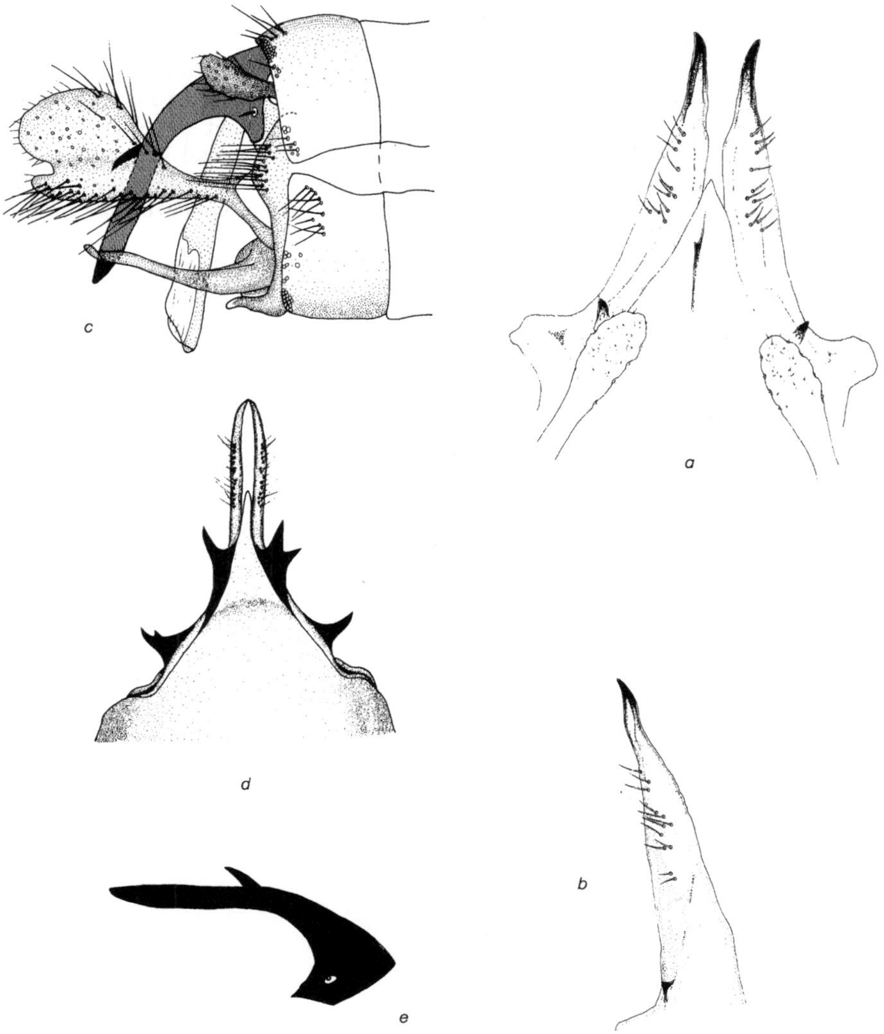

Fig. 11. *Sericostoma siculum* McL.: ♂ — Specimen from Sicily: a = Segment X with lateral spiniform processes, superior appendages, from above; b = lateral spiniform process, side view; specimen from *Basilicata*: c = lateral view of genital armature; d = segment X with lateral spiniform processes, from above; e = lateral spiniform process, side view.

ever, have three large black teeth very protruding and pointing outwards, simple or furcate; in some specimens there is also a dorsal tooth (Fig. 11: a, b, c). The map of the geographical distribution (Fig. 18: 1) and the extreme variations in the notches in the peninsular forms lead us to suppose that there is a Calabrian-Lucanian population derived from the Sicilian one. We also maintain that the *S. cianficconii* species from the Marches, Umbria, the Apuanian Mountains and Romagna have a phyletic origin from *S. siculum*, from which it is nevertheless conspicuously different, above all from the lyre-shaped form of the superior branch of the inferior appendage and in its smaller size (Fig. 12).

10. *Sericostoma subaequale* McL. (KEMPNY, 1900; KIMMINS, 1957; McLACHLAN, 1874-1880, 1898; MARCUZZI, 1956; THIENEMANN, 1904, 1905; ULMER, 1927).

The spines of segment X are furcate and of equal length. Some specimens have a tooth about halfway from the inferior edge of the superior branch (Fig. 13: a). At the base of both the appendages of segment X there are teeth and dark crests, very irregularly placed: here at the top and only on one side, there on both appendages but differently shaped (Fig. 13: a, b, c).

Wing expanse 28-32 mm, length of body 10-11 mm.

We have been able to study different specimens collected by Hartig in the Dolomites (Fig. 18: 2). The validity of this species, already separated by McLACHLAN from *S. pedemontanum* and then sanctioned according to KIMMINS' systematic rules in 1957 (KIMMINS, 1957), is also in our opinion quite unquestionable.

11. *Sericostoma timidum* HAG. (MORETTI, 1937).

This has furcate appendages of segment X, very close together and dorsally turned downwards (Fig. 14). For this reason, it is similar to *S. turbatum*, but the markedly

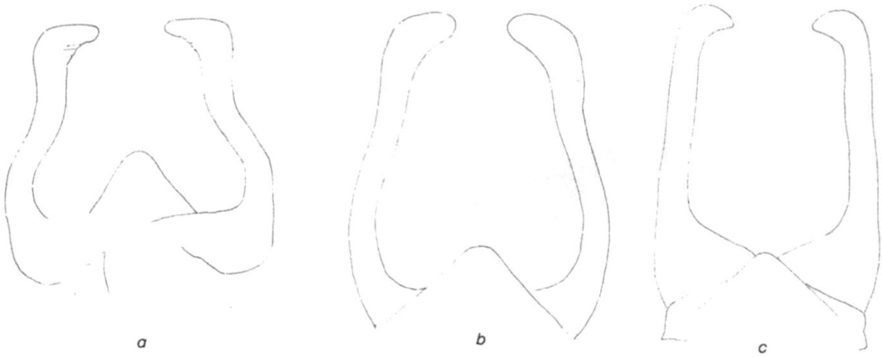

Fig. 12. Ventral branches of inferior appendage and ventral process of segment IX. a = *Sericostoma cianficconii* MORET. specimen from Marches; b = *S. siculum* McL.; , c = *S. cianficconii* MORET. specimen from the Apuanian Mountains and Tuscany-Emilian Apennines.

annulated antennae make this insect a good species. (Fig. 15).

Wing expanse 25 mm.

It has been diagnosed by one of us from a specimen in the Moretti collection coming from Trentino (Fig. 18: 3).

12. *Sericostoma turbatum* McL. (McLACHLAN, 1874-1880; MORETTI, 1937; NAVAS, 1932, 1933).

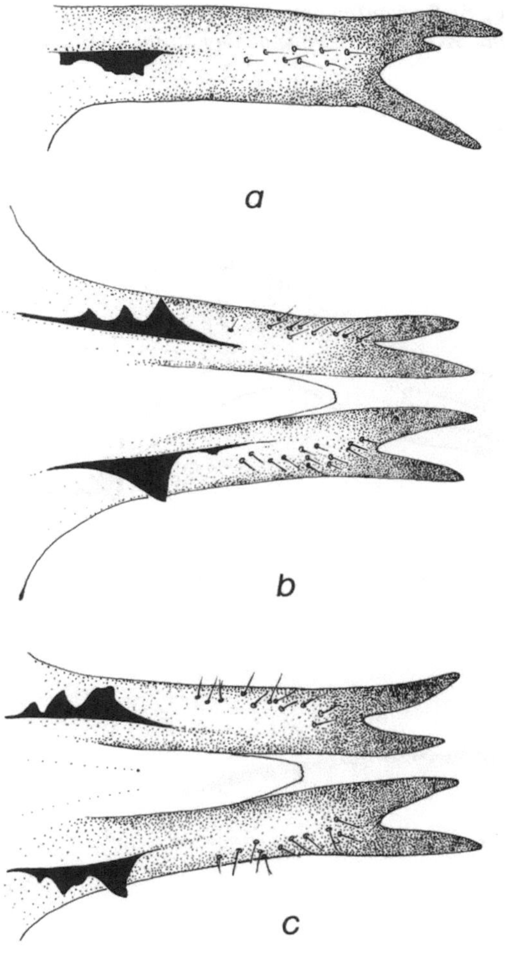

Fig. 13. *Sericostoma subaequale* McL.: ♂ — a, b, c = lateral spiniform processes of segment X, side view (specimen from the Dolomites).

As regards the shape of the appendages of segment X we are unable to discern any clear diagnostic differences from *S. timidum.* However, the lack of any annulations in the antennae must not be neglected from the systematic point of view. Perhaps this is a subspecies rather than a species.

It has been found in Piedmont by NAVAS and in Trentino (classified by MORETTI). (Fig. 18: 4).

To conclude, therefore, there are in Italy today 12 species of *Sericostoma* (Tab. 1). Of these, 8: *S. cianficconii* MORET., *S. clypeatum* HAG., *S. galeatum* RAMB., *S. italicum* MORET., *S. maclachlanianum* COSTA, *S. siculum* McL.,

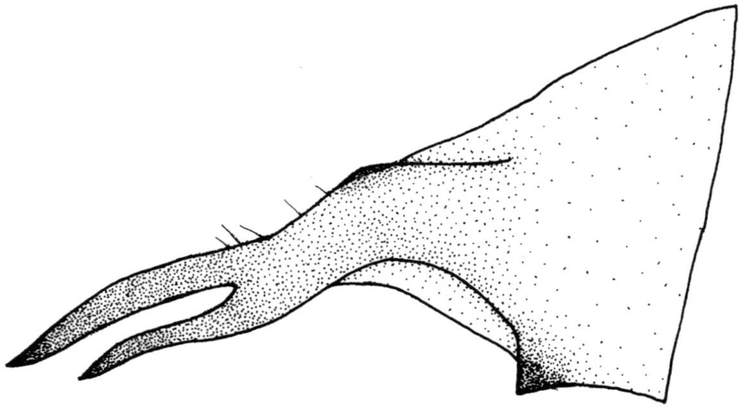

Fig. 14. *Sericostoma timidum* Hag.: ♂ — Lateral spiniform process of segment X, side view.

Fig. 15. *Sericostoma timidum* Hag.: ♂ — Antenna annulations.

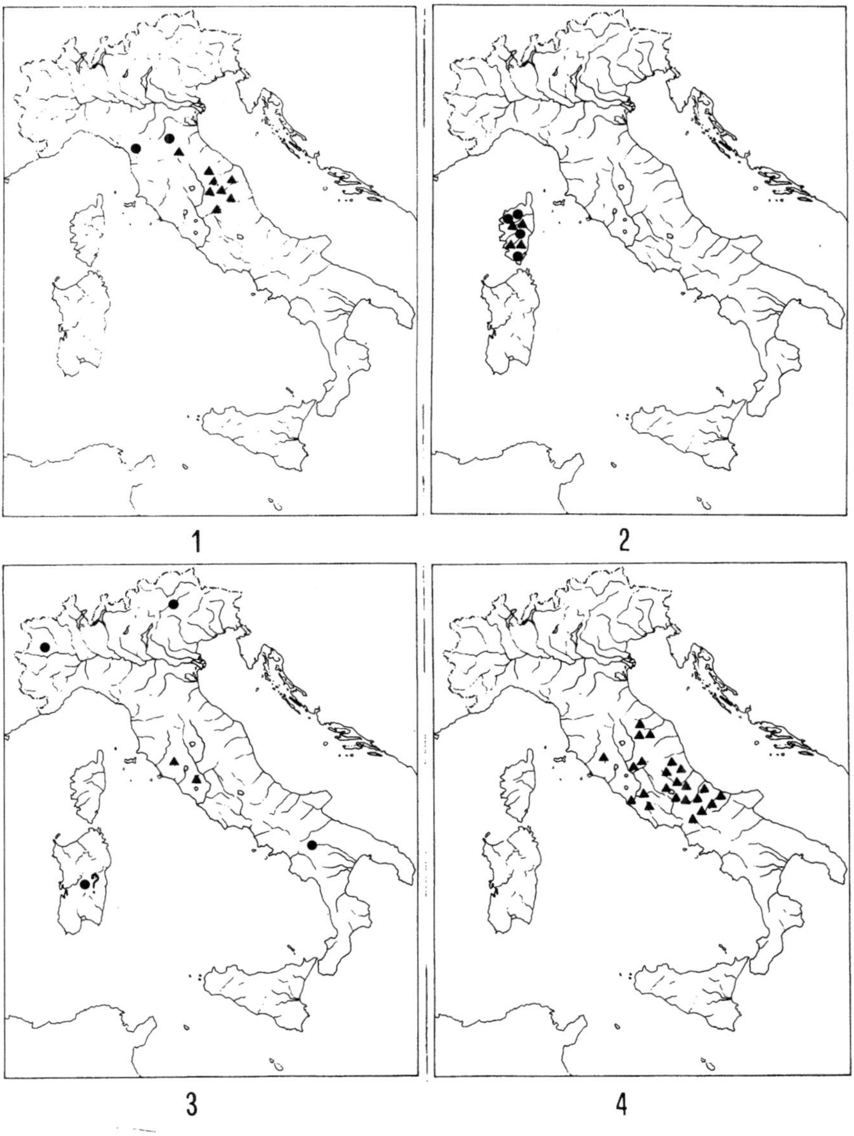

Fig. 16. Distribution area in Italy of: 1 — *Sericostoma cianficconii* MORET.; 2 — *S. clypeatum* HAG.; 3 — *S. galeatum* RAMB.; 4 — *S. italicum* MORET.
● = Findings already given in bibliography; ▲ = new findings.

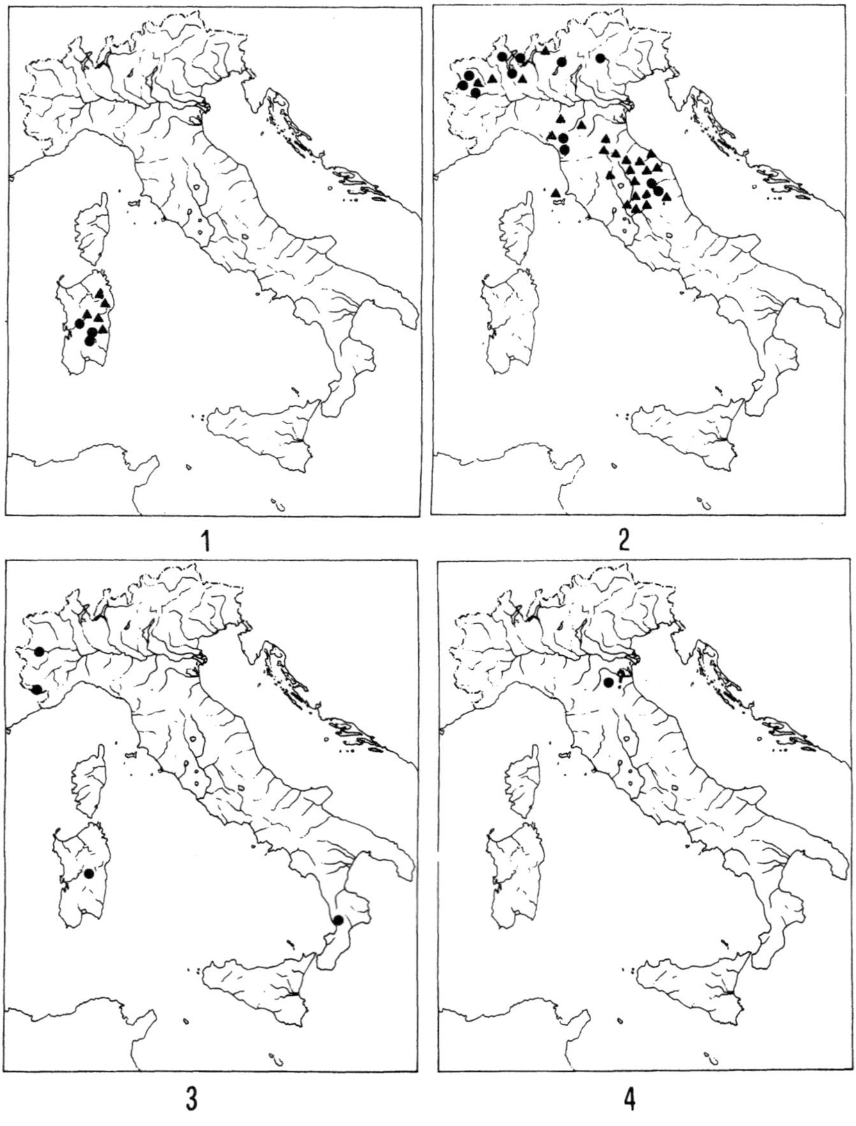

Fig. 17. Distribution area of: 1 — *Sericostoma maclachlanianum* COSTA; 2 — *S. pedemontanum* McL.; 3 — *S. personatum* SPENCE; 4 — *S. romanicum* NAV. (idem Fig. 16)

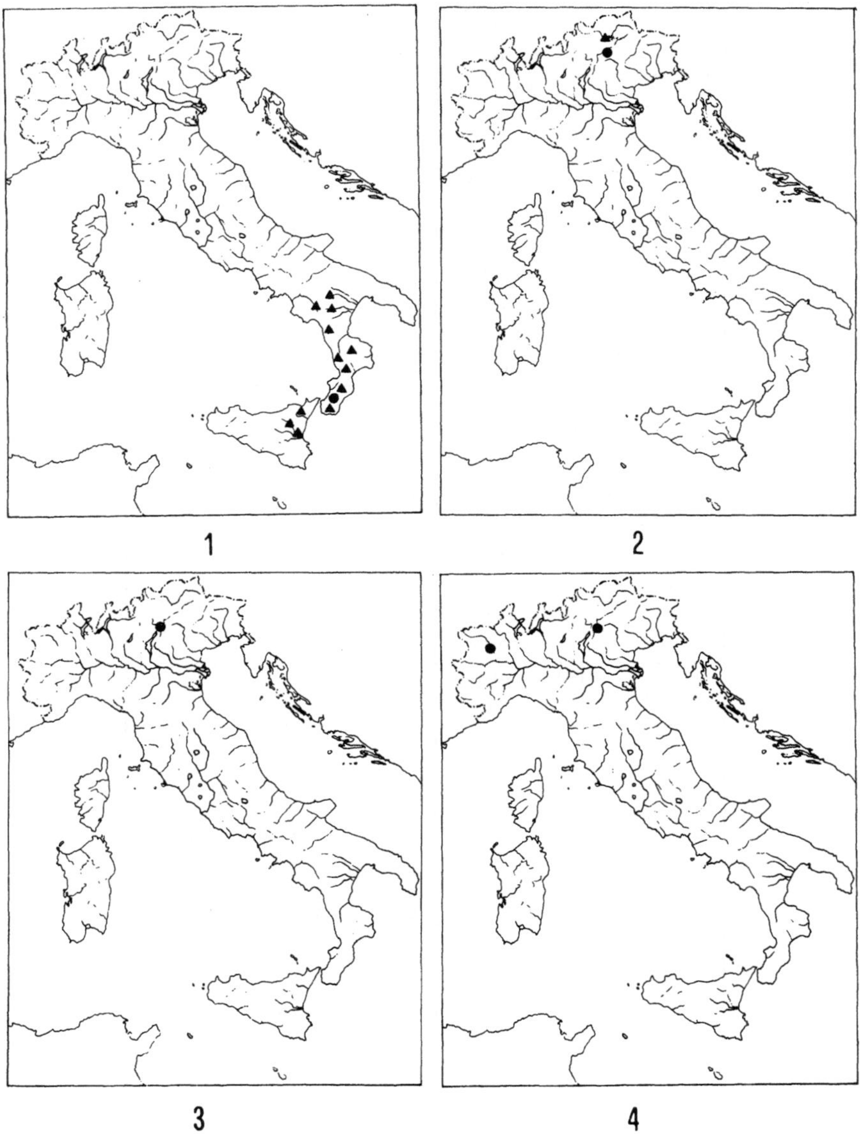

Fig. 18. Distribution area of: 1 — *Sericostoma siculum* McL.; 2 — *S. subaequale* McL.; 3 — *S. timidum* Hag.; 4 — *S. turbatum* McL. (idem Fig. 16).

Tab. 1. *Sericostoma* LATR. — Species found in Italy up to 1977.

	REGIONS and ISLANDS / TAXA		Piedmont	Valle d'Aosta	Liguria	Lombardy	Trentino-Alto Adige	Venetia	Friuli-Venezia Giulia	Emilia-Romagna	Tuscany	Apuanian Mountains	Isle of Elba	Umbria	Marches	Latium	Abruzzo	Molise	Campania	Apulia	Basilicata	Calabria	Corsica	Sardinia	Sicily
1	*Sericostoma*	*cianficconii* MORET.								+		+		+	+										
2	”	*clypeatum* HAGEN																					+		
3	”	*galeatum* RAMB.	+				+				+					+					+			+	
4	’	*italicum* MORET.									+			+		+	+	+	+						
5	”	*maclachlanianum* COSTA																						+	
6	”	*pedemontanum* McL.	+	+		+	+			+	+	+	+	+	+										
7	”	*personatum* SPENCE	+		+																	+		+?	
8	”	*romanicum* NAVAS								+?															
9	”	*siculum* McL.								⋮											+	+			+
10	”	*subaequale* McL.					+			⋮															
11	”	*timidum* HAG.					+			↓															
12	”	*turbatum* McL.	+				+			+															

S. subaequale McL., *S. timidum* Hag. can be considered good species, as has been shown in the figures of the ♂ genitalia. Four species still leave one somewhat perplexed as to their validity. The first problem to be solved is the separation of *S. personatum* SPENCE and *S. pedemontanum* McL..

We are convinced that it is necessary to separate the two species which are clearly divided geographically (*S. pedemontanum* along the Pre-Alps and the Apennines; *S. personatum* along the Alps. They have visibly different characters, which are also very variable in both species.

We are not convinced of *S. romanicum* NAV., judging by the drawing and the description and we feel it should be synonymous with *S. turbatum*, if not with *S. pedemontanum.*

If one takes into consideration the non-annulated antennae of the *S. turbatum* McL., this could be considered at least as a subspecies of *S. timidum* or vice versa.

As regards the distribution of *S. siculum* we consider that this has spread in the peninsula very probably from Sicily and has reached Tuscany, the Marches and Romagna with isolated populations, by now already with specific characters.

References

ANGELIER, E. 1959. Les eaux douces de Corse et leur peuplement. Vie et Milieu. Paris suppl. 8: 19.

BERLAND, L. & Mosely, M.E. 1963. Catalogue des Trichoptères de France. Ann. Soc. ent. France 105: 137-138.

BOTOSANEANU, L. 1967. Trichoptera. In: ILLIES, Limnofauna Europea: 308, Stuttgart (Fischer).

BOTOSANEANU, L. & Schmid, F. 1973. Les Trichoptères du Muséum d'Histoire naturelle de Genève. Rev. suisse Zool. 80: 251-254.

BRAUER, F. 1876. Festschr. zool. bot. Ges.: 284.

COSTA, A. 1869. Ann. Mus. Zool. Napoli 5 (1865): 13.

——. 1871. Ann. Mus. Zool. Napoli 6 (1866): 15.

——. 1884. Atti Acc. Napoli (2) 1 no. 9: 7, 32, 52.

——. 1884. Nota intorno i Neurotteri della Sardegna. Rend. Acc. Napoli 23: 21.

——. 1885. Diagnosi di nuovi Artropodi di Sardegna. Bull. Soc. ent. Ital. 17: 242-243.

ESBEN-PETERSEN, P. 1912. Addition to the Knowledge of the Neuropterous insect fauna of Corsica. Ent. Meddel. (2) 4: 349.

——. 1913. Addition to the Knowledge of the Neuropterous insect fauna of Corsica. Ent. Meddel. 10: 25.

——. 1934. Fl. og Fauna: 20-97.

FELBER, J. 1908. Die Trichopteren von Basel und Umgebung. Arch. Naturg. 74: 8, 14, 33, 75, 80, 82, 85.

FISCHER, F.C.J. 1970. Trichopterorum Catalogus 11: 218-254.

GRIFFINI, A. 1897. Imenotteri, Neurotteri, Pseudoneurotteri, Ortotteri e Rincoti Italiani. Hoepli Milano: 181-182.

HAGEN, H.A. 1859. Die Phryganiden Pictet's. Stettin. ent. Zeit. 20: 147-148.

——. 1860. Révision critique des Phryganides décrites par M. Rambur d'après l'examen des individus types. Ann. Soc. ent. Belg. 4: 75.

——. 1860. Examen des Nevrotères (non Odonates) recueillis en Sicile par M.E. BELLIER DE LA CHAVIGNERIE. Ann. Soc. ent. France (3) 8: 746.

HAGEN, H.A. 1864. Phryganidarum synopsis synonymica. Verh. Zool. bot. Ges. 14: 880.
——. 1864. Névroptères (non Odonates) de la Corse, recueillis par M.E. BELLIER DE LA CHAVIGNERIE en 1860 et 1861. Ann. Soc. ent. France (4) 4: 43.
KEMPNY, P. 1900. Verh. zool. bot. Ges. 50: 258.
KIMMINS, D.E. 1957. Lectotypes of Trichoptera from McLACHLAN collection. Bull. Brit. Mus. Ent. 6: 124.
KLAPALEK, F. 1917. Über die von Herrn. Prof. A. HETSCHKO in Korsika gesammelten Neuropteroiden nebst Bemerkungen über einige ungenügend bekannte Arten. Wien. Ent. Zeit. 36: 193.
KOLENATI, F. 1859. Wien. ent. Mschr. 3: 20.
——. 1859. Genera et species Trichopterorum Praga 2: 162, 179, 289.
McLACHLAN, R. 1869. Notes additionelles sur les Phryganides décrites par le Dr. M. RAMBUR. Ann. Soc. ent. Belg. 13: 9.
——. 1870. Cat. Br. Neur. : 31.
——. 1875. Descriptions de plusieurs Névroptères Planipennes et Trichoptères nouveaux de l'île de Célèbes et de quelques espèces nouvelles de Dipseudopsis, avec considérations sur le genre. Tijdschr. ent. 18: 26-27.
——. 1874-1880. A monographic revision and synopsis of the Trichoptera of the European fauna. London I. V. Voorst., (1884) First additional supplement.
——. 1898. Some new species of Trichoptera belonging to the European fauna, with notes on others. Ent. Mon. Mag. 34: 49-50.
MALICKY, H. 1971. Trichopteren aus Italien. Entom. Zeit. 23: 257-265.
MARCUZZI, G. 1956. Fauna delle Dolomiti. Mem. Cl. Sci. matem. nat. Venezia. 31: 201.
MARTIN, R. 1893. Feuille jeun. Natural. 23: 57.
MEYER-DÜR, L.R. 1875. Die Neuropteren — Fauna der Schweiz bis auf heutige Erfahrung. Mitt. Schweiz. ent. Ges. 4: 200-401.
MORETTI, G.P. 1937. Tricotteri della Venezia Tridentina 1921-1935. Studi sui Tricotteri. IX. Studi Trentini Sci. Nat. 18: 67, 73.
——. 1939. Studi sui Tricotteri. XI. Alcuni Tricotteri dell'Instituto di Entomologia della R. Universita di Bologna. 11: 91-93.
——. 1940. Studi sui Tricotteri. XIII. I Tricotteri della Sardegna. Mem. Soc. Entom. Ital. 19: 288.
——. 1949. Valutazione biologica del Fiume Potenza come esponente delle acque fluviali delle Marche. Verh. Int. Ver. Limnol. 10: 337.
——. 1950. Tricotteri fonticolo dell'Appennino Umbro Marchigiano. Boll. Soc. Eustachiana. 4: 207.
——. 1952. Bilancio ecologico di una raccolta di Tricotteri delle Marche, Umbria ed Abruzzo. (Studi sui Tricotteri XXII). Boll. Zool. Torino. 19: 256, 260, 261, 262, 264, 266, 267.
——. & MICHELETTI, P.A. 1952. Facies primaverile delle biocenosi zoofile del fiume Potenza. Boll. Pesca. Piscic. Idrobiol. 6: 143, 174-175.
——. 1955. Sulla presenza dei foderi dei Tricotteri e dei ditteri Tanitars sui fondi del Lago Maggiore. Mem. Ist. Ital. Idrobiol. suppl. 8: 212-215.
——. & VIGANO', A. 1959. L'habitat e la biologia di *Helicopsyche sperata* McL. in Toscana. Boll. Zool. 26: 585.
——. & VIGANO', A. 1961. Fisionomia di una raccolta tricotterologica compiuta in Toscana. Att. Acc. Naz. It. Ent. 8: 258.
——. & GIANOTTI, F.S. 1967. Tricotteri presenti in un altro impianto di troticoltura dell'Umbria. Riv. Idrobiol. (1) 6: 109-111.

MORETTI, G.P. & CIANFICCONI, F., GIANOTTI, F.S., PIRISINU, Q., VIGANO', A. 1970. Informazioni sui Tricotteri delle Apuane. Soc. Ital. Biog. 1: 492-509.
MORTON, K.J. 1934. Notes on some Odonata, Trichoptera and Neuroptera collected in Corsica. Ent. monthly Mag. 70: 5.
MOSELY, M.E. 1930. Corsican Trichoptera. Eos. Rev. Espanola Ent. 6: 181.
——. 1932. Corsican Trichoptera and Neuroptera (S.L.) 1931. Eos. Rev. Espanola Ent. 8: 178.
NAVAS, L. 1930. Insetti della Romagna. Boll. Soc. Entom. Ital. 62: 149-151.
——. 1932. Alcuni insetti del Museo di Zoologia della R. Universita di Torino. Boll. Mus. Zool. Anat. Comp. Univ. Torino. 42 (3): 38.
——. 1934. Insetti Neurotteri ed affini del Piemonte. Mem. Soc. ent. Ital. 12 (1933): 162.
——. 1933. Lambillionea 33: 25.
NOCENTINI, A.M. 1963. Strutture differenziali della fauna macrobentonica litorale del Lago Maggiore. Mem. Ist. Ital. Idrobiol. 16: 221, 222, 244, 261, 267.
RAMBUR, P. 1842. Histoire naturelle des insectes Nevroptères. Roret, Paris: 496.
RIS, F. 1889. Trichopteren des Kantons Tessin und angrenzender Gebiete. Mitt. schweiz. Ent. Gesellsch. 8: 120-144.
ROSTOCK, M. 1874. SB. Ges Isis Dresden (1873): 21.
——. 1888. Neur. Germ. : 54.
SCHMID, F. 1947. Sur quelques Trichoptères suisses nouveaux ou peu connus. Mitt. Schweiz. Ent. Gesell. 20: 335.
SECONDARI, A. 1950. La presenza nelle Marche del *Sericostoma siculum* McL. (Insetti-Tricotteri). Boll. Soc. Eustachiana 43: 47-50.
THIENEMANN, A. 1904. *Ptilocolepus granulatus* Pt., eine Übergangsform von den Rhyacophiliden zu den Hydroptiliden Allg. Z. Ent. 9: 212.
——. 1905. Z. Ferdinand. Innsbrück (3) 49: 390.
ULMER, G. 1905. Z. Insbiol. 1: 24.
——. 1907. Trichoptera. In: Genera Insectorum. 60: 83.
——. 1907. Trichoptera. Catalogues de Collections zoologiques du Baron Edm de Selys Longchamps (1) 6: 31.
——. 1909. Trichoptera in Die Süsswasserfauna Deutschlands. : 197-201 G. Fischer, Jena.
——. 1927. Trichoptera. Tierwelt Mitteleuropas 6: XV, 45.
VIGANO', A. 1958. Aliquote tricotterologiche nel tubó digerente dei pesci di acqua dolce. Riv. Biol. 50: 388.
ZANGHERI, P. 1966. Repertorio sistematico e topografico della flora e fauna vivente e fossile della Romagna. Mus. Civ. St. Nat. Verona 2: 828.

Discussion

NIELSEN: As shown by yourself and CRICHTON, the male maxillary palp in this genus is a scent organ. The alleged specific difference between *pedemontanum* and *personatum* may be a difference in physiological state.

MORETTI: Such a difference might be possible but we have not yet considered it because so far we have found only *pedemontanum* in the Apennines. *S. flavicorne* males sent to me by MALICKY always show well-developed and protruding maxillary palps.

DENIS: Dans le département de l'Isère, j'ai trouvé une importante population de *S. galeatum.* Si les males ont la teinte uniforme typique des Sericostomatidae, les femelles sont soit de teinte uniforme sombre (elles ont alors des oeufs verts), soit bicolores avec des taches blanches importantes (elles ont des oeufs jaunes). Enfin il existe toute une série de types intermediaires entre ces deux formes.

MORETTI: Tout ce que vous m'avez fait savoir maintenant est très interessant et je crois qu'il est bon de nous tenir bien en contact sur cet argument.

MARLIER: Je ne trouve pas de différences stables entre *personatum* et *pedemontanum*. Les larves ne seraient pas du tout différents. Ce sont peut-être des sous-espèces.

MORETTI: La question est vraiment complexe; je prends note de ce que vous justement dites, mais en Italie la question de deux sous-espèces est géographiquement et aussi taxonomiquement à approfondir à travers des recherches supplémentaires.

Note. After the Symposium, Dr. P.C. BARNARD of the British Museum (Natural History), kindly sent us for examination the *Sericostoma hamatum* that McLACHLAN believed might have come from North Italy, and that KIMMINS considered a synonym of *S. galeatum*. There is no doubt, in our opinion, that this is truly *S. galeatum*.

In the specimens of *S. galeatum* from south east France, kindly sent by Dr. C. DENIS, the apex of the appendages of segment X, has a pronounced outward curve. This character is not present in the Italian specimens from Latium and Tuscany, which are also smaller.

Proc. of the 2nd Int. Symp. on Trichoptera, 1977, Junk, The Hague

Les larves et nymphes des Trichoptères des Seychelles

G. MARLIER

Abstract

The immature instars of 6 species of Trichoptera collected in the Seychelles Islands by Prof. F. STARMÜHLNER's (1974) and by the author's (1976) expeditions are described.

Of great interest is the discovery of the larva of two unknown species; the first one attributed to *Hughscottiella* (ULMER) is characterized by a most elongated and retractile prothorax; the second one, attributed to *Leptodermatopteryx*, is more classical and rather similar to a beraeid larva.

Introduction

Les Trichoptères des Iles Seychelles sont connus depuis longtemps et, chose curieuse, par les résultats d'une seule expédition scientifique: la Percy-Sladen Trust Expedition, conduite par M. STANLEY GARDINER, en 1905, dans les Iles de l'Océan Indien du Sud-Ouest. Les récoltes entomologiques de cette expédition furent faites par H. SCOTT, qui séjourna huit mois dans les Iles. Il ne récolta des Trichoptères que sur les îles granitiques de Mahé et de Silhouette et ne captura que des exemplaires adultes.

Les matériaux rapportés, qui comprenaient 81 exemplaires furent étudiés par G. ULMER et publiés en 1910.

Plusieurs expéditions scientifiques eurent lieu depuis lors et de nombreux entomologistes de talent séjournèrent aux Seychelles depuis le début du siècle mais, apparemment, ne furent pas intéressés par la maigre mais intéressante faune des Trichoptères.

En 1974, le laboratoire de zoologie de l'Université de Vienne, organisa une expédition dans cet archipel, dans le but d'y étudier la faune des eaux douces. Cette expédition se faisait dans le cadre d'une vaste enquête sur les faunes dulcicoles des îles océaniques, qui avait déjà donné lieu à l'exploration d'autres îles de l'Océan Indien.

Le Professeur F. STARMÜHLNER qui la dirigeait, rapporta des Seychelles un matériel d'environ 352 larves et nymphes de Trichoptères qu'il confia au Dr. H. MALICKY. En 1976 nous fîmes nous aussi une brève mission dans les mêmes îles d'où nous rapportâmes 188 adultes et environ 1400 larves et nymphes.

Mr. le Dr. MALICKY nous confia à son tour le matériel de la mission STAR-

MÜHLNER ce qui nous mit en mesure d'étudier une importante collection faite dans une île de 148 km^2. Tous les insectes recueillis le furent en eau courante, en majeure partie torrentueuse. Les adultes furent capturés soit à la main soit au piège à rayon U.V.

Liste des espèces
Six espèces nouvelles pour la Science avaient été signalée par G. ULMER en 1910
Hydroptilidae: *Petrotrichia palpalis* ULMER n.g. n.sp.
Polycentropodidae: *Cyrnodes scotti* ULMER n.g. n.sp.
Ecnomidae: *Ecnomus insularis* n.sp.
Hydropsychidae: *Hydromanicus seychellensis* ULMER n.sp.
Odontoceridae: *Leptodermatopteryx tenuis* ULMER n.g. n.sp.
Hughscottiella auricapilla ULMER n.g. n.sp.
L'étude du matériel larvaire et nymphal aboutit aux résultats suivants: *Petrotrichia* est un Helicopsychidae et non un Hydroptilidae, ce qui explique l'anomalie relevée par ULMER dans la structure des palpes maxillaires. Une deuxième espèce d'*Helicopsyche* fut trouvée, souvent dans les mêmes stations que la première.

Hydromanicus seychellensis, *Ecnomus insularis* sont communs à l'état larvaire dans les eaux des Seychelles et nos deux collections en renfermaient beaucoup. *Cyrnodes scotti* ne fut plus retrouvé à aucun des stades, peut-être a-t-il disparu depuis le début du siècle.

Aucun adulte ni aucune nymphe d'Odontoceridae ou d'autre famille de Trichoptères à fourreau ne fut découvert. Deux espèces de larves à fourreau conique furent retrouvées, l'une en nombre réduit, l'autre en deux exemplaires (par F. STARMÜHLNER) on peut supposer qu'elles correspondent aux deux 'Odontoceridae' signalés par ULMER, mais on n'en a aucune preuve.

Description des larves recueillies
Famille: Ecnomidae
1. *Ecnomus insularis* ULMER
Espèce très répandue dans les cours d'eau de Mahé depuis les sources (bassin Gd. Saint Louis, Bassin Grand Anse, Riv. Baie Lazare) jusqu'aux embouchures près des mangroves: riv. Grand Anse, Sèche, Caïman, Anse aux Poules Bleues, Cascade etc. ... sur tous les versants et à toutes les altitudes explorées. Rarement en grands nombres.

Construit des galeries de soie dans les anfractuosités des pierres et des écorces comme les autres *Ecnomus* de tous les continents. La grande différence que présente cette espèce avec le plus grand nombre de ses congénères est d'ordre écologique. Elle fréquente en effet non des lacs ou des cours d'eau calmes mais des torrents de montagne.

Larve
Très semblable à toutes celles du genre.

Longueur: 9,5 mm au maximum; tête comprise 5,3 fois dans la longueur totale. Coloration variable, les larves plus claires ayant les sclérites céphaliques et thoraciques jaunes, sans aucun dessin. Exemplaires très colorés présentant les caractères suivants: tête jaune ambrée avec la moitié antérieure du frontoclypéus, jusqu'aux bras antérieurs du tentorium, gris-noirâtre, le bord postérieur de cette zone échancrée anguleusement vers l'avant; dans les angles antérieurs de cette zone sombre, 3-4 taches claires arrondies; moitié postérieure du frontoclypéus entièrement claire, bandes furcales à partir des bras antérieurs du tentorium, largement noires, de même que les tempes et la suture épicraniale; 3-4 taches rondes claires sur les tempes; zone périoculaire blanche, bord occipital rembruni, face ventrale de la tête claire sauf une zone noire au bord occipital. Pronotum entièrement clair sauf le bord postérieur qui forme un bourrelet noirâtre et se prolonge à la face ventrale jusqu'à la ligne médiane, derrière l'épimère. Mésonotum jaune transparent avec une marge noire très fine sur les côtés, après le tiers antérieur, et au bord postérieur, ainsi qu'un trait noir oblique partant des angles antérieurs jusqu'à un point situé à la moitié de la longueur de segment et à la moitié de la largeur de chaque demi-sclérite. Métanotum semblable un peu plus pâle.

Zones colorées de la tête, surtout les tempes, à microsculpture polygonale très étirée transversalement.

Pleure prothoracique à trochantin long, aigu, presque de la longueur de la coxa, fusionné à sa base avec l'épisternum; épimère très long entourant la coxa vers l'arrière. Ligne latérale abdominale formée de très longs poils fins sur plusieurs rangs, sur les segments II à VII.

Patte antérieure un peu plus courte et plus épaisse que les deux autres, à fémur un peu dilaté, coxa portant deux fortes soies noires antérieures, une antéroexterne et une courte, interne; fémur armé de deux longues soies au bord ventral; tibia avec deux longues soies terminales ventrales, une courte médio-ventrale et deux dorsales terminales; tarse un peu plus court que le tibia, portant une brosse ventrale de poils serrés, deux soies médio-ventrales de part et d'autre de l'arête et deux soies dorsales terminales; griffe aussi longue que le tarse, aiguë, faiblement courbée avec un long éperon inséré un peu au-dessus de la base, et une soie au pied de l'éperon.

Patte intermédiaire à fémur armé de deux longues soies calcariformes ventrales, tibia court (17/27 du fémur), tarse encore plus court (11/27), griffe (10/27 du fémur) portant un éperon supra-basilaire atteignant les 2/3 de la longueur de la griffe; patte postérieure encore plus longue à griffe fine, presque droite, armée d'un éperon supra-basilaire semblable à celui de la patte intermédiaire.

Appendices terminaux (pygopodes) à griffe terminale coudée à angle droit, sans dent dorsale mais avec des dents internes croissantes de la base à l'extrémité, les basales extrêmement petites, le nombre total des dents dépassant 65.

Nymphe

Longueur: 5 mm (♀), largeur maximum au niveau du métathorax, régulièrement rétrécie vers l'arrière en fuseau, extrémité tronquée.

Antennes allongées le long du corps et ramenées avec les ailes vers la face ventrale, atteignant chez la femelle l'extrémité du VIIe segment; composées de 38 articles.

Palpes maxillaires divergents jusqu'au milieu des coxas I puis repliés perpendiculairement vers la ligne médio-ventrale où ils se croisent de toute la longueur du 5e article. Fourreaux alaires antérieurs atteignant l'extrémité du Ve segment.

Clypéus avec 2 longues soies à la base, de chaque côté, dirigées vers l'avant et le milieu et se croisant avec celles du côté opposé. Labre orné de 3 groupes de soie de chaque côté: un groupe basal latéral de 2 soies noires dirigées en oblique vers l'avant, un groupe antéro-latéral de 3 soies dirigées vers le bas et un groupe médian de 2 soies redressées verticalement, crochues au bout, beaucoup plus longues que les précédentes, très proches de la ligne médiane. Mandibules très fines, très aiguës, sans dents.

Pas de branchies, une ligne latérale de longs poils sur les segments VI à VIII.

Appareil d'accrochage peu développé, formé de petites plaques présegmentales arrondies à dents postérieures nettes; les plaques postsegmentales (sur V), plus écartées, allongées transversalement; dents répondant au schéma suivant:

	Pré	Post
III	5	—
IV	6	—
V	3-4	8
VI	6	—
VII	5-6	—
VIII	5	—

Franges de poils natatoires présentes sur les articles 1 à 4 des tarses intermédiaires.

Lobes terminaux de l'abdomen arrondis au bout, à base large, brusquement rétrécis dans leur premier tiers, leur dernier tiers hérissé, à la face ventrale et à l'extrémité, de soies noires très longues et raides.

Stations de récolte: Mission G. MARLIER: toutes les rivières de Mahé. Mission F. STARMÜHLNER Sta 9b: Riv. Grand St. Louis, 8.II.74; Sta $25c^3$: N.W. Küste; Riv. Cascade: 19.II.74: 2 larves.

Famille: Hydropsychidae

2. *Hydromanicus seychellensis* ULMER

Espèce très largement répandue dans tous les cours d'eau de l'île de Mahé à toutes les altitudes, parfois même dans les eaux faiblement polluées comme celles de la rivière Jouanis. *H. seychellensis* a également été trouvée dans l'île de Praslin (Ruisseau baie St. Anne) sous forme d'une exuvie nymphale, et dans la rivière de la Vallée de mai sous forme de très nombreuses larves. En tout 552 larves et nymphes furent récoltées dans toutes les rivières de Mahé (20 stations différentes) de même que 19 adultes. L'expédition STARMÜHLNER récolta également environ 200 larves et nymphes à Mahé.

Les larves de *H. seychellensis* se rencontrent dans les mêmes conditions que les *Hydropsyche* en Europe et construisent des retraites et des filets-pièges tout à fait semblables: le filet d'une larve très mûre était plus ou moins circulaire avec des mailles carrées de 133 μ de côté et un diamètre total de 30 à 36 mailles. La coque nymphale est faite de petites pierres et de grains de sable grossier et la soie en est assez résistante et rigide.

Larve
Taille très variable: maximum observé 15 mm de longueur totale, dont 2 mm de longueur de tête. Aspect tout à fait semblable à celui d'une larve d'*Hydropsyche.*
Coloration de la tête: d'un brun fauve assez uniforme, parfois uniformément claire: le plus souvent le bord antérieur du fronto-clypéus surtout les angles latéraux, deux taches arrondies aux insertions des bras tentoriaux, les tempes plus faiblement et la moitié apicale du fronto-clypéus rembrunis; parfois aussi un point brun au sommet de l'échancrure occipitale, prolongé ou non sur la suture épicraniale. Face ventrale de la tête assez claire, la zone stridulante large, foncée, à 42-45 stries.
Coloration du thorax: prothorax brun uniforme avec une bordure noire étroite et régulière en arrière, large dans l'angle postérieur, amincie à l'angle antérieur, devenant une simple ligne noire en avant; mésothorax souvent un peu plus pâle, avec un bord noir non sinueux assez épais sur les côtés, élargi encore dans l'angle antérieur, aminci au bord postérieur où il se dilate au milieu en une tache concave vers l'avant avec deux pointes aiguës; métathorax à tache médio-postérieure presque carrée, très petite, finement échancrée en arrière, une fine marge noire latérale, un peu dilatée en avant.
Morphologie et chaetotaxie: Submentum à bord antérieur concave, à pointe postérieure assez courte, de proportions suivantes: largeur totale: 7; hauteur au milieu: 2.4; largeur des 'bras latéraux: 1 (rapport largeur bras/largeur totale: 0,14) Postgula minuscule; pas de pilosité à la face inférieure de la tête.
Frontoclypéus à bord antérieur concave, festonné régulièrement (15-17 festons sur toute la largeur: dans les creux entre les festons, des groupes de soies plumeuses insérées sur la base des festons; surface du fronto clypéus presque lisse, avec des microtriches très fins et espacés devenant plus denses sur les bords de l'apotome, surtout dans les angles antérieurs et au sommet; reste de la face dorsale densément couvert de poils courts, aigus ou tronqués, très aigus surtout dans l'angle occipital, beaucoup plus longs autour des yeux.

Pronotum orné de façon semblable de poils épais assez courts, entremêlés de poils plus fins et plus courts, les poils épais non dilatés en massue à l'extrémité. Mésonotum à bord antérieur renforcé, portant 2-3 rangées de poils longs, obtus, le reste du sclérite couvert de poils courts, renflés près de la base, tronqués au bout et mêlés de poils plus fins et courbés. Métanotum moins sclérifié, à pilosité analogue mais plus fine, les poils épais étant beaucoup plus courts; entre ceux-ci, le tégument présente une microsculpture en petits peignes de denticules courbés vers l'avant; (chez *Hydropsyche angustipennis*, la microsculpture ne présente pas de petits peignes réguliers et les poils épais sont élargis au bout, renflés en massue ou fendus en écailles).

Sclérites pairs du prosternum très allongés transversalement, étirés en pointe et à bord postérieur irrégulier, sclérites pairs du mésosternum grands, en triangle à angle antérieur très obtus, à base postérieure concave.

Pleures prothoraciques formées d'un ensemble épisterno-épiméral grand, subrectangulaire et d'un trochantin distinct, fourchu, à branche dorsale conique aiguë, aussi longue que la ventrale, portant quelques fines soies à la base des faces externe et dorsale. Au mésothorax épisternum petit, trapézoïdal, à base dorsale longue, base ventrale courte et bord antérieur très oblique, postérieur presque vertical, épimère grand, en croissant à concavité dorsale, bord dorsal non épaissi, de couleur jaune. Au métathorax l'épisternum est un peu plus grand.

Patte antérieure: coxa avec les deux crêtes antérieures ornées de deux longues soies noires et de poils aigus assez épais, face externe ornée de poils épais et aigus se raréfiant lorsqu'on s'éloigne vers la face postérieure et remplacés par des poils longs et fins, ceux-ci entremêlés de soies plumeuses rigides bisériées beaucoup plus courtes et plus épaisses, face ventrale avec seulement ces soies plumeuses; arête ventrale garnie de poils normaux denses et plus longs; face interne à pilosité rare et simple.

Trochanter et fémur à sétosité noire, dense, surtout ventrale, sans soies plumeuses (pas même celles qui se trouvent sur l'arête ventrale des fémurs chez *Hydropsyche angustipennis*); tibia et tarse courts, fort épineux; griffe longue, peu courbée.

Patte intermédiaire à pilosité moins forte mais à coxa ornée de soies plumeuses externes, ventrales, de quelques soies noires raides et courtes au bord antérieur et de quelques soies plumeuses près du bord ventral; reste de la patte à poils aigus denses souvent mêlés de poils très fins; griffe à éperon interne très gros, valant 1/3 de la griffe.

Patte postérieure à coxa portant quelques soies plumeuses très rares et très fines vers l'arête ventrale.

Segments abdominaux couverts de poils fins ou assez épais dirigés dans tous les sens mais surtout vers l'arrière, très denses; parmi ces poils, mais plus rares, de grosses écailles plates et brunes disposées grossièrement en rangées transversale (pas plus de 8 à 10 sur une rangée et de 3-4 rangées par segment), entre ces phanères, le tégument offre un recouvrement de microdenticules très courts et très épais, ceux-ci surtout denses sur le milieu des segments III à IV devenant de plus en plus petits et rares vers l'arrière.

Sclérites du segment VIII rapprochés, triangulaires à angles émoussés, portant 6-8 soies raides noires; sclérite médio-ventraux du segment IX grands, trapézoïdaux à grande base médiane, et côté antérieur très oblique, portant une dizaine de soies raides noires, sclérites latéraux du même segment très petits, plus ou moins carrés avec 1-3 longues soies, sclérites dorsaux très écartés, arrondis en avant, droits en arrière plus ou moins en forme de demi-cercle, ornés d'une forte soie noire externe et de quelques soies postérieures.

Griffes des appendices anaux du type habituel, courbées en un angle presque droit avec une longue soie dorsale, atteignant l'extrémité de la griffe.

Branchies semblables à celles des Hydropsyche; une paire mésosternale de 18 à 20 rameaux, une paire latérale et une branchie impaire médiane sur le métasternum, chacune de 16 rameaux. Sur les segments abdominaux, troncs branchiaux latéraux juxtaposés le plus médian un peu antérieur, filaments insérés tout autour du tronc sur le plus médian, du côté externe sur le plus externe. Tronc externe simple sur segment I, double sur segments II à VII; tronc 'médian' disparaissant sur le segment VII (La présence de branchies sur le segment VII fait différer *H. seychellensis* de *H. flavoguttatus* Albda). Nombre de filaments sur les troncs branchiaux:

	Touffe plus médiane	touffe plus externe
I	14	18
II	15	15 + 16
III	15	16 + 15
IV	9 — 10	10 + 14
V	10	13 + 12
VI	9	5 + 8
VII	0	7 + 7

'Branchies ≫ coniques. (Zipfelkiemen) très petites sur les segments III-VII au nombre de 1, 2, 2, 2, 1, d'observation difficile.

Nymphe

La nymphe est très semblable à celle des *Hydropsyche*. Longueur de l'exuvie étalée: 13 mm (ex. de Praslin) .

Labre arrondi ou presque, couvert de poils, non échancré au bord antérieur Mandibules à bord externe légèrement coudé vers le milieu, la gauche armée de 5 dents, l'apicale et les deux basales fortes subégales, les deux intermédiaires plus petites, la droite avec 4 dents subégales, leur tranchant très peu crénelé entre ces dents.

Appareil d'accrochage à plaques petites et arrondies présegmentales sur les segments III à VIII, leurs denticules disposés au bord des plaques en une série arrondie suivant le tableau ci-joint, plaques postsegmentales à peine plus grandes ovales, écartées, à 2 rangées de denticules, sur le segment III et à 1 rangée sur IV, les denticules des plaques s'accroissant de l'avant à l'arrière de l'abdomen.

Schéma des denticules

Segment	*Présegmentales*	*Postsegmentales*
III	7	24-26
IV	6-7	8-12
V	5-7	—
VI	7-9	
VII	7-8	
VIII	6-9	

Pilosité dorsale longue sur les segments IV et V, courte sur I à III, très éparse et très courte sur VI à VIII.

Branchies de la forme habituelle disposées comme chez la larve sur les segments II à VII plus ou moins suivant le schéma suivant:

	Médiane	Externe
II	13	16
III	12	16
IV	13	14
V	9	16
VI	5	10
VII	7	11

Appendices terminaux formés de 2 bâtonnets épais, à base large, à zone médiane rétrécie, légèrement aplatis à l'extrémité, celle-ci échancrée, sinueuse et porteuse de spinules à sa marge postérieure; (non fourchue comme celle des *Hydropsyche*)

Des franges natatoires sur les articles 1 à 4 des tarses intermédiaires.

Stations de récolte: dans toutes les rivières de Mahé et de Praslin, à toutes les altitudes (G. MARLIER)

Mission F. STARMUHLNER: Sta 2: Mamelles Riv.: 5.II.74: 1 larve; Sta 3 Cape Riv., 5.II.74: 18 larves; Sta 9: Riv. Gd. St. Louis, 8.II.74: 16 larves; Sta 10 Riv. Gd. St. Louis, 8.II.74: 3 larves; Sta 14: Praslin, Cascade Riv. 12.II.74: 32 larves; Sta 15: 2 larves; Sta 16: Jasmine Riv. 14.II.74: 23 larves et nymphes; Sta 18 et 19: Gd. Anse: 15.II.74: 40 larves et nymphes; Sta 21: Gd. Anse, 17.II.74: 5 larves et nymphes; Sta 22: Athonas Riv., W. Küste: 18.II.74: 10 larves; Sta 23: Desert R. 19.II.74: 8 larves; Sta 25: R. Cascade (W): 13.II.74: 23 larves et nymphes.

Famille: Helicopsychidae

Les exemplaires de la collection Scott que G. ULMER (1909) décrivit comme Hydroptilidae et dont il fit un nouveau genre *Petrotrichia* et une espèce spéciale *P. palpalis*, appartiennent en fait à la famille des Helicopsychidae, ainsi que les récoltes récentes d'adultes et de larves l'ont prouvé.

De plus, deux espèces ont été trouvées ensemble qui se reconnaissent bien à tous les stades. La plus commune est *Helicopsyche palpalis* (ULMER), la seconde est nouvelle. Nous en décrivons les stades jeunes ci-après.

3. *Helicopsyche palpalis* (ULMER)

Larve

Dimensions: Longueur 2 mm.

Coloration: face supérieure de la tête jaune brunâtre sale à taches ovales ou arrondies plus claires, aire oculaire noire; moitié basale des mandibules noire, moitié distale brun clair, pronotum jaune sale.

Tête arrondie, aussi large que longue, à face antérieure aplatie mais sans carène, frontoclypéus élargi en avant, échancré au niveau des bras tentoriaux antérieurs, à peine élargi en arrière et à sommet très peu aigu; face ventrale à suture gulaire effacée mais à partie antérieure un peu plus colorée, à genae jaune clair marquées de 3 ou 4 grosses taches brunâtres.

Sclérites médians du submentum (A. NIELSEN, 1942) longs, courbes vers l'arrière minces et aigus à l'extrémité; palpes maxillaires très courts, émoussés. Lobe labial dépassant en avant les maxilles, conique, arrondi au bout.

Labre presque deux fois plus large que sa longueur au milieu (1,77 x) largement échancré en avant, garni sur tout son bord antérieur de nombreux poils serrés et courts.

Mandibules à moitié basilaire très distincte de la distale, face interne offrant une longue brosse basale et un pinceau apical de bâtonnets recourbés vers la ligne médiane, la brosse moins développée à droite qu'à gauche de même que le pinceau apical, bâtonnets et soies ciliés plumeux.

Antennes extrêmement courtes, réduites aux bâtonnets sensoriels, insérées à mi-distance entre l'oeil et le bord de la tête.

Pronotum à angles antérieurs arrondis, à bord postérieur mal défini, sinueux avec une tache brun foncé dans les angles postérieurs où s'articule la pleure, bord antérieur avec 10 soies de chaque côté, une rangée pré-médiane de 6 longues soies, les marges latérales, surtout en arrière, couvertes de poils fins et courts.

Mésonotum moins sclérifié que le pronotum, montrant des marbrures plus claires surtout au milieu et dans les angles postérieurs, le sclérite divisé longitudinalement et élargi en avant, angles postérieurs arrondis, bord postérieur irregulier, offrant une petite plaque sclérifiée distincte de part et d'autre de la ligne médiane, en forme de goutte renflée vers l'avant, atteignant 14 à 18% de la longueur du mésonotum; au centre de chaque demi-sclérite, une tache ronde foncée.

Métathorax portant un sclérite assez grand, plus ou moins triangulaire médian, peu sclérifié et incolore, puis une paire de plaques postérieures allongées plus latérales et enfin une paire de petites pièces sclérifées allongées, juste au-dessus des, pleures.

Propleure articulée juste à l'angle postérieur du pronotum par une côte d'articulation oblique presque verticale; en arrière, épimère membraneux mais soutenu par une côte chitinisée transparente assez aiguë dirigée vers l'arrière, une baguette sternale rejoignant son homotype sur la ligne médio-ventrale prolongeant les angles postérieurs du pronotum; ces deux formations invisibles sur l'animal frais. Trochantin long, aigu, dirigé vers le bas, le bord ventral (postérieur) très convexe, le dorsal rectiligne, terminé par une forte soie aiguë aussi longue que la moitié du trochantin.

Mésopleure peu sclérifiée, renforcée par une côte d'articulation foncée. Métapleure sclérifiée et rembrunie seulement en avant de la côte d'articulation.

Tubercules latéraux du 1er segment abdominal revêtus d'un sclérite non coloré mais couvert de 14 rangées de soies raides, doubles, chaque rangée comportant environ 25 paires de soies, les plus longues atteignant 20 microns, sclérite se prolongeant vers l'arrière par une baguette transparente assez rigide passant derrière le tubercule dorsal; celui-ci lisse et rond.

Pattes jaune-grisâtre clair, croissant d'avant en arrière, portant des griffes émoussées peu allongées, courbées, armées d'une soie basale assez courte. Au stade de maturité complète il n'y a pas de longue soie-pinceau sur les coxa postèrieures (BOTOSANEANU, 1956) mais bien sur les fémurs II et III, sur le trochanter III et sur la coxa II; il y a en outre de courts poils plumeux sur les trochanters, 10 aux antérieurs, 6 environ aux intermédiaires et aux postérieurs.

Pas de branchies visibles sur les larves fixées. Ligne latérale remplacée par une rangée de tubercules portant des soies simples (ou bifides) aiguës de 20 μ de long insérées sur les segments III à VIII: 6 à 7 soies simples sur III à VIII (rarement 5); 12 à 14 bifides sur VIII; les soies simples ont leurs bords serrulés mais pas aussi barbelés que ceux de *H. bacescui* ORGH. et BOTOS.

Sixième segment avec une ceinture de très petites soies serrées dirigées vers l'avant.

Neuvième segment dépourvu de sclérite dorsal, orné de 2 groupes de 4 soies dorsales fines (4 μ).

Appendices anaux (pygopodes) très courts, à base large et bien sclérifiée, griffe de la forme usuelle des *Helicopsyche*, à pointe principale portant seulement une crête formée de 8 ou 9 épines décroissantes; cette crête recourbée vers la ligne médiane de telle sorte que les épines semblent former une couronne autour de la griffe.

Fourreau

Le fourreau est du type déprimé, analogue à une coquille de Zonitidae aux stades larvaire final et nymphal, il forme 1,75 tour; il a un diamètre transversal de 1,27 mm et un diamètre longitudinal de 1,55 mm, (grains de sable saillants exclus); l'orifice antérieur est très oblique, presque dans le plan de la dernière spire, l'orifice postérieur béant est déporté à droite du diamètre antéro-postérieur, le rapport de la hauteur totale au diamètre maximum de la base est de 0,45 le fourreau est composé de grains de sable fort semblables entre eux et il est très peu rugueux; le tour de spire externe présente quelques grains un peu plus gros en saillie. Le fourreau nymphal est obturé par une membrane ovale (rapport des 2 axes 5/4) avec une zone circulaire de 280 μ de diamètre, presque centrale un peu déplacée vers la périphérie, cette zone forme une grille à mailles irrégulières plus ou moins arrondies ou ovalaires, de 14 à 21 μ de diamètre environ.

Nymphe

Corps courbé sur lui-même, antennes un peu plus longues que le corps, de 44 articles (♂) à premier article long et renflé avant son extrémité où il porte un gros tubercule. Labre petit, arrondi, convexe portant 3 paires de soies sur une rangée antérieure et 8 paires sur une rangée postérieure, ces soies aiguës et non crochues à l'extrémité ni, apparemment, spatulées. Mandibules courtes, assez courbées, à tranchant très faiblement serrulé, à base renflée portant 2 longues soies dorsales.

Palpes maxillaires gros, de 3 articles chez le mâle, courbés en demi-cercle trois fois plus longs que les mandibules; palpes labiaux plus courts, n'atteignant pas l'extrémité des palpes maxillaires lorsque ceux-ci sont en position courbée.

Pattes antérieures et intermédiaires ornées d'une double rangée de poils natatoires sur les articles 1-4 des tarses.

Appareil d'accrochage formé de plaques présegmentales petites, arrondies, insérées sur les baguettes sclérifiées des tergites des segments III à VII avec respectivement 2, 1, 1, 3 crochets très petits; une paire de plaques postsegmentales carrées sur V, armées de 2 crochets plus forts: sur les pleures des segments III, IV, V une protubérance plus ou moins sphérique, noire, hérissée de tubercules pointus placés en rangées plus ou moins concentriques; tous les tergites, de II à VII, couverts de minuscules spinules, souvent doubles, dirigées vers l'arrière, VIII et IX lisses.

Appendices terminaux en cornes aiguës recourbées vers le haut et l'extérieur à bord interne portant des soies à fortes embases, 3 avant le milieu et 2 juste dans la courbure avant la pointe.

Stations de récolte: North-East point: 1 ♂ adulte; Riv. Grand St. Louis, de 550 m à 250 m d'altitude nombreux adultes, nymphes et larves (550 ex) du 30-IX au 1.IX. 76; Riv. Mare aux Cochons et Cascade (Ouest) à 300 m d'altitude: 65 larves et nymphes, à 50 m d'altitude; Riv. Grand-Anse sous la Cascade aval de Salazie: 1 adulte ♀ et 1 larve; Riv. Grand-Bois 27.X.76, 4 larves; Riv. Cascade (Côté E) région des sources (450-500 m) 7.X.76: 4 larves; Riv. du Cap à la route de l'Anse Boileau 19.X.76: 3 larves et 2 adultes le 10.X.76.

Mission STARMÜHLNER: Aff. 7b: Riv. Rochon, 6.II.74: 8 larves; Sta 22 N.W. Küste: 18.II.74: 21 larves; Sta. 25c Cascade Riv. (W), 19.II.74: 1 fourreau.

4. *Helicopsyche kantilali* MARLIER & MALICKY.

Larve (avant dernier stade)

Dimensions: longueur 3,5 mm

Tête à face antérieure très aplatie et même concave, cet aplatissement limité par une carène bien nette passant devant les yeux, qui sont très latéraux et à mi-distance entre la base des mandibules et le vertex.

Sclérites céphaliques jaune clair sans macules grises ou brunes, face inférieure presque blanche, sans gula différenciée mais avec une prégula jaunâtre en demi-lune; autres sclérites jaunâtres; pronotum rembruni dans la moitié postérieure et à bord postérieur moins irrégulier que chez *H. palpalis* (ULMER). Mésonotum en deux parties en trapèze avec une suture médiodorsale, un peu plus sombre en avant qu'en arrière; à bord postérieur arrondi régulièrement, le petit sclérite médio-postérieur minuscule, allongé, à bords parallèles valant 1/6 de la longueur du mésonotum.

Métanotum à sclérites latéraux externes très petits, à sclérite médian grand et coloré tant en avant qu'en arrière.

Sclérites couvrant les tubercules du 1er segment abdominal hyalins portant les mêmes épines doubles que chez les autres *Helicopsyche*, les plus longues de ces épines atteignant 27 μ. Une soie princeau à la face interne du fémur II ainsi que sur la coxa, le trochanter et le fémur de la patte III.

Ligne latérale d'épines semblable à l'espèce précédente. Pas de branchies visibles à l'état fixé.
Griffes des crochets anaux avec 12 épines dorsales.

Fourreau
Le fourreau de *H. kantilali* est très différent de celui de *H. palpalis* ULMER. Sa forme est haute, hélicoïde, à sommet un peu asymétrique, construit en sable assez grossier mais avec le dernier tour particulièrement hérissé de grains très gros surtout du quartz, faisant fortement saillie à la surface.
La hauteur maximum de la 'coquille' rapportée au diamètre maximum (longitudinal) de la base varie de 0,75 à 0,91. Il y a 2,5 tours de spire.
Le fourreau nymphal est obturé à la partie antérieure par une membrane ovale, étirée parallèlement à la 'columelle' de la 'coquille' et présentant un orifice grillagé très excentrique comportant une trentaine de mailles.

Nymphe
Antennes de 31 articles (♀) ou 35 articles (♂). Premier article moins renflé que chez *H. palpalis*, clypéus granuleux, labre arrondi, un peu prolongé en bec au milieu, avec 5 paires de longues soies disposées comme dans l'espèce précédente. Mandibules semblables à celles de *H. palpalis* mais à serrulation pratiquement effacée.

Palpes maxillaires et labiaux très courts, les premiers divergents, 2, 24 fois plus longs que les mandibules.

Appareil d'accrochage semblable à celui de *H. palpalis* mais à crochets un peu plus nombreux et plus forts, les présegmentaux au nombre de 3-4, 3-4, 3-4 et 4 les postsegmentaux au nombre de 4.

Face dorsale des segments abdominaux à spinules groupées en petits peignes; ceux-ci très faibles mais présents également sur le tergite VIII.

Appendices terminaux à bord interne plus fortement boursoufflé par l'embase de la soie médiane, qui est unique et celle de l'unique soie subterminale.
Stations de récolte: Rivière Grand-St. Louis: 28 larves et 17 adultes; Riv. Grand-Anse à Salazie: 1 adulte et 11 larves; Riv. du Cap (route Anse Boileau) 7 adultes.
Mission STARMÜHLNER: Sta 22b N.W. Küste: 17 larves, Sta 23 b: Desert Riv. 19-2-74: 63 larves.

Famille: Odontoceridae
5. *Hughscottiella auricapilla* ULMER
Aucun matériel nymphal ni adulte n'a été recueilli par F. STARMÜHLNER ni par nous-même, l'attribution spécifique doit donc être considérée comme provisoire.

Larve
Larve campodéiforme de 11 mm de longueur (le plus grand exemplaire recueilli).

Tête très allongée, élargie vers les 3/4 de la longueur, convexe dorsalement et concave ventralement, à taches oculaires antérieures. Proportions de la tête: largeur du bord antérieur: 1; longueur totale, 3,5; largeur maximum 1,4. Face dorsale de la tête sans suture épicraniale ni clypéo-frontale, face ventrale sans suture gulaire, échancrée en triangle par l'orifice occipital; zone prégulaire un peu épaissie et plus colorée.

Antennes uniarticulées, insérées presque à l'aplomb de la base des mandibules. Mandibules fortes à deux échancrures très courbées; la gauche à tranchant dorsal portant 2 dents internes médianes inégales et à tranchant ventral très concave, portant outre la dent terminale une autre, très distale; mandibule droite armée de 4 dents apicales; les 2 mandibules dépourvues de brosses.

Labre deux fois plus large que long non échancré mais un peu crénelé au bord antérieur qui porte 4 soies; courbées, renflées; juste en arrière, dans chacun des angles, 1 groupe de 3 soies; tormae petits.

Un grand sclérite submental impair rectangulaire flanqué latéralement par les 'cardos' triangulaires et très larges; palpes maxillaires longs, coniques, de 4 articles dépassant fortement le lobe labial médian.

Tous les sclérites céphaliques comme thoraciques à chaetotaxie extrêmement rare et fine. Soie supraoculaire 9 (NIELSEN, 1942) très mince, soie 12 courte; groupe de soies 15, 16 et 17, très peu développé, situé en arrière de l'oeil.

Prothorax modifié de manière extraordinaire, il semble formé de 3 anneaux successifs, le premier portant la paire de pattes antérieures; les deux suivants recouverts chacun d'un notum qui les enveloppe, ne laissant qu'un espace membraneux à la face ventrale; ces trois anneaux suivis d'un 'intersegment' membraneux aussi long que l'ensemble du pronotum parcouru par 8 rubans sclérifiés longitudinaux puissants.

Tout le prothorax, à l'exception de l'extrême bord antérieur peut se rétracter dans le mésothorax, les pattes antérieures étant à peine visibles; premier arceau du pronotum sans suture longitudinale, deuxième et troisième divisés en deux 'valves' latérales; angles postérieurs de chacun des anneaux non prolongés vers la ligne médioventrale; prosternum représenté par un sclérite pentagonal à sommet antérieur, séparant les bases des coxas.

Propleures à trochantin petit, aigu et relevé vers l'avant non soudé à l'épisternum; celui-ci très petit, soudé à l'épimère et partiellement caché sous le bord du pronotum; épimère un peu plus large, rectangulaire.

Mésonotum offrant au quart de sa longueur une suture séparant un anneau antérieur tout à fait sclérifié, mobile sur la partie postérieure.

Anneau antérieur à bord oral prolongé vers l'avant et la face ventrale, divisé au milieu par une suture longitudinale, orné de 6 paires de soies raides de chaque côté au bord antérieur; un peu en arrière 2 longues soies obliques vers la ligne médiane et encore plus en arrière près du bord latéral, un groupe de 3 soies.

Partie postérieure du mésonotum plate au-dessus et convexe latéralement, divisée par une suture longitudinale médiane, angles antérieurs arrondis, séparés par une suture du reste du sclérite, cette zone arrondie ornée d'un bouquet de 7 soies raides, bord latéral longé par un bourrelet prolongeant la suture de l'angle antérieur jusqu'au bord postérieur, face ventrale occupée par un sclérite portant à la face interne la furca, très sclérifiée, de couleur foncée.

Métanotum occupé par un grand sclérite à bord antérieur droit, à bord postérieur convexe, angle antérieur portant 4 longues soies raides; hors de ce sclérite, le segment porte dans ses angles antérieurs de chaque côté un groupe de 9 soies, en arrière 2 groupes de 3 soies alignées transversalement, face ventrale portant vers l'arrière une rangée transversale de 16 soies dirigées vers l'avant.

Premier segment abdominal sans gibbosités mais offrant au milieu de la face dorsale un sclérite en demi-lune convexe vers l'arrière; sur les côtes, à l'emplacement des bosses latérales, un sclérite cordiforme à pointe antérieure; en avant de ce sclérite, face ventrale portant une rangée transversale de 10 soies moins épaisses que les métasternales, dirigées vers l'avant. Les sclérites latéraux sont hérissés d'un feutrage de spinules à doubles pointes dirigées vers l'avant affaiblies en petites soies au centre. Sur le sclérite dorsal les spinules sont plus longues et recourbées en crochets à leur extrémité.

Une ligne latérale de poils très fins, courts, s'étend du milieu du segment III au bout du VIII.

Tergite VIII sans sclérite mais avec 2 longues soies, tergite IX sans sclérite, avec 3 fortes soies de chaque côté et 1 ou 2 plus petites.

Pygopodes courts à segment b sclérifié dorsalement et griffe terminale armée d'une forte dent dorsale.

Branchies isolées, insérées en 6 rangées au bord antérieur des segments II à VIII suivant le schéma suivant:

	Dorsales	Latérales	Ventrales
II	I	I	I
III	I	I	I
IV	I	I	I
V	I	I	I
VI	I	—	I
VII	I	—	I
VIII	I	—	I

Fourreau

Fourreau de grains de sable grossier, de couleur souvent claire, alignés en ceintures successives plus ou moins régulières; de forme conique, courbé à orifice antérieur droit, orifice postérieur légèrement refermé au bord dorsal par des grains de sable laissant une fente ventrale concave vers le haut à angles très arrondis. Longueur maximum observée Sta. 25: 15 mm; diamètre antérieur 4,3 mm; diamètre postérieur: 2 mm. Stations de récolte: Riv. Grand St. Louis: 51 larves; Riv. Islette; Plateau Mare aux Cochons (et Riv. Cascade W.): 17 larves, Riv. Cap: 1 larve; Riv. Gd. Anse: 1 fourreau vide.

Mission STARMÜHLNER: Affl. Riv. Grand Bois à Casse Dent, 6.II.74 (480 m): 4 larves.

6. *Leptodermatopteryx tenuis* ULMER
L'attribution des deux autres larves à fourreau non hélicoïde présentes dans la collection STARMÜHLNER, et dont nous n'avons vu qu'un exemplaire, est aussi peu certaine que la précédente. Aucun matériel nymphal ni adulte n'a été recueilli.

Larve
Longueur: 4,35 mm
Larve de type éruciforme, ayant la tête et le pronotum seuls sclérifiés.
Coloration jaune d'ambre sans taches, une ligne noire horizontale sur les coxas II et III et une autre, perpendiculaire à la première sur la suture épisternum-épimère; bord externe du pronotum très étroitement rembruni; un trait noir-épais sur le sclérite *b* des appendices anaux.

Tête ronde, assez aplatie antérieurement, avec une carène marquée entre la base des mandibules et le bord supérieur de l'oeil, celui-ci arrondi, situé peu avant le milieu de la tête, entouré d'une zone plus pâle, face inférieure de la tête avec une lunule antérieure plus brune. Orifice occipital grand et oblique.

Pronotum à bord antérieur un peu concave à angles antérieurs étirés en deux pointes qui prolongent une carène oblique rejoignant en arrière le centre de chaque demi-sclérite; bords latéraux convexes, bord postérieur mal délimité à angles éffacés. Pas de corne prosternale.

Mésonotum membraneux portant une rangée transversale de soies dirigées vers l'avant, le reste du segment à pilosité fine et éparse.

Métanotum membraneux, avec la même rangée transversale de soies. Propleure large, à trochantin trapézoidal très obtus en avant.

Patte antérieure courte, à coxa conique à crête antérieure tranchante; fémur court et large, à arêtes dorsale et ventrale parallèles presque jusqu'à l'extrémité puis rétréci assez brusquement, formant pince avec l'arête ventrale du tibia; celui-ci court, élargi au bout, terminé par un fort éperon interne, griffe longue et forte, à courte épine basale, arête ventrale du fémur hérissée de soies, arête dorsale et face externe également couvertes de soies.

Patte intermédiaire plus longue, mince, à tibia dépourvu d'éperon terminal; à longue griffe portant un fort éperon empodial. Patte postérieure encore plus longue à longues soies dorsales sur le tibia et le tarse, à longue griffe courbée porteuse d'une soie basale aiguë.
Pas de branchies visibles.

Premier segment abdominal à protubérance dorsale effacée, les latérales aplaties couvertes d'un sclérite hyalin allongé armé de spinules.
Ligne latérale très basse.

Neuvième segment sans sclérite coloré avec de fines soies, 2 soies latérodorsales assez antérieures, un groupe de 3 petites, latérales, une rangée postérieure transversale de 14 soies.
Appendices anaux avec le segment *b* portant 4 fortes soies postérieures mais pas de sclérite; sclérite du segment *c* triangulaire, jaune très pâle avec une bande noire

dorsale; base de la griffe assez renflée, incolore, incomplètement divisée par une fente ventrale étroite; griffe proprement dite brune, aiguë, recourbée en angle aigu, porteuse d'un crochet dorsal fort et pointu; 2 soies sur le sclérite basal, très fines, une très longue soie épaisse insérée au bord dorsal de ce sclérite, dépassant largement la griffe en arrière, 3 fois aussi longue que l'appendice; à la face interne de la griffe, un peigne de 6 fortes soies noires très longues, dirigées vers l'arrière et le bas, un peu distal par rapport à 2 soies noires situées dans le même alignement.

Fourreau

Le fourreau de sable presque lisse, courbé, conique, à orifice postérieur rétréci par une membrane avec un trou circulaire central; longueur 3,5 mm; diamètre antérieur: 0,97 mm; diamètre postérieur: 0,52 mm.

Stations de capture: les deux exemplaires capturés par la mission F. Starmühlner l'ont été à Mahé à la station 4b le 6.II.74, affluent Riv. Grand Bois près de Casse Dent.

Remerciements

Je dois remercier ici le Professeur F. STARMÜHLNER qui m'a autorisé à étudier le riche matériel recueilli par lui et le Dr. H. MALICKY qui a bien voulu s'en dessaisir pour me le confier. Les illustrations de la larve de *Hughscottiella* sont l'oeuvre de l'artiste Carlo CHAPELLE. Les autres illustrations sont dues pour une part à mon épouse, pour la seconde à Madame G. STROOBANTS. A toutes ces personnes, j'adresse mes vifs remerciements.

Bibliographie

BOTOSANEANU, L. 1956. Le développement postembryonnaire, la biologie et la position systématique d'un des Trichoptères les plus intéressants de la faune européenne: *Helicopsyche bacescui* ORGHIDAN et BOTOSANEANU Vestn. Ceskosl. Zool. Spol. 20: 285-312.

NIELSEN, A. 1942. Uber die Entwicklung und Biologie der Trichopteren mit besonderer Berücksichtigung der Quelltrichopteren Himmerlands. Arch. Hydrobiol. Suppl. XVII: 255-621.

ULMER, G. 1910. Scientific Results of the Percy-Sladen Trust Expedition 1905. III Trichoptera. Trans. Linn Soc. 14: 41-54.

——. 1955. Köcherfliegen von der Sunda-Inseln, Teil II. Arch. Hydrobiol. Suppl. XXI: 408-608.

Discussion

WINTERBOURN: How does the probable *Leptodermatopteryx* feed and what does it eat?

MARLIER: In fact this seems most probably to be the larva of *Hughscottiella*. It is carnivorous and seems to feed on chironomids as well as on other caddis larvae.

MORETTI: Moi, je voudrais savoir si vous avez fait des observations d'anatomie musculaire sur le curieux thorax et sur une possible presence de particuliers organes sensoriels en correspondance de cette zone du corps très modifiée.

MARLIER: Non, mais Dr BARLET (Liège) a étudié la musculature du prothorax et a pu démontrer la nature prothoracique des trois anneaux antérieurs et la nature intersegmentale du tube rétractile, le mésothorax ne commençant qu'en arrière de ce dernier.

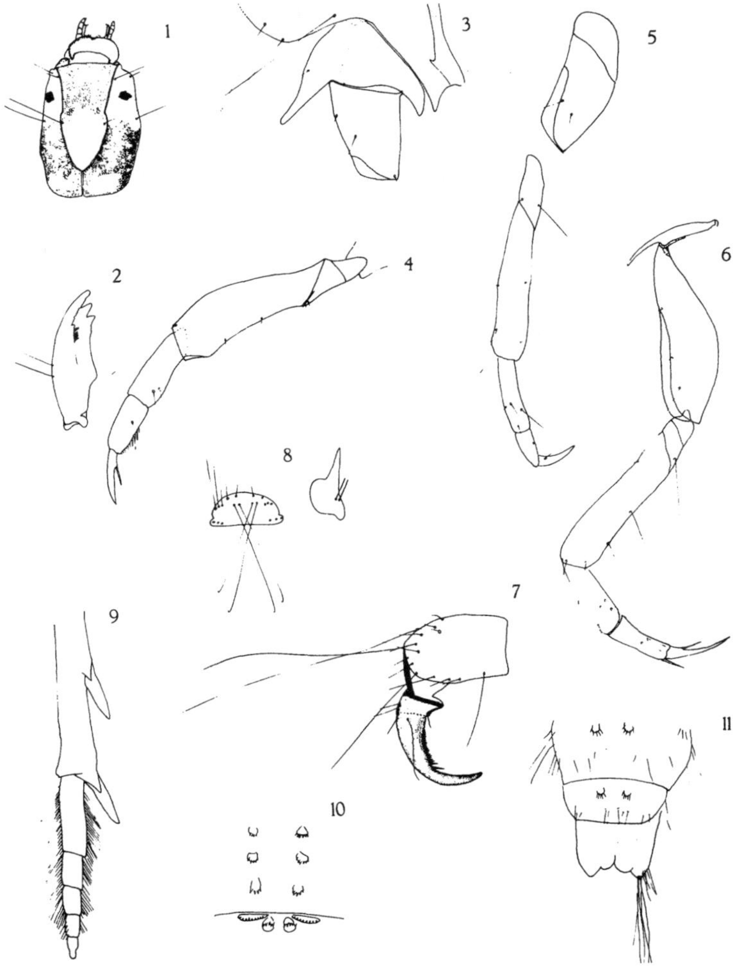

Pl. I: *Ecnomus insularis* ULMER, larve (1-7) et nymphe (8-11); 1. tête, 2. mandibule gauche, 3. pleure et coxa antérieure, 4. patte antérieure, 5. patte intermédiaire, 6. patte postérieure, 7. griffe des pygopodes, 8. labre et mandibule, 9. patte intermédiaire, 10. appareil d'accrochage III-VI, 11. extrémité de l'abdomen à partir du septième segment.

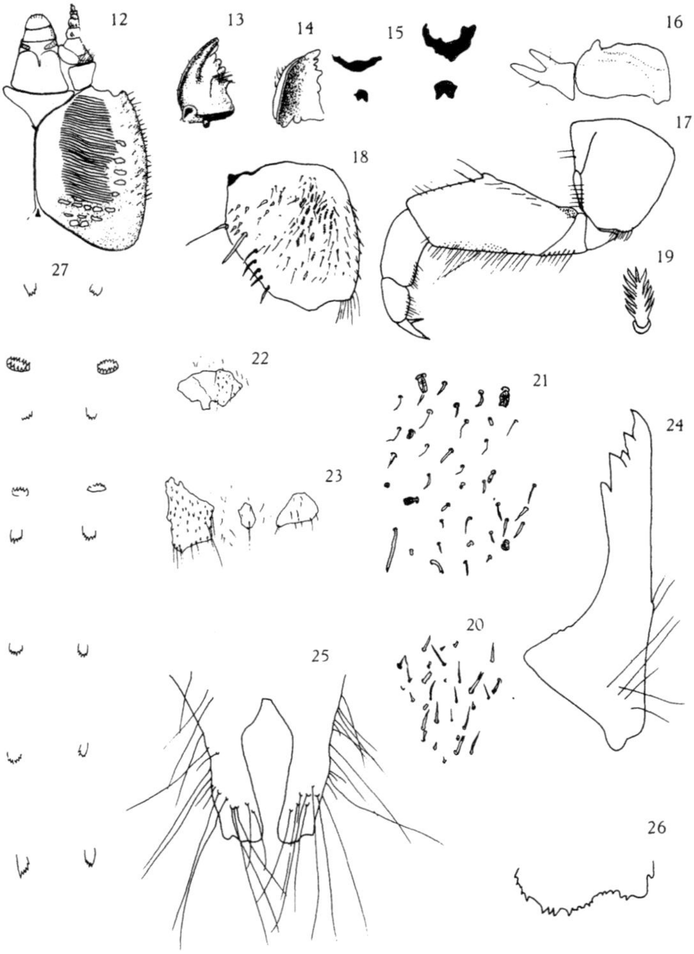

Pl. II: *Hydromanicus seychellensis* ULMER, larve (12-23) et nymphe (24-26); 12. face ventrale de la tête, 13. mandibule gauche (dorsal), 14. mandibule droite (ventral), 15. taches thoraciques: variations, en haut: mésonotum en bas: métanotum, 16. pleure antérieure, 17. patte antérieure, 18. coxa antérieure, 19. 1 soie plumeuse de la coxa, 20. revêtement du sclérite métathoracique, 21. revêtement de l'abdomen (l'avant de la larve est dirigé vers le haut dans les deux figures), 22. sclérite ventral du segment VIII, 23. de gauche à droite sclérites ventral, latéral et dorsal du segment IX, 24. mandibule, 25. appendices terminaux, 26. extrémité des appendices terminaux, 27. appareil d'accrochage.

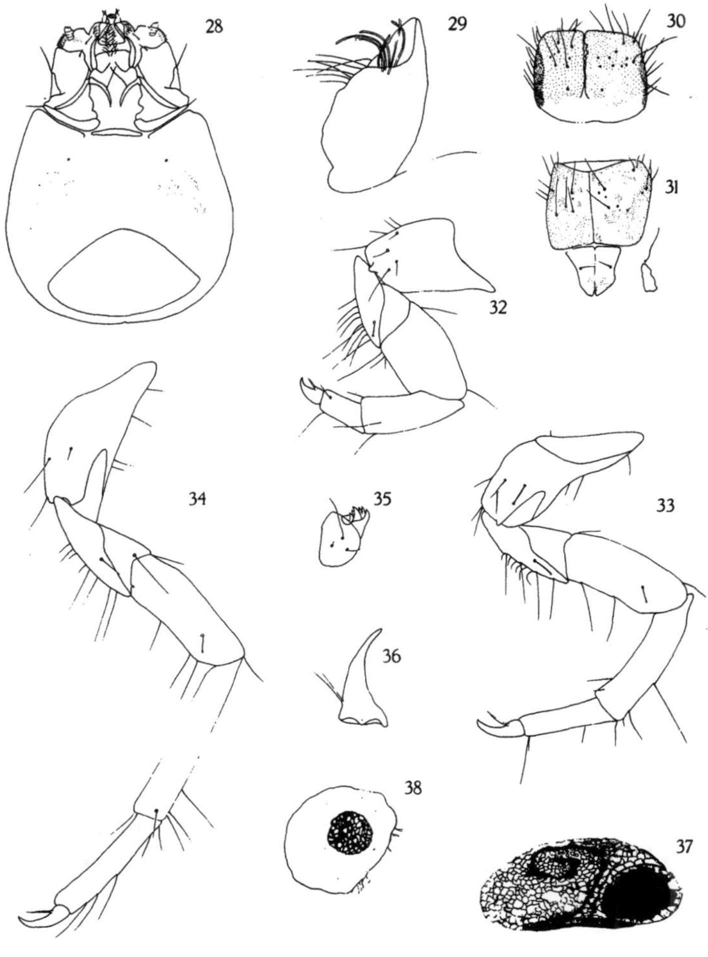

Pl. III: *Helicopsyche palpalis* (ULMER) larve 28-35, nymphe 36-38; 28. tête, face ventrale, 29. mandibule, 30. pronotum, 31. méso et métanotum, 32. patte antérieure, 33 et 34. pattes intermédiaire et postérieure, 35. griffe des pygopodes, 36. mandibule, 37. fourreau nymphal, 38. membrane du fourreau.

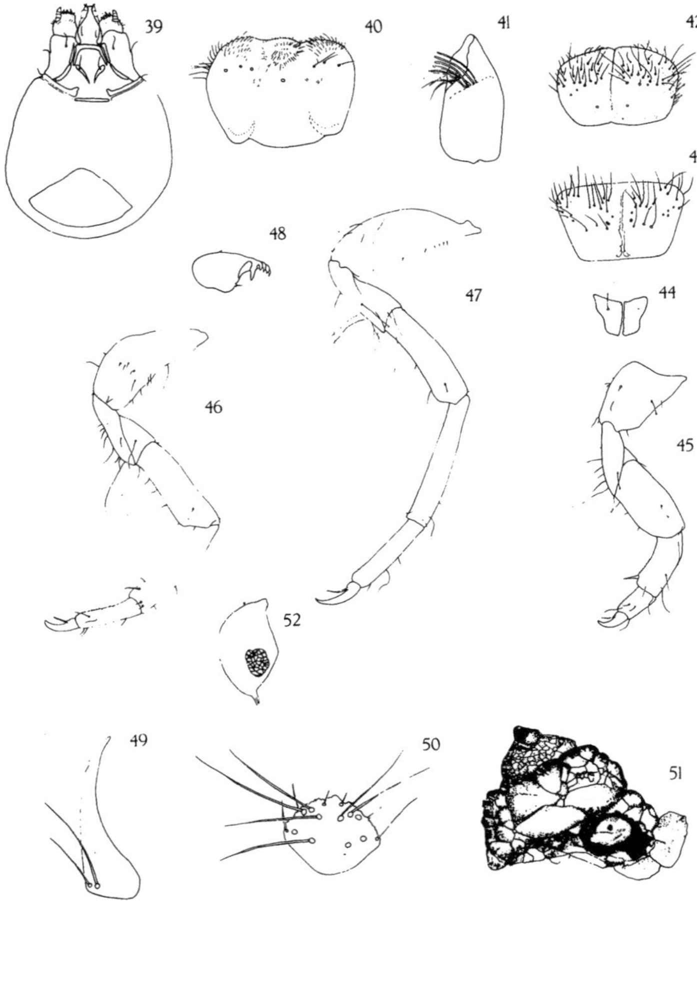

Pl. IV: *Helicopsyche kantilali* MARLIER & MALICKY larve (39-48), nymphe (49-50) et fourreau (51-52); 39. tête, face ventrale, 40. labre, 41. mandibule, 42, 43, 44. pro-méso-métanotum, 45, 46, 47. pattes antérieure, intermédiaire et postérieure, 48. griffe des pygopodes, 49. mandibule, 50. labre, 51. fourreau larvaire, 52. opercule du fourreau nymphal.

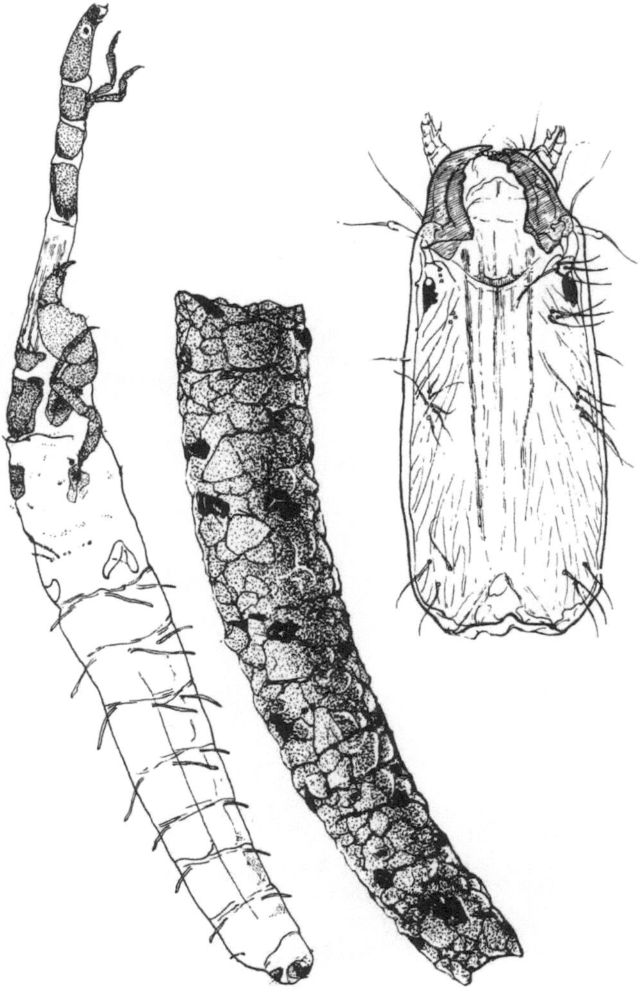

Pl. V: *Hughscottiella auricapilla* ULMER (?) larve, fourreau et tête.

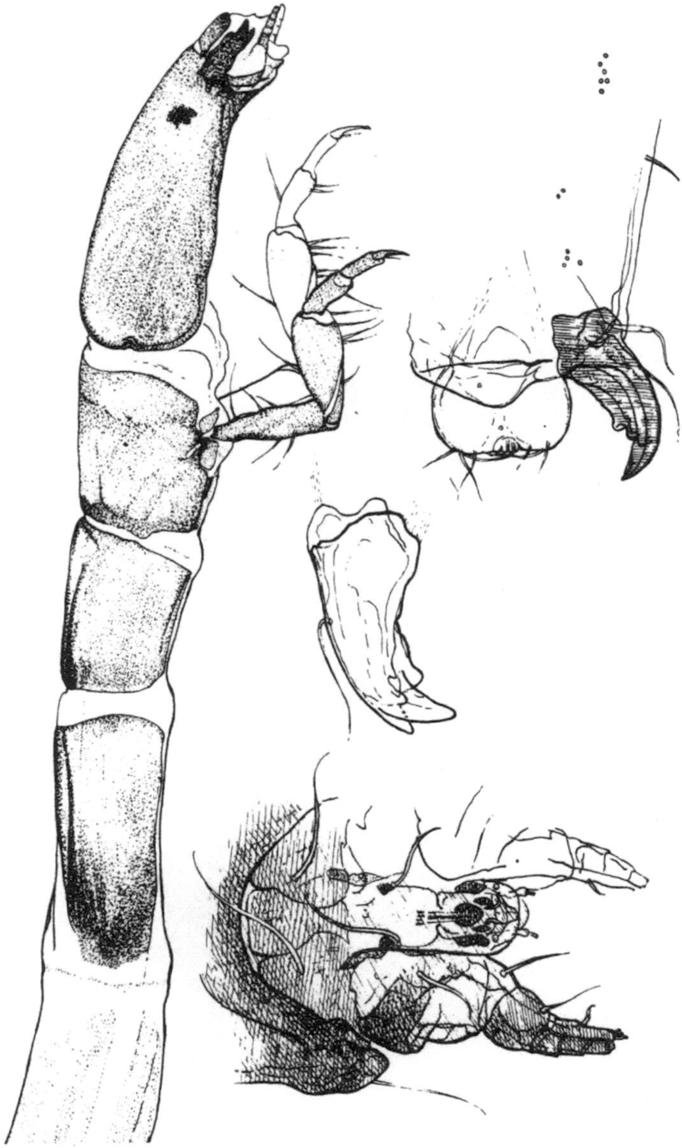

Pl. VI: *Hughscottiella auricapilla* ULMER tête et prothorax; mandibules; maxillo-labium.

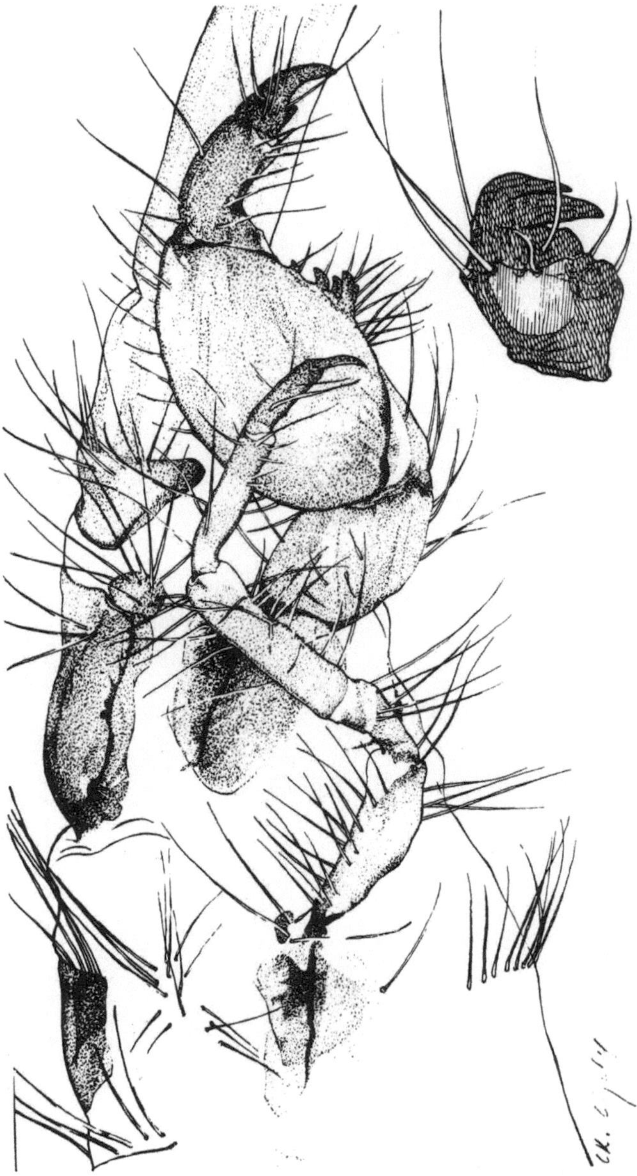

Pl. VII: *Hughscottiella auricapilla*, larve méso- et métathorax, griffe des pygopodes.

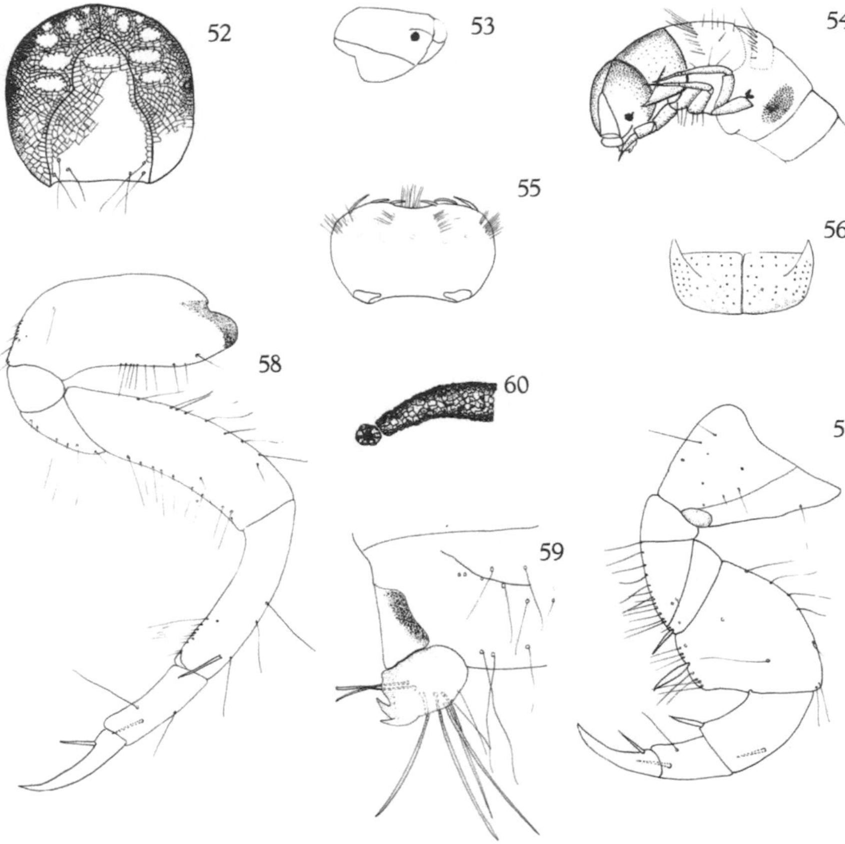

Pl. VIII: *Leptodermatopteryx tenuis* ULMER (?) larve; 52. tête, vue dorsale, 53. tête, vue latérale, 54. partie antérieure du corps, 55. labre (ventral), 56. pronotum, 57. patte antérieure, 58. patte intermédiaire, 59. extrémité de l'abdomen et pygopode gauche, 60. fourreau.

Proc. of the 2nd Int. Symp. on Trichoptera, 1977, Junk, The Hague

The food and occurrence of larval Rhyacophilidae and Polycentropodidae in two New Zealand rivers

M.J. WINTERBOURN

Abstract

The food of late instar larvae of five species of Rhyacophilidae and one species of Polycentropodidae was investigated in two lowland rivers in New Zealand. Some life history information was also obtained. Final instar larvae of all species were present in most months although those of the three most common rhyacophilids were most abundant in late spring and early summer. Pupae were also taken in many months. Larvae of all species were primarily carnivorous although a considerable amount of detritus had been ingested by some individuals. Prey species composition tended to reflect prey abundance and availability with a predominance of chironomid larvae (Orthocladiinae) taken in one river and chironomid larvae and mayfly nymphs in the other. Oligochaetes were important foods of *Psilochorema bidens* and *Polyplectropus puerilis* but were not seen in the guts of *Hydrobiosis* species. Gut contents of *Neurochorema confusum* were predominantly arthropod fragments which may have been parts of trichopteran pupae. The conditions which allow two ecologically similar species of *Hydrobiosis* to coexist are discussed.

Introduction

The Rhyacophilidae is the largest trichopteran family in New Zealand with 58 described species. All belong to the subfamily Hydrobiosinae and are included in nine genera (WISE, 1973). The final instar larvae of 20 species have been described by McFARLANE (1951) who also outlined the distribution of rhyacophilid larvae in streams and rivers forming part of the large Waimakariri River system on the eastern side of the Southern Alps (McFARLANE, 1938). More recently, CROSBY (1975) has investigated the food of *Hydrobiosis parumbripennis* McFARLANE larvae in a coastal stream in the South Island and MICHAELIS (1974) obtained some information on the life history and habitat of *Psilochorema tautoru* McFARLANE in a cold stenothermal spring. Apart from these studies and some brief notes on occurrence and natural history in more general works, nothing has been published on the biology of the New Zealand Rhyacophilidae.

The Polycentropodidae are represented in New Zealand by two genera, *Polyplectropus* and *Plectrocnemia*. Six species of *Polyplectropus* have been described, of which *P. puerilis* (McLACHLAN) is probably the most common. Its larva has been described by COWLEY (1975), and MICHAELIS (1974) considered its life history briefly in her study of a spring. No studies have been made on the food or feeding of New Zealand polycentropodids.

The aim of the present study was to investigate the species assemblages, foods and temporal occurrence of larval and pupal Rhyacophilidae and Polycentropodidae in two contrasting lowland rivers.

Study areas

Both study areas were sited on rivers which flow eastwards across the Canterbury Plains in the South Island of New Zealand and derive much of their flow from subterranean aquifers. The Kaiapoi River is a tributary of the lower Waimakariri River and arises from springs several kilometres northwest of the city of Christchurch. Above the sampling site it flows through farm land and a fish hatchery is present near its source. The sampling site was a riffle 4-6 m wide and varying in depth from 0-60 cm. Mean current speed in midstream during the study was about 30 cm/sec. The river bed consisted of compacted gravel, sand and silt with occasional larger stones resting on the surface.

The Selwyn is a major river arising in the foothills of the Southern Alps and emptying into coastal Lake Ellesmere about 30 km south of Christchurch. Field work was carried out at Chamberlain's Ford where the river flows through arable and pastoral farmland. Here the river bed is up to 50 m wide in places but except during times of severe flooding, flow is restricted to a narrow channel. At the sampling site, the channel was about 5 m wide and up to 25 cm deep with a mean flow of about 1 m/sec. The river bed was composed of rounded pebbles, gravel and some larger stones packed loosely to a depth of 10-15 cm above a compacted base. Little sand or silt was present. Further details of the site and the invertebrate fauna occurring there are given by WINTERBOURN (1974).

Water temperature measured above the substratum between 0900 and 1100 h on each sampling day is shown in Fig. 1. An annual temperature range of 7.5°C was recorded in the Kaiapoi River whereas it was 11.5°C in the Selwyn where both upper and lower temperatures were more extreme.

Methods

Benthic samples were taken from both rivers over a period of 13 months. Five samples were taken each month from the Kaiapoi with a 0.1 m^2 Surber sampler (mesh size 0.5 mm) by disturbing bed materials to a depth of about 5 cm. Three of the samples were taken along a transect across the river in water less than 10 cm deep and the other two were taken in water 30-40 cm deep at the head of the riffle where larger stones were found. Additional non-quantitative collections of fauna were made with a 0.5 mm mesh hand net. Six Surber samples were obtained from the Selwyn River each month. They were taken along a diagonal transect across the stream to a depth of 10-15 cm, i.e. the bottom of the layer of loose stones. All

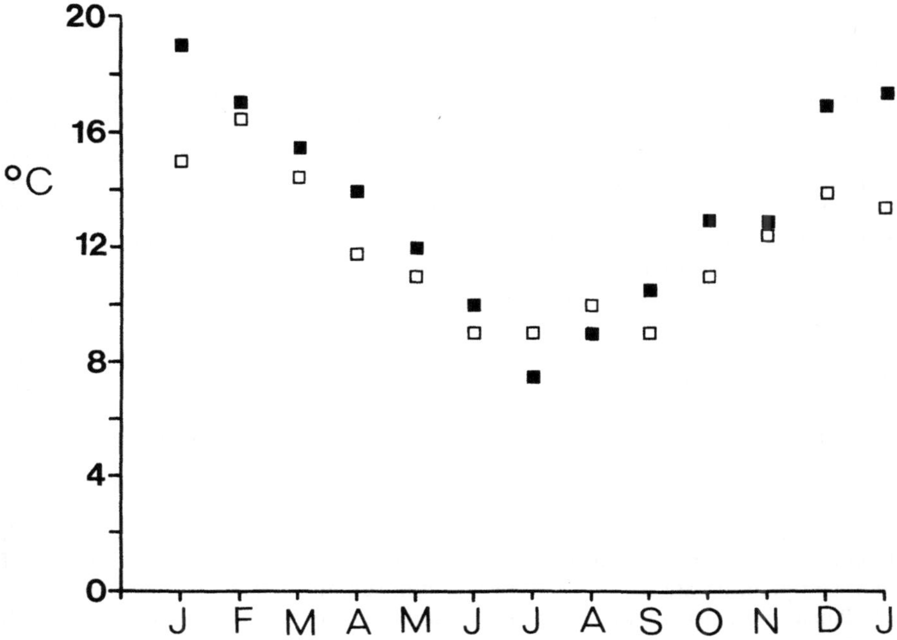

Fig. 1. Water temperature recorded in the Selwyn (■) and Kaiapoi (□) Rivers between 0900 and 1100 h on each sampling day.

collections were preserved immediately in 10% formalin.

Late instar rhyacophilid larvae were identified from the descriptions given by McFARLANE (1951) but early instars of *Hydrobiosis* species could not always be identified since their head and pronotal markings are not well developed. Pupae were identified from the descriptions of genitalia in MOSELY and KIMMINS (1953).

Head widths of all larvae collected were measured across the eyes with a linear eyepiece graticule at X 25 magnification. Gut contents of final and penultimate instar larvae were dissected from larvae, teased out on microscope slides and mounted in lactophenol — PVA containing the stain Lignin Pink. Slides were examined under a microscope and all animal fragments were identified as accurately as possible. The presence of detritus, sediment and algae was noted. As the number of larvae of each species collected each month varied considerably and was often small, most individuals used for gut analysis were obtained when larvae were most abundant. Therefore, no information on seasonal changes in food taken is provided.

Results

The fauna

Four species of Rhyacophilidae and one polycentropodid occurred in both rivers and a fifth rhyacophilid, *Hydrobiosis clavigera* McFARLANE was found only in the Selwyn (Table 1). *Psilochorema bidens* McFARLANE was the most abundant species in both rivers while *Hydrobiosis* was the only genus represented by more than one species. Most larvae taken were final or penultimate instars and with the exception of two second instar *Polyplectropus puerilis*, no first or second instar individuals were seen. It is possible that small larvae occur mainly in non-riffle habitats or they may simply have been missed by the 0.5 mm mesh net.

Head widths of late instar larvae of all species are shown in Fig. 2. No intraspecific differences were found between rivers. Mean head widths of final instar rhyacophilids ranged from 0.52 to 1.44 mm. Two of the congeneric species were similar in size but the third, *Hydrobiosis clavigera* was much smaller. *P. puerilis* was larger than the rhyacophilids and possessed the greatest size range within instars.

Temporal occurrence of late larvae and pupae.

Final instar larvae of *Psilochorema bidens* were found in both rivers in all months. In the Kaiapoi River where the population was larger, peak numbers of final and penultimate instar larvae occurred in summer (December-February) (Fig. 3). Pupae were found in either the Selwyn or the Kaiapoi in all months with maximum numbers in summer.

Final instar larvae of *Neurochorema confusum* (McLACHLAN) were taken from the Kaiapoi River from August to February with largest numbers (17/m^2) recorded in January. Although fewer larvae were collected from the Selwyn ($<7/m^2$), final instars were seen in all months except April and October. Pupae occurred in all seasons of the year.

Table 1. The species of Rhyacophilidae and Polycentropodidae taken from the Kaiapoi and Selwyn Rivers and the relative abundance of their final instar larvae in Surber samples during the study.

	Percentage	
	Kaiapoi	Selwyn
RHYACOPHILIDAE		
Hydrobiosis parumbripennis	15.9	16.2
H. umbripennis	13.6	6.8
H. clavigera	absent	1.4
Psilochorema bidens	54.3	45.9
Neurochorema confusum	7.7	24.3
POLYCENTROPODIDAE		
Polyplectropus puerilis	8.5	5.4

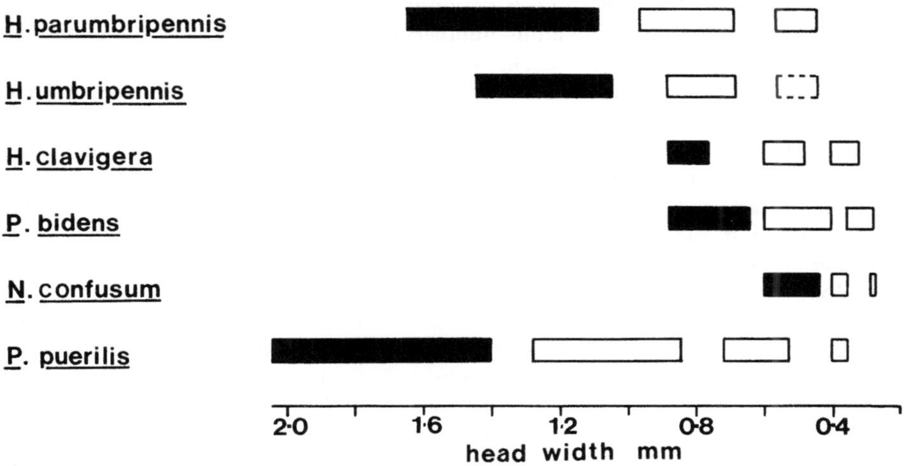

Fig. 2. The range of head capsule widths found in the last three instars of five rhyacophilid species and in the last four instars of *Polyplectropus puerilis*. The final instar is shaded.

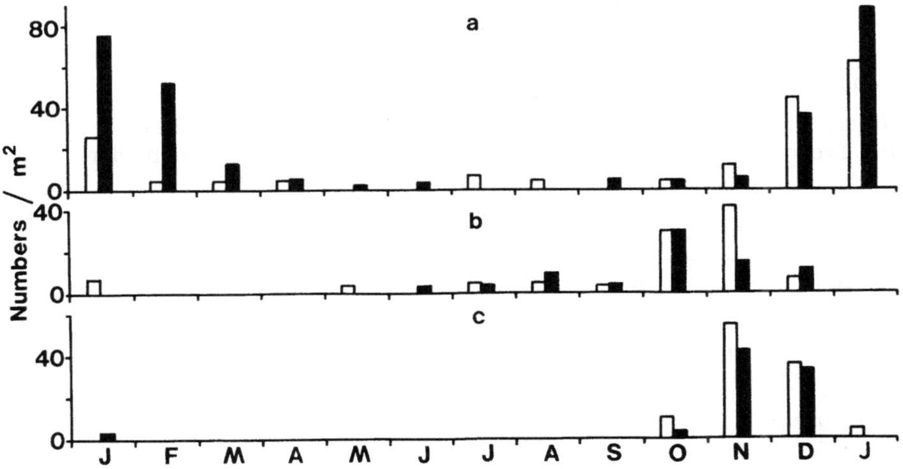

Fig. 3. Numbers of final and penultimate instar larvae of three species of Rhyacophilidae taken in Surber samples from the Kaiapoi River each month. Open bars, instar F-1; shaded bars, instar F. a, *Psilochorema bidens*; b, *Hydrobiosis umbripennis*; c, *H. parumbripennis*.

Some final instars of *Hydrobiosis parumbripennis* occurred in the Kaiapoi River from June to January with maximum numbers in late spring and early summer (Fig. 3). Before October they were taken only in non-quantitative net samples which explains their absence prior to that month in Fig. 3. Pupae were taken from August to January. In contrast, most final instar larvae ($12/m^2$) and pupae were found in the Selwyn River in late autumn and winter (May-July) although some final instar larvae were taken in most months.

Final instars and pupae of *Hydrobiosis umbripennis* were obtained from the Kaiapoi River in the same months as those of *H. parumbripennis*. No pupae were collected from the Selwyn River but small numbers ($<8/m^2$) of final instar larvae were found in April, May, June, December and January.

Hydrobiosis clavigera was not found in the Kaiapoi River and was the least common rhyacophilid in the Selwyn where final instar larvae were taken only in December and January. In a third Canterbury river, the Glentui, where *H. clavigera* is more common, final instar larvae have been found in all months except August and November and pupae are known to occur in summer and winter (author's unpublished records). It seems likely therefore, that more intensive collecting would have revealed final instar larvae in the Selwyn in more months especially as pupae were found in August as well as January and February.

Polyplectropus puerilis was rarely abundant on riffles in either river and no pupae were found. They may have been more abundant in slower water which D.R. COWLEY (pers. comm.) considers is their most usual habitat. Some final instar larvae were collected from riffles in all months except September and October with maximum numbers ($30/m^2$) in the Kaiapoi River in March.

Food of larvae

Results of gut content analyses made on final and penultimate instar larvae from the two rivers are summarized in Tables 2 and 3. All species were predominantly

Table 2. Numbers of caddis larvae from the Kaiapoi River containing different food items. Data from all months and instars F and F-1 are combined.

Species	n	detritus	Chironomidae	Oligochaeta	Others	sediment	empty
H. parumbripennis	43	0	30	0	9*	0	10
H. umbripennis	56	0	47	0	5**	0	9
P. bidens	51	15	27	10	28***	0	0
N. confusum	12	9	1	0	3	0	0
P. puerilis	40	6	12	14	21****	5	0

* includes *Deleatidium, Hydrobiosis* and Amphipoda
** all unidentified arthropod fragments
*** includes *Deleatidium*, Copepoda, *Potamopyrgus*
**** includes *Deleatidium*, Copepoda, Acari

Table 3. Numbers of caddis larvae from the Selwyn River containing different food items. Data from all months and instars F and F-1 are combined.

Species	n	detritus	*Deleatidium*	Chironomidae	Trichoptera	Acari	Unidentified Arthropoda
H. parumbripennis	34	20	7	6	5	0	6
H. umbripennis	20	10	2	12	1	0	4
P. bidens	33	20	6	6	5	3	9
N. confusum	18	5	0	3	0	0	11
P. puerilis	6	1	0	1	1	0	3

Table 4. Numbers of chironomid larvae found in guts of final instar caddis larvae which had eaten Chironomidae

Species	n	Numbers of chironomid larvae	
		mean	range
H. parumbripennis	17	5.4	1–14
H. umbripennis	37	7.6	1–40
P. bidens	17	1.6	1–4
P. puerilis	10	1.0	1–1

carnivorous although detritus occurred in the guts of some individuals particularly those from the Selwyn River. In the Kaiapoi River chironomid larvae (mostly Orthocladiinae but also a few Tanytarsini and Diamesinae) were numerically the dominant prey of all rhyacophilid species except *Neurochorema confusum*. They occurred in 53, 70 and 84% of the dissected *Psilochorema bidens, Hydrobiosis parumbripennis* and *H. umbripennis* respectively. Large numbers of chironomids had been eaten by some *Hydrobiosis* larvae (Table 4) and other taxa constituted only 2 and 5% of the total prey items in all *H. parumbripennis* and *H. umbripennis* respectively. In contrast, other prey made up about 50% of the total prey items in *P. bidens*. They included mayfly nymphs (*Deleatidium spp.*) and copepods; oligochaete chaetae were found in 20% of the *P. bidens* larvae and a radula of the gastropod, *Potamopyrgus antipodarum* was found in one individual.

In the Selwyn River, *Deleatidium* nymphs were more important and chironomid larvae less important prey of rhyacophilids than in the Kaiapoi. Nevertheless, mean numbers of chironomids in the guts of final instar caddis larvae were similar to those for the Kaiapoi River. Detritus had been ingested by numerous larvae of all species. As in the Kaiapoi River, *Psilochorema bidens* exhibited the most catholic diet which included a variety of aquatic insects and mites. All species contained some insect fragments which could not be identified accurately but may have included pieces of caddis larvae and pupae. In particular they were found in 61% of larvae of *Neurochorema confusum* which according to McFARLANE (1938) preys on pupae of *Aoteapsyche colonica* (Hydropsychidae) a species that was abundant in the study area.

Larvae of *Polyplectropus puerilis* build a loose retreat from silk, bits of stick and gravel tied together and attached to a stone. A roughly cone-shaped net extends upstream and outwards from the retreat and may reach up to 10 cm in diameter. According to D.R. COWLEY (pers. comm.) the silk threads of the net appear to be sticky and often collect large amounts of silt and other debris. Larvae from the Kaiapoi River were principally carnivores although some had ingested fine detrital particles and inorganic sediments. Oligochaetes appeared to be an important food as chaetae were seen in 35% of the larvae examined. Chironomid larvae occurred in less than one third of the *P. puerilis* dissected and never in numbers greater than one per larva. Other arthropods eaten included *Deleatidium* spp., copepods and a mite.

Discussion

With one exception the same species of Polycentropodidae and Rhyacophilidae were found in both rivers despite marked differences in the nature of the substrate and to a lesser extent water temperature, flow and potential prey. In common with many invertebrate predators, larval population densities were relatively low during much of the year. Because sampling was confined to riffles and was carried out with a 0.5 mm mesh net, detailed information on population size structure and life histories was not obtained. However, it was found that some final instar larvae of all species were present during much of the year and pupae occurred in many months. A similar situation has been described by MICHAELIS (1974) for *Polyplectropus puerilis* and *Psilochorema tautoru* in a cold, constant temperature spring and CROSBY (1975) found that because most larval instars of *Hydrobiosis parumbripennis* were present each month from March to November (the duration of his study) no distinct life history pattern could be discerned. Because first instar larvae occurred at all times in his study, it seems probable that eggs were hatching during much of the year.

The flight periods of Trichoptera in a forested stream in northern New Zealand have been investigated by NORRIE (1969) who obtained some rhyacophilids including species of *Hydrobiosis*, *Psilochorema* and *Neurochorema* in all months of the year. Of the rhyacophilid species considered in the present study, only *H. parumbripennis* and *N. confusum* were taken by Norrie, the former every month of the year except July and the latter in all months. *Polyplectropus puerilis* also occurred at his light trapping site and was taken in eight months while two other polycentropodids also had flight periods exceeding six months. Although flight periods and presumeably emergence periods may be long, the Kaiapoi River data on occurrence of late larvae and pupae of *P. bidens*, *H. umbripennis* and *H. parumbripennis* indicate that there are seasonal emergence peaks. Such peaks may be less pronounced at lower latitudes at least in some species, since Norrie took *N. confusum* in large and comparable numbers in five non-sequential months.

Polycentropodid larvae build silken tubes, bags, shelters or trap nets and are thought to be almost exclusively carnivorous (ANDERSON, 1976). *Polyplectropus puerilis* conforms with this description although it ingests fine particles of detritus and sediment as well as animal prey. Whether it utilizes detrital particles or their surface microflora as food is not known. The frequent occurrence of oligochaetes and sediments in *P. puerilis* guts suggests that its feeding nets are positioned beneath the surface of the river bed rather than on surface stones as in most New Zealand Hydropsychidae. The relative paucity of chironomids in the guts of predators supports this contention as the former were most numerous at the surface of the river bed.

The free living larvae of most rhyacophilids also are thought to be carnivorous although THUT (1969) and MECOM (1972) showed that some species feed on algae, living plant tissue and detritus and CROSBY (1975) found that the early instars of *Hydrobiosis parumbripennis* were detritivores whereas the last two instars were carnivores. Late instar rhyacophilid larvae in the Kaiapoi River were predominantly chironomid feeders whereas mayfly nymphs as well as chironomid larvae were important prey in the Selwyn. Since these were the most abundant invertebrate taxa in the two rivers it is probable that many caddis larvae were essentially opportunist feeders. Nevertheless, some differences in gut contents were found between species. Thus a wider spectrum of materials which included considerable amounts of detritus and oligochaetes was found in *Psilochorema bidens* than in the *Hydrobiosis* species. This suggests that *P. bidens* may feed over a greater depth range within the substratum than the *Hydrobiosis* species which appear to be essentially surface feeders. An apparently greater degree of food specialization was shown by *Neurochorema confusum* which according to McFARLANE (1938) feeds on the pupae of Hydropsychidae.

The findings of this study make it interesting to compare the ecological niches of the two common *Hydrobiosis* species in the light of the competitive exclusion principle. Larvae of both species occurred together with peak abundance of late instars in late spring and early summer, they attained almost the same size and utilized the same food supply. No clear differences in ecology have been demonstrated yet they appear to coexist successfully. This suggests that shared resources were not in short supply at the population densities found and that competitive interactions, if they occur, are not harmful. One mechanism which could be operating to reduce contacts between species and thus prevent competition by interference is the possession of different foraging times as found in lycosid spiders (ENDERS, 1976). However, a more important feature which may permit the occurrence of ecologically similar aquatic larvae is the nature of the life cycle with its non-aquatic reproductive stage. The number of eggs laid in a section of river each year, and subsequently the number of larvae which develop there may have little relation to the size of the larval population of the previous generation but may be influenced primarily by events which occur out of the water. Consequently, larval populations could be expected to fluctuate irregularly in size from year to year. If so, one might expect to find a continually shifting equilibrium between potentially competitive species which occur together in low numbers with neither being favoured long enough to establish dominance to the extent that the other is excluded. KERST & ANDERSON (1975) postulated that a situation of this kind might help account for the coexistence of congeneric plecopteran species in an Oregon stream. Equally it could apply to the *Hydrobiosis* species considered here especially as they both probably have long flight and hatching periods which permit the continual recruitment of larvae during much of the year.

Acknowledgement

I thank Dr. DON COWLEY of the University of Auckland who kindly provided me with information on the habitat and nets of *Polyplectropus* larvae.

References

ANDERSON, N.H. 1976. The distribution and biology of the Oregon Trichoptera. Oregon Agric. exp. Sta. Techn. Bull. 134: 152 pp.

COWLEY, D.R. 1975. Systematic studies on the immature stages of New Zealand Trichoptera (caddisflies). Ph.D. thesis, University of Auckland, New Zealand.

CROSBY, T.K. 1975. Food of the New Zealand trichopterans *Hydrobiosis parumbripennis* McFARLANE and *Hydropsyche colonica* McLACHLAN. Freshwat. Biol. 5: 105-114.

ENDERS, F. 1976. Size, food-finding, and Dyar's constant. Env. Ent. 5: 1-10.

KERST, C.D. & ANDERSON, N.H. 1975. The Plecoptera community of a small stream in Oregon, U.S.A. Freshwat. Biol. 5: 189-203.

McFARLANE, A.G. 1938. Life histories and biology of New Zealand Rhyacophilidae (order Trichoptera). M.Sc. thesis, University of Canterbury, New Zealand.

——. 1951. Caddis fly larvae (Trichoptera) of the family Rhyacophilidae. Rec. Cant. Mus. 5: 267-289.

MECOM, J.O. 1972. Feeding habits of Trichoptera in a mountain stream. Oikos 23: 401-407.

MICHAELIS, F.B. 1974. The ecology of Waikoropupu Springs. Ph.D. thesis, University of Canterbury, New Zealand.

MOSELY, M.E. & KIMMINS, D.E. 1953. The Trichoptera (caddis-flies) of Australia and New Zealand. Br. Mus. (nat. Hist.), London.

NORRIE, P.H. 1969. The flight activity of Ephemeroptera and Trichoptera in a Waitakere stream. M.Sc. thesis, University of Auckland, New Zealand.

THUT, R.N. 1969. Feeding habits of larvae of seven *Rhyacophila* (Trichoptera: Rhyacophilidae) species with notes on other life-history features. Ann. ent. Soc. Am. 62: 894-898.

WINTERBOURN M.J. 1974. The life histories, trophic relations and production of *Stenoperla prasina* (Plecoptera) and *Deleatidium* sp. (Ephemeroptera) in a New Zealand river. Freshwat. Biol. 4: 507-524.

WISE, K.A.J. 1973. A list and bibliography of the aquatic and water-associated insects of New Zealand. Rec. Auck. Inst. Mus. 10: 143-187.

Discussion

ANDERSON: Would you comment further on the large amount of detritus consumed by rhyacophilids in the rapid-flowing Selwyn River?

WINTERBOURN: This is difficult to explain especially as the smooth stones of the Selwyn River bed appeared to trap less detritus than the Kaiapoi. Perhaps it is largely gut contents of the large prey organisms, e.g. mayfly nymphs ingested in the Selwyn River.

MORETTI: Did you find gregarines in the gut contents?

WINTERBOURN: No gregarines were found in caddis guts.

WALLACE: Could the occurrence of the molluscan radula be explained by the caddis larva scavenging the decaying body of the snail?

WINTERBOURN: Yes.

WALLACE: I have noticed final instar larvae of *Athripsodes albifrons, A. bilineatus*, and *A. commutatus* frequently biting into pupal cases of polycentropids and hydropsychids to feed on the enclosed pupa.

Proc. of the 2nd Int. Symp. on Trichoptera, 1977, Junk, The Hague

Atriplectididae, a new caddisfly family (Trichoptera: Atriplectididae)

A. NEBOISS

Abstract

A new family Atriplectididae is proposed for the Australian caddisfly genus *Atriplectides* MOSELY, which so far is represented by only one species, *Atriplectides dubius* MOSELY. The new family is related to the family Odontoceridae, but differs in larval, pupal and adult characteristics. It is suggested that the genus *Hughscottiella* ULMER from Seychelles might also belong to the new family. Together with the family diagnosis the description of the larva and pupa of *Atriplectides dubius* is also given. The larva possesses an unusual capability to extend and retract the pronotum.

Introduction

While investigating the benthic fauna at Great Lake, Tasmania, Mr. W. FULTON of Zoology Department, University of Tasmania, observed rather unusual slender caddisfly larvae. These larvae occurred in benthic samples and occasionally were observed extending their pronota. The larvae were present in samples from many localities, and quite obviously represented a common species. They were found living within the upper layer of the lake sediments. Some time later Mr. FULTON successfully bred several adults, which together with previously collected larvae, were sent to the present author and identified as *Atriplectides dubius* MOSELY. It is interesting to comment that the type specimen was also collected at Great Lake, Tasmania.

Family Atriplectididae fam. nov.
Type Genus *Atriplectides* Mosely, 1936

Atriplectides dubius MOSELY.
MOSELY, 1936a: 120; 1936b: 421, 424; MOSELY & KIMMINS, 1953: 168; FISCHER 1965: 71; 1972: 60; NEBOISS 1977: 113.

Description

Larva (Figs. 1-17): length of final instar larva with extended pronotum up to 18 mm. Colour of specimens preserved in alcohol creamy-white; all sclerotized parts yellowish brown.

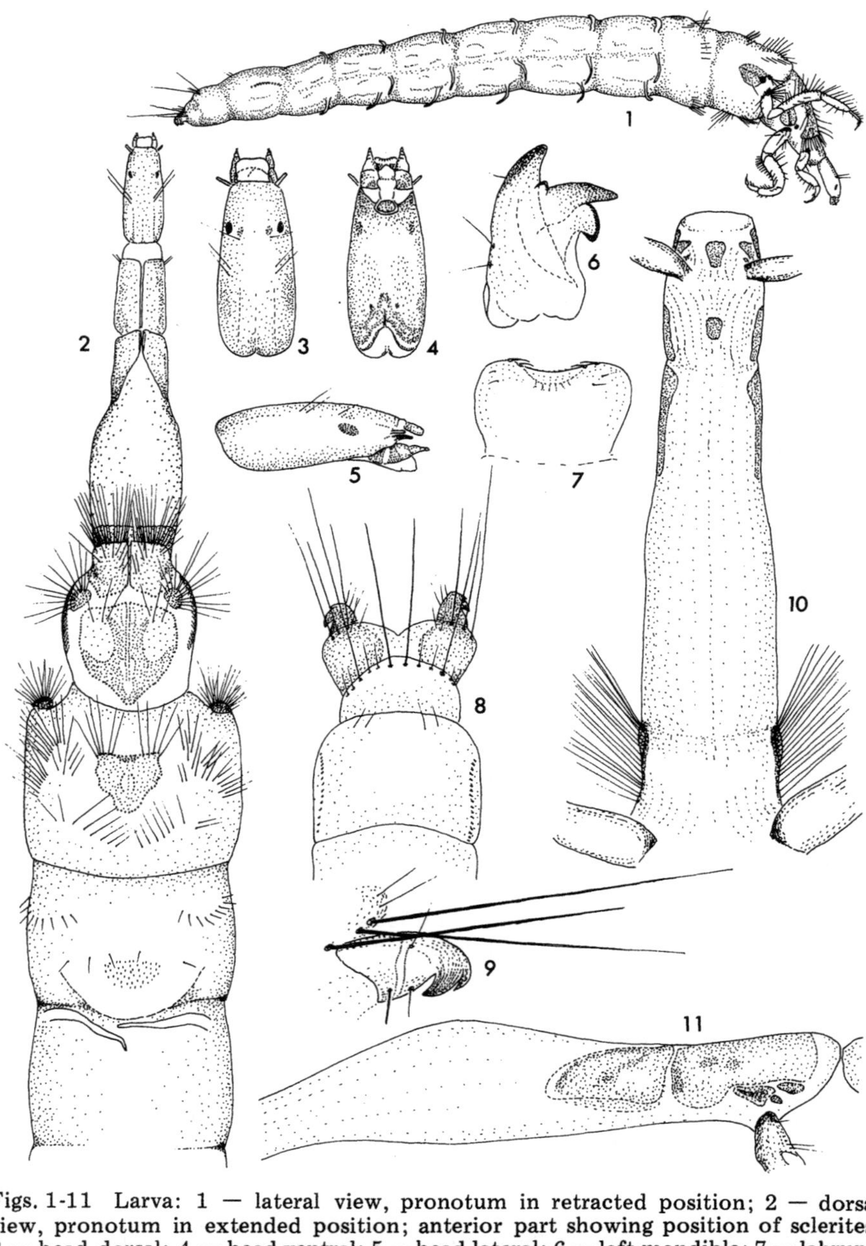

Figs. 1-11 Larva: 1 — lateral view, pronotum in retracted position; 2 — dorsal view, pronotum in extended position; anterior part showing position of sclerites; 3 — head dorsal; 4 — head ventral; 5 — head lateral; 6 — left mandible; 7 — labrum; 8 — abdomen, terminal segments dorsal; 9 — anal claw, oblique lateral view; 10 — pronotum ventral; 11 — pronotum lateral.

Head (Fig. 3) narrow, elongate, almost cylindrical, without visible sutures; dorsal surface smooth, with one fine seta behind the dark spot above the eye and two fine, rather inconspicous setae a short distance behind eyes; ventral surface partially covered with sparse, short spicules, visible only under higher (100x) magnification. Antennae situated at the antero-lateral angles, near the base of mandibles. Eyes black, elongate. Labrum (Fig. 7) pale-yellowish, wider than long, anterior margin concave mesally, setae fine, lateral margins curved. Mandibles (Fig. 6) without mesal brushes, cutting edges developed into separate teeth, apices of which are directed forwards, inwards and backwards respectively.
Pronotum (Fig. 10) very slender; two pairs of dorsal sclerites situated on the anterior half; posterior half without sclerites, integument soft, retractable into mesonotum; two small mesoventral sclerites — one just anterior and the other posterior to the fore-legs.
Mesonotum (Fig. 15), anterior sclerites with ventrally extended antero-lateral angles, a row of long, anteriorly directed setae along the anterior margin; small, somewhat rounded dorsolateral sclerite on either side just anterior to the posterior shield-like dorsal sclerite; a pair of angular, pale-yellowish sclerites ventrally.

Metanotum with small, shield-like dorsal sclerite (Fig. 2); sternum without sclerites. A large lateral sclerite immediately posterior to the base of mid- and hind-legs.

Legs relatively long (Figs. 12-14), especially hind legs; abundant chaetotaxy on femora and tibiae of mid- and hind-legs. Trochantin very small. Femur and tibia of fore-leg each with a short distal spine; claw dark brown, slightly curved, basal seta about half the length of the claw. Femur of mid-leg stout, slightly curved, tibia and tarsus short, rather stout, claw curved, dark brown, basal seta long, reaching almost to the apex of the claw. Hind-legs slender, claw curved, more slender than the ones of fore- and mid-legs, basal seta less than half the length of the claw.

Abdomen gradually tapered caudad, lateral line indistinct. Gills filiform, tapered apically, three pairs (dorsal, lateral and ventral) at the anterior end of segments 2-7 (lateral gills absent on segment 7). First abdominal segment with a wide, low, slightly sclerotized dorsal hump; laterally only very lightly sclerotized area. Specimens killed and preserved while still in their cases, have the humps somewhat flattened, and do not show the true size. Abdominal segment 8 with two pairs of small dorso-caudal setae; laterally a comb of 12-18 bifid spicules. Tergite of abdominal segment 9 with a pair of long, stout and dark mesal and lateral bristles; a pair of short, dark setae between the long bristles; a pair of short pale setae laterad of the long lateral bristles. Abdominal segment 10 with sclerotized areas yellowish-brown; anal claw (Fig. 9) with dorsal tooth; three long, stout, dark bristles, and two smaller setae arising from the sclerotized plate above the base of each anal claw.

Case up to 22 mm long, anteriorly about 3 mm wide, slender, slightly curved, at the beginning squarish or triangular, with distinct angular ridges, later becoming more rounded; constructed of small sandgrains, intermixed, and outside usually covered with fine mud particles; inside lined with a layer of silken threads. Fully grown larva apparently extends the case anteriorly, slightly narrowing the anterior

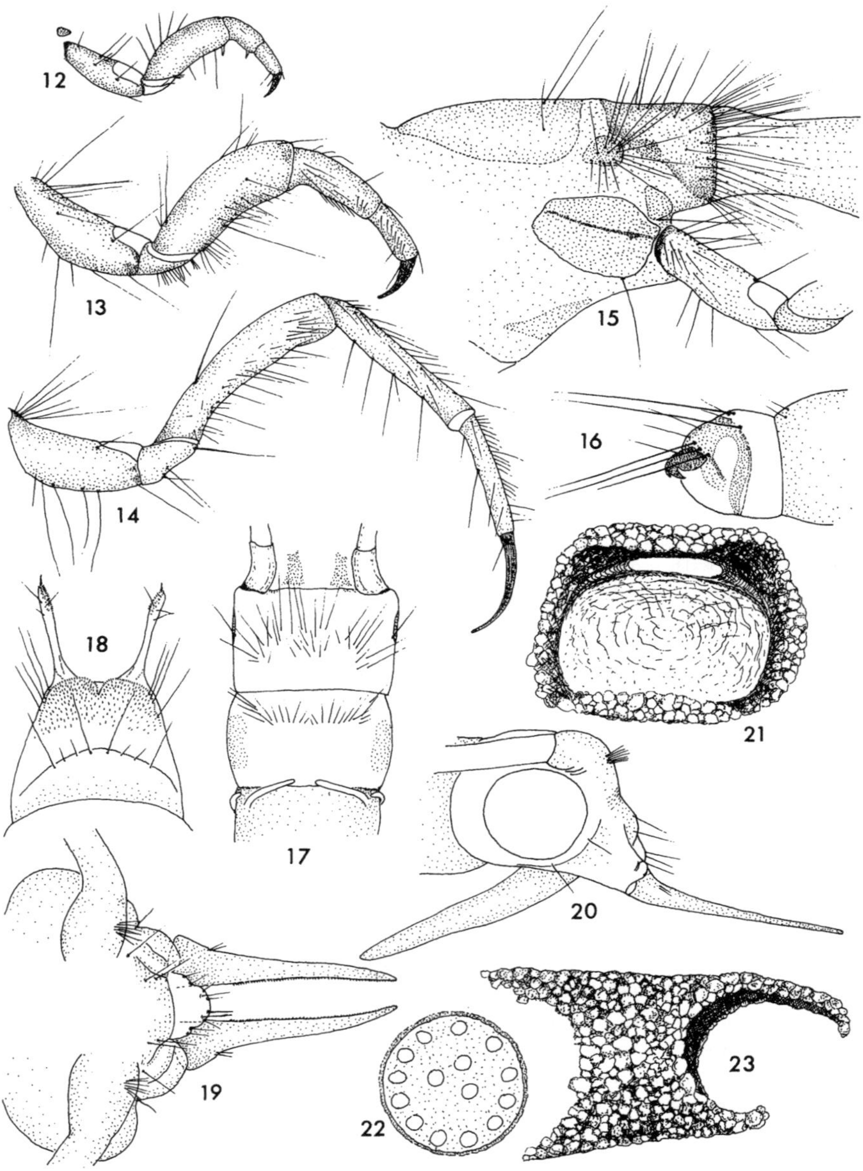

Figs. 12-17 larva; 18-20 pupa; 21-23 case: 12 — fore-leg, with trochantin; 13 — mid-leg; 14 — hind-leg; 15 — mesonotum lateral; 16 — abdomen, terminal segments lateral; 17 — metasternum and first two abdominal segments ventral; 18 — terminal segments, dorsal; 19 — head dorsal; 20 — head lateral; 21 — anterior closure; 22 — posterior closure; 23 — anterior end of pupal case lateral.

end, thickening the margins and constructing a definite dorsal and ventral lip (Fig. 23). The case is shortened posteriorly before pupation. Anterior end closed by a convex silken membrane (Fig. 21), with a dorsal slit-like opening situated close against the dorsal wall of the case; the posterior end is sealed by a firmly constructed perforated closure (Fig. 22).

Pupa: Length 11-12 mm. Mandibles longer than head (Figs. 19-20), slender, gradually tapering apically, almost straight, only distal third slightly curved inwards, serrate along the entire inner margin. Basal antennal segment with a small group of setae; a pair of setae on each side of the frons just above the labrum, one in front of the eye, and another below that on the antero-lateral margin of the head. All setae on labrum about the same length. A curved row of small triangular denticles along the caudal margin of the first abdominal tergite. Gills on segments 2-7 filiform, tapered apically, three pairs on each segment in similar position to those in larvae, dorsal and lateral gills apparently absent on segment 7. Hookplates anteriorly on segments 2-7 small, oval, with single, caudally directed hook; the posterior hookplates on segment 5 larger, somewhat squarish, usually with 2, but sometimes 3 anteriorly directed hooks. The last abdominal tergite covered with short, chitinous spicules. Anal processes straight, rod-like, slightly divergent, apical third covered with short setae on inner surface, and with three dorsal bristles on each process; apex terminates with a single spine.

Material: Tasmania, Great Lake, 6 larvae Swan Bay, 17 May, 1973; 2 larvae — Cramps Bay, 11 Oct., 1973; 6 larvae — Brandum Bay, 14 Sept., 1973; 1 larva, 2 ♀ pupae — Brandum Bay, 27 Jan. 1975; 3 ♀ hatched with pupal skins and cases, Great Lake, Dec. 1975. All specimens collected by W. FULTON.

2 larvae and 3 pupal cases, Main dam, Waratah, 3 Nov. 1960, (collector unknown). All specimens deposited in the National Museum of Victoria collection.

Discussion

The family Odontoceridae, in which the genus *Atriplectides* was placed by MOSELY & KIMMINS (1953), is rather small, with a total of about fifteen genera ascribed to it. Not all of these genera agree fully with the diagnostic characteristics of the family. The Australian genus *Caloca* has already been transferred to a new family Calocidae by ROSS (1967). A number of structural differences found between the larvae, pupae and adults of typical odontocerids (FLINT, 1969) and the genus *Atriplectides* are considered as sufficient to establish a separate family Atriplectididae.

Family diagnosis: Adults — ocelli absent. Antennae long, exceeding the length of the anterior wing; basal segment short, enlarged, somewhat bulbous, shorter than the vertical diameter of eye, touching, or nearly touching each other at the base. Maxillary palpi long, stout, 5-segmented, and similar in both sexes; segments 4 and 5 distinctly thinner than segments 1-3. Anterior wings long, narrow; discoidal cell closed, short, base of the cell situated at the distal third of the wing; median cell absent; thyridial cell long. Posterior wing with discoidal cell closed, short; a row of strong macrotrichia along the basal half of the costal margin, extending distally as far as the base of discoidal cell. Pronotum with one pair of transversally elongate warts. Mesoscutum without warts, but with two parallel rows of setiferous punctures. Scutellum short, dome-shaped, wart large, occupying most of the sclerite. Legs slender. Spurs 2:4:4, covered with fine setae.

Larva: Head cylindrical, labrum wider than long, mandibles with strong teeth, mesal brushes absent. Pronotum with two pairs of dorsal sclerites. Mesonotum with two large dorsal sclerites. Metanotum with one small median sclerite. Lateral sclerites immediately posterior of the base of mid- and hind-legs. Dorsal spacing hump present. Gills filiform. Anal claw with dorsal tooth.

Pupa: Mandibles gradually tapering apically; inner margin serrate. Gills filiform.

The family Odontoceridae differs in adults by having a pair of rounded warts on mesoscutum; shorter and broader anterior wings in which the base of discoidal cell is situated proximally from the middle of the wing.

In larvae the head is shorter and broader with longer and narrower labrum, mandibles without teeth, differently arranged notal sclerites, branched gills and the lack of dorsal tooth on anal claw. The pupa also differs by having branched gills.

In attempting to diagnose the family Odontoceridae, certain major differences were noted between some of the included genera. It is outside the scope of this paper to analyze all these differences, but comparing the Australian genus *Atriplectides* with others, a remarkable similarity was noticed with adults of the genus *Hughscottiella* ULMER from Seychelles. Further suggestions on their likely relationship came from Dr. MALICKY (in litt.), who had received an unknown larva from Seychelles resembling that of *Atriplectides dubius* and which might belong to *Hughscottiella* is transferred to the new family Atriplectididae. The position of the Australian larvae was discussed with the present author at a previous meeting (pers. comm., 1975).

Based on the similarities in adults, as described by ULMER (1910), particularly in wing venation and the structure of male genitalia, it is proposed that the genus *Hughscottiella* is transferred to the new family Atriplectididae. The position of the second Seychelles genus *Leptodermatopteryx*, which shows affinities with *Hughscottiella*, remains unchanged until further material can be studied. The position of the genus *Marilia* Fr. MÜLLER also remains uncertain, as at least the Australian species placed in this genus lack the mesonotal warts found in typical odontocerids.

Acknowledgements

The author expresses sincere thanks to Mr. W. FULTON, for providing the material for this study, to Dr. H. MALICKY of Biologische Station, Lunz am See, Austria, to Prof. G. MARLIER of the Institut Royal des Sciences Naturelles de Belgique, Brussels, and to Dr. T. NEW of La Trobe University, Bundoora, for helpful discussions and constructive criticism.

References

FISCHER, F.C.J. 1965. Trichopterorum Catalogus VI: 71; Amsterdam: Ned. Ent. Veren.
—. 1972. Ibid. XIV: 60.
FLINT, O.S. 1969. Studies of neotropical caddis flies, VIII: The immature stages of *Barypenthus claudens* (Trichoptera: Odontoceridae). Proc. ent. Soc. Wash. 71: 24-28.
MOSELY, M.E. 1936a. A revision of the Triplectidinae, a subfamily of the Leptoceridae (Trichoptera). Trans. R. ent. Soc. Lond. 85: 91-129.
—. 1936b. Tasmanian Trichoptera or Caddis-flies. Proc. Zool. Soc. Lond. 1936: 395-424.
—. & KIMMINS, D.E. 1953. 'The Trichoptera (Caddis-flies) of Australia and New Zealand'. London: Br. Mus. (nat. Hist.).
NEBOISS, A. 1977. A taxonomic and zoogeographic study of Tasmanian caddisflies (Insecta: Trichoptera). Mem. natn. Mus. Vict. 38: 1-208.
ROSS, H.H. 1967. The evolution and past dispersal of the Trichoptera. A. Rev. Ent. 12: 169-206.
ULMER, G. 1910. Trichoptera of the Seychelles. Trans. Linn. Soc. Lond. 14: 41-54.

Proc. of the 2nd Int. Symp. on Trichoptera, 1977, Junk, The Hague

First results of Trichoptera collecting with light traps at Vila do Conde (Portugal)

L.S.W. TERRA

Abstract

Two light traps have been operated every third night from early May 1976 to late May 1977 in the grounds of the Portugese Freshwater Biological Station at Vila do Conde. The total number of species collected so far is 28. Two species, *Tinodes assimilis* and *Hydropsyche lobata*, have very long flight periods, being collected almost all the year round, while *Tinodes waeneri* shows also a long flight period with two well-defined peaks. Other species have shorter flight periods, which are especially short for *Agapetus laniger* and *Oecetis alexanderi*. The light traps are still in operation.

Introduction

The first records of Portugese Trichoptera were presented by McLACHLAN (1874-80) based on collections made in 1880 by the Rev. A.E. EATON. In McLACHLAN's work, 57 species were listed with certainty.

My own collections of caddis flies date from 1972, and 98 species have been found so far. Of these species, 54 had not been recorded by McLACHLAN so the total number known from Portugal is now 111. Eleven of the species I collected were new to science and they have been described by MALICKY (1975, 1976) and by MALICKY & KUMANSKI (1974).

My collections cover all the territory of Portugal, but they were more frequent to the north of the River Tagus. Collecting was with a hand-net, and occasionally with a light trap. Light traps were used in the grounds of the Portugese Freshwater Biological Station at Vila do Conde, where systematic trapping began in 1975. Only the catches for 1976 and part of 1977 are presented in this paper.

Methods and sites of collections

The light trap used consisted of a mercury vapour lamp placed over a bowl of water to which detergent had been added, thus reducing the surface tension so that the caddisflies sank in the water. This method was suggested by H. MALICKY.

Light traps were first used occasionally at the end of 1974. They were then used systematically in 1975, but the frequency of operation was not convenient, so in 1976 and 1977 they were operated every third night, as in the observations of CRICHTON (1960).

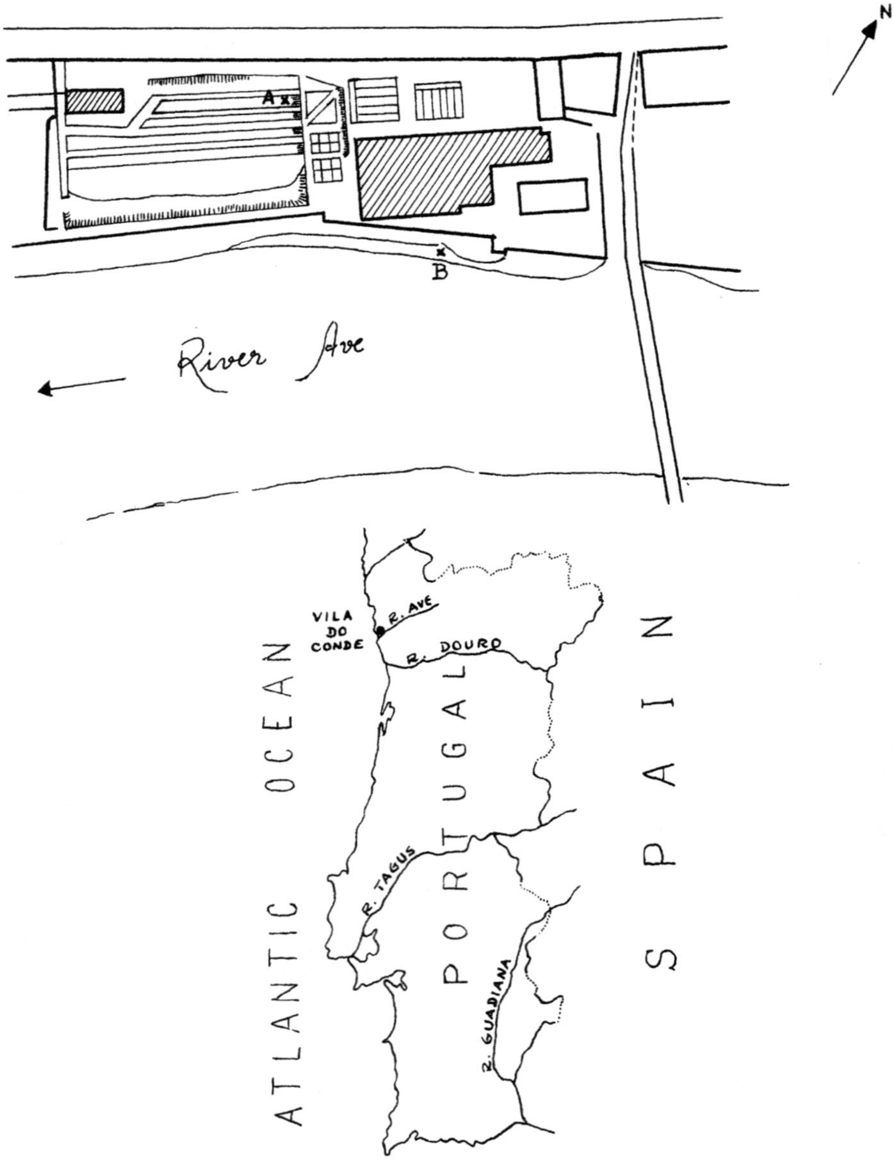

Fig. 1. Above: plan of the grounds of the Portugese Freshwater Biological Station showing sites of light traps A and B. Below: map of Portugal, to show the position of Vila do Conde.

One light trap was placed in the middle of some fish ponds and drainage channels, where there was a good flow of water and the cover of some trees. This light was operated every third night from 2 March 1976, with no interruption except by accident. A second light trap was placed on the bank of the river in a clearing among some bushes and a few small trees, and was operated on the same nights as the first trap from 1 May 1976. The two lights were not visible from each other, and their positions can be seen in Figure 1.

The situation of Vila do Conde, on the right bank of the River Ave, 3 km from the Atlantic, is also shown in the map of Portugal (Figure 1). The river here has a mean width of 70 m, and a depth varying from 0.9 to 1.5 m, with the bed consisting mainly of sand. The water is fairly polluted, especially by the effluents of

Table 1. Total catches of 13 species of Trichoptera in light traps at Vila do Conde in 1976 and 1977.
A — light trap at fish ponds; B — light trap on river bank

	1976 2 Mar. — 27 Dec. A	 B	1977 11 Jan. — 31 May A	 B
Rhyacophila munda McL.	2	4	20	25
Agapetus laniger (PICT.)	8	14	0	0
Polycentropus telifer McL.	83	398	2	37
Ecnomus deceptor McL.	47	1501	0	1
Tinodes assimilis McL.	657	310	174	106
Tinodes waeneri (L.)	907	187	149	166
Hydropsyche lobata McL.	468	580	181	337
Limnephilus guardarramicus SCHM.	14	4	3	3
Limnephilus marmoratus CURT.	19	13	6	2
Ceraclea sobradieli (NAV.)	27	165	9	21
Mystacides azurea (L.)	97	348	0	12
Oecetis alexanderi KUMANSKI	3	11	0	0
Oecetis testacea (CURT.)	9	7	0	0
	2341	3542	544	710

textile industries which usually are not submitted to any treatment before discharge into the river. The highest concentration of these industries is some 20 km upstream.

Further collections were made by hand from December 1976 to February 1977 on the outside northern walls of the Station building, close to which is a small concrete channel with water flowing into some ponds. It is only 30 x 30 cm in section and a few metres in length, and is not represented in the figure.

Results

The total number of species collected was 28, but 11 of these were caught only

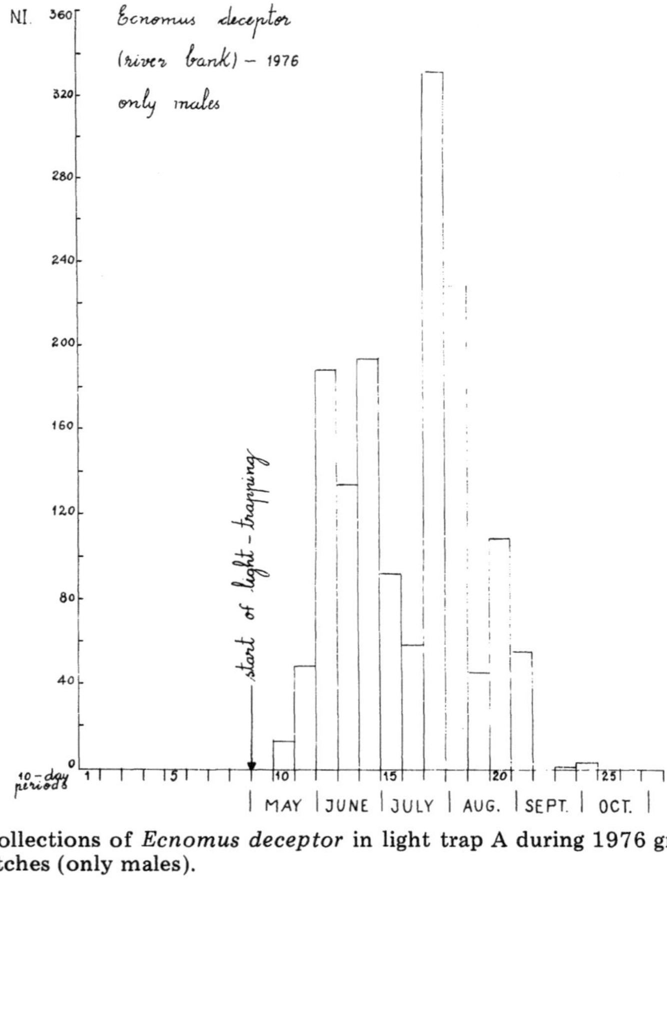

Fig. 2. Collections of *Ecnomus deceptor* in light trap A during 1976 grouped into 10-day catches (only males).

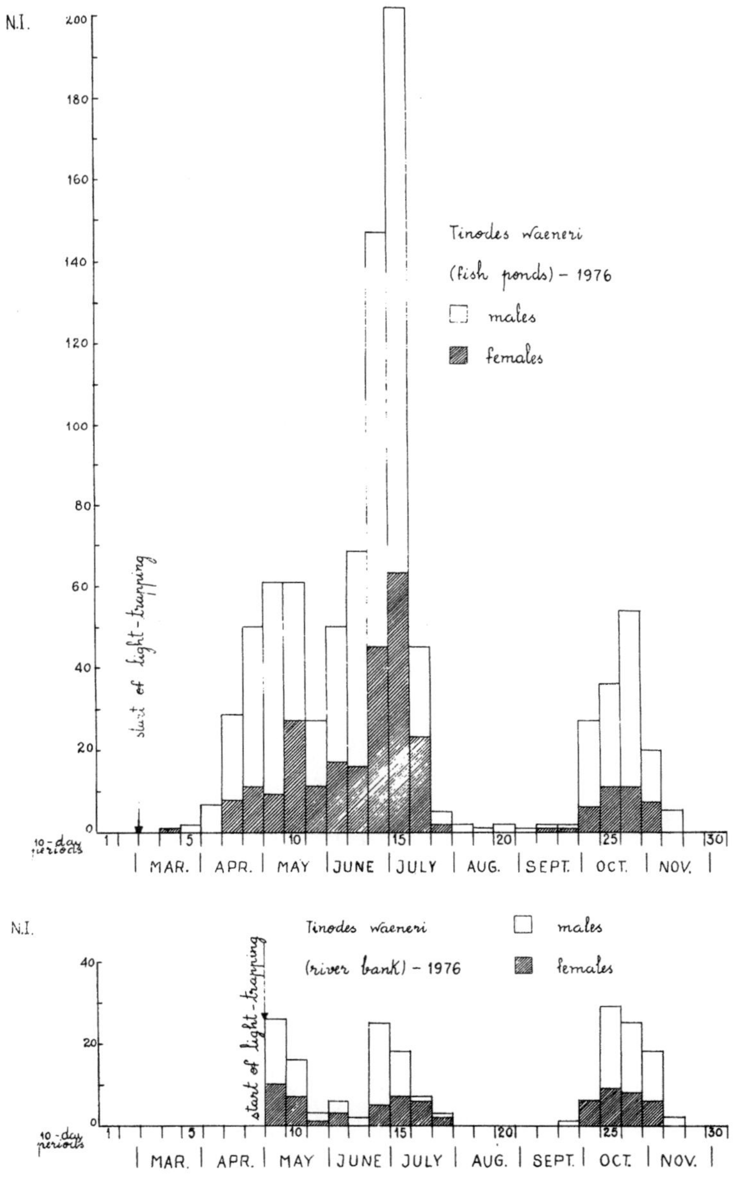

Fig. 3. Collections of *Tinodes waeneri* during 1976 in light trap A (above) and light trap B (below) grouped into 10-day catches.

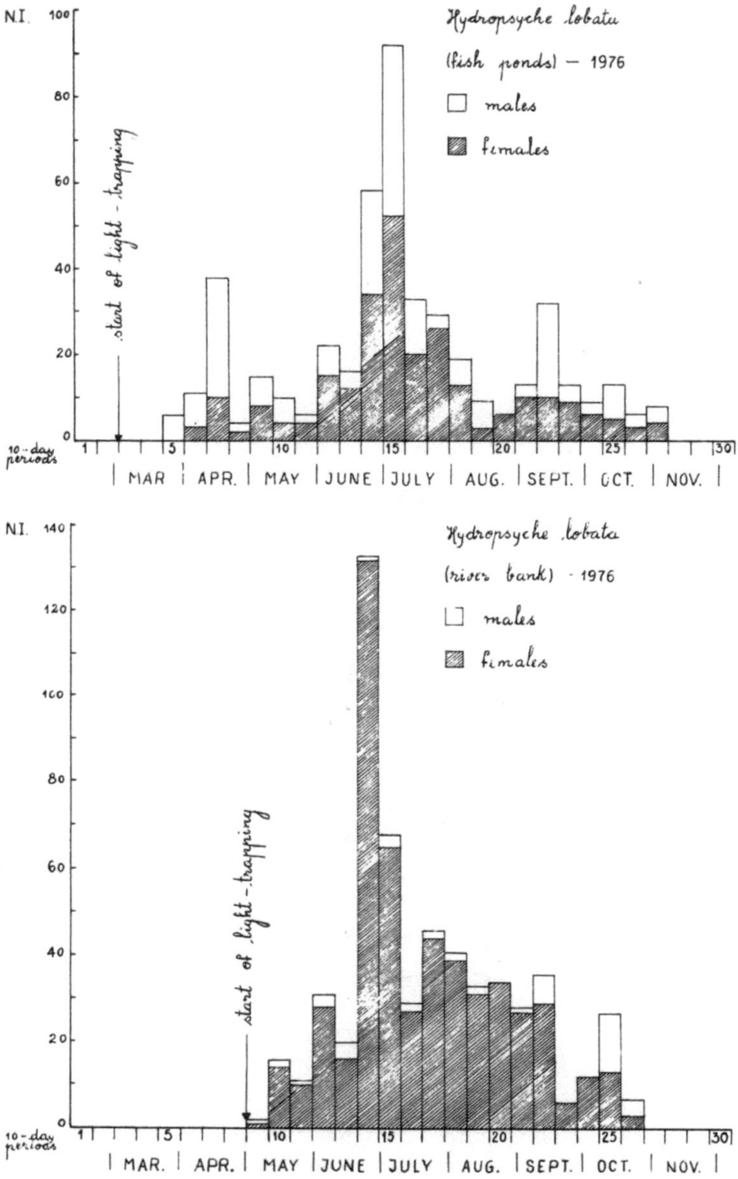

Fig. 4. Collections of *Hydropsyche lobata* during 1976 in light trap A (above) and light trap B (below) grouped into 10-day catches.

occasionally and a few were represented by single specimens during the whole period of trapping.

The catches of 13 species are listed in table 1 for the periods 2 March to 27 December 1976, and 11 January to 31 May 1977. In each year the catches from the light trap at the fish ponds are given in column A, and those from the trap on the river bank in column B.

During the period 21 December 1976 to 1 March 1977 regular collections of *Tinodes assimilis* were made by hand, usually every third day at 10.00 a.m. on the outside northern walls of the Station building. Some insects were caught on each visit and the total was 336 specimens, which compares with only 29 caught in both light traps on the same dates. Air temperatures at 10.00 a.m. on these days varied between 3.5 and 16°C, while the water temperature in the channel feeding the ponds was almost constant at 12°C.

Discussion

The most numerous species in 1976 were *Ecnomus deceptor*, *Tinodes waeneri*, *T. assimilis* and *Hydropsyche lobata*, all with more than 500 specimens caught in at least one of the traps. A second group of species, *Polycentropus telifer*, *Mystacides azurea* and *Ceraclea sobradieli* were represented by 150 to 400 specimens each in one of the traps.

Ecnomus deceptor was represented by 1501 specimens in the trap on the river bank, and by only 47 in the other trap, which is evidence that it does not fly far from the river. It had a long flight period, from 11 May to 8 October, with a peak at the end of July (Fig. 2).

Tinodes assimilis was remarkable for its long flight period, being caught every night from 2 March to 21 December 1976 and again from 21 January to 31 May 1977. Even during the months December to February, when few were caught in light traps it was found in reasonable numbers on the outside walls of the buildings. Its larvae were found in the swiftly-flowing water in the small channel feeding the ponds. It was more abundant in the fish ponds trap and peak numbers were in June and July.

Tinodes waeneri also had a long flight period, from March to early November in 1976, and showed one well-defined peak in early July, and a smaller one in late October (Fig. 3). This is similar to the pattern recorded in England, where it was suggested that it had two generations in the year (JONES, 1976a, 1976b).

Hydropsyche lobata is another species with a long flight period, from late March to the end of October (Fig. 4). This figure also gives the numbers of males and females; females were 55% in trap A and 90% in trap B. *Polycentropus telifer* was more abundant in trap B, and had a flight period from April to November, with no well-defined peak. *Mystacides azurea* was caught from May to mid-November, and *Ceraclea sobradieli* from April to mid-October.

The Limnephilidae were not common; the most numerous species were *Limne-*

philus marmoratus and *L. guadarramicus*. The former was collected from May to November, and the latter from March (from January in 1977) to August, but there was no clear indication, from the small numbers caught, of two flight periods and a summer diapause (CRICHTON, 1976). Only 4 specimens of *L. lunatus* CURT. were caught, in trap B in 1976. There were five males of *Halesus radiatus* (CURT.), during March 1976, but at other places a few specimens were caught in October. A few *Micropterna fissa* (McL.) were collected in October 1974, and there were also a few *L. hirsutus* (PICT.) and *Mesophylax aspersus* (RAMB.), and a single *Grammotaulius submaculatus* (RAMB.).

Other species represented by very few or single specimens were *Rhyacophila adjuncta* McL., *Cyrnus cintranus* McL., *Polycentropus kingi* McL., *Psychomyia ctenophora* McL., *Ecnomus tenellus* (RAMB.), *Lepidostoma hirtum* (F.), *Ceraclea albimaculata* (McL.), *Adicella reducta* (McL.), and *Calamoceras marsupus* BRAUER.

Acknowledgements

I am indebted to Dr Hans MALICKY for his great help, especially in identification. I am much indebted to my wife for drawing graphs. I must also thank Mr Americo QUADROS for graphs and plans and for recording temperatures, and Mr Domingos LEMOS for dealing with the operation of the light traps and separating the caddisflies from other insects.

References

BOTOSANEANU, L. 1972. Observations sur quelques Trichoptères du Portugal. Ciência Biológica (Portugal) 1: 19-23.

CRICHTON, M.I. 1960. A study of captures of Trichoptera in a light trap near Reading, Berkshire. Trans. R. ent. Soc. Lond. 112: 319-344.

— 1976. The interpretation of light trap catches of Trichoptera from the Rothamsted Insect Survey. Proc. 1st int. Sym. Trich. 1974: 147-158.

JONES, N.V. 1976a. The Trichoptera of the stony shore of a lake, with particular reference to *Tinodes waeneri* (L.) (Psychomyiidae). Proc. 1st int. Sym. Trich. 1974: 117-130.

— 1976b. Studies on the eggs, larvae and pupae of *Tinodes waeneri* (L.). ibid. 131-143.

McLACHLAN, R. 1874-80. A monographic revision and synopsis of the Trichoptera of the European fauna. (Reprint) Hampton: Classey.

MALICKY, H. 1975. Fuenfzehn neue Mediterrane Koecherfliegen. Mitt. ent. Ges. Basel 25: 81-96.

— 1976. Beschreibung von 22 neuen Westpaläarktischen Köcherfliegen (Trichoptera). Z. Arb. Gem. öst. Ent. 27: 89-104.

— & KUMANSKY, K. 1974. Neun neue Köcherfliegen aus Südeuropa (Trichoptera). Ent. Z. 84: 9-20.

Proc. of the 2nd Int. Symp. on Trichoptera, 1977, Junk, The Hague

Some characteristics of the Yugoslav fauna of Trichoptera

MARA MARINKOVIĆ-GOSPODNETIĆ

Abstract

This paper reports on some zoogeographical characteristics of the Yugoslav fauna of Trichoptera. The specificity of the Dinarides fauna is emphasized, as well as the affinity of the entire Balkan fauna of Trichoptera.

Introduction

Yugoslavia occupies a diversified area of the north-western and central part of the Balkan peninsula. The old mountain massif, the Rhodope, rises in the south-eastern part of the Balkan peninsula; its northern part penetrates as a wedge between the younger mountains, the Dinarides in the west, and the Carpathian and Balkan mountain range in the north-east. The Dinarides run parallel to the Adriatic and Ionian Seas. In many respects the fauna of the Balkan peninsula differs from that of the north; there are several reasons for this. It is certain that the Balkan peninsula witnessed many changes during Tertiary times and the effect of these changes is of particular interest. In addition to the relicts of the Tertiary fauna, many immigrants inhabit the Balkan peninsula. The new species, having their origin in several Balkan centres, contribute also to the specificity of the fauna.

The first investigations (KLAPALEK, 1898, 1900, 1902) have already shown that many species belonging to the Balkan fauna of Trichoptera are unknown in other parts of Europe. This fauna, however, has remained insufficiently known for several decades until the appearance of the papers of RADOVANOVIĆ (1935), BOTOSANEANU (1960), MARINKOVIĆ-GOSPODNETIĆ (1966, 1973, 1975), KUMANSKI (1968, 1971), OBR (1969) and MALICKY (1974). In the meantime, BOTOSANEANU has examined the fauna of Roumania. All these papers are a contribution not only to our knowledge of the fauna of different parts of the Balkans but also to a clear view of the characteristics of the fauna of the whole Balkan peninsula. Having reached this point it is possible to get an idea of the position of the fauna of Yugoslavia in the Balkan fauna of Trichoptera as a whole.

Some zoogeographic characteristics of the Yugoslav fauna of Trichoptera

The first characteristic to be mentioned is the presence of the relatively great number of endemic species. Some of them inhabit the entire Balkan peninsula, and a few also the southern Carpathian mountains, but the greatest number of species are restricted to smaller areas of distribution. Many of them are endemic species of the Dinarides, sometimes of a single river system. The species of the genus *Drusus* demonstrate a striking example of endemism. In the Dinarides and western Macedonia this genus is represented by 17 species and subspecies. Among them, only *Drusus discolor* and *Drusus biguttatus* extend their range beyond the Balkan peninsula. All other species are limited to small parts of the Dinarides. The largest area is that of the *D. schmidi*, which occurs in almost all river systems of the Dinarides. In these localities, it has been found with some other species of *Drusus* (Fig. 1). All *Drusus* species of the group *bosnicus*, however, are allopatric. Their area of distribution is extremely restricted; sometimes, they do not inhabit all the springs of the tributaries of the river system (*D. bosnicus* lives only in the right tributaries of the river Bosna). Comparative morphological studies have shown (MARINKOVIĆ-GOS-

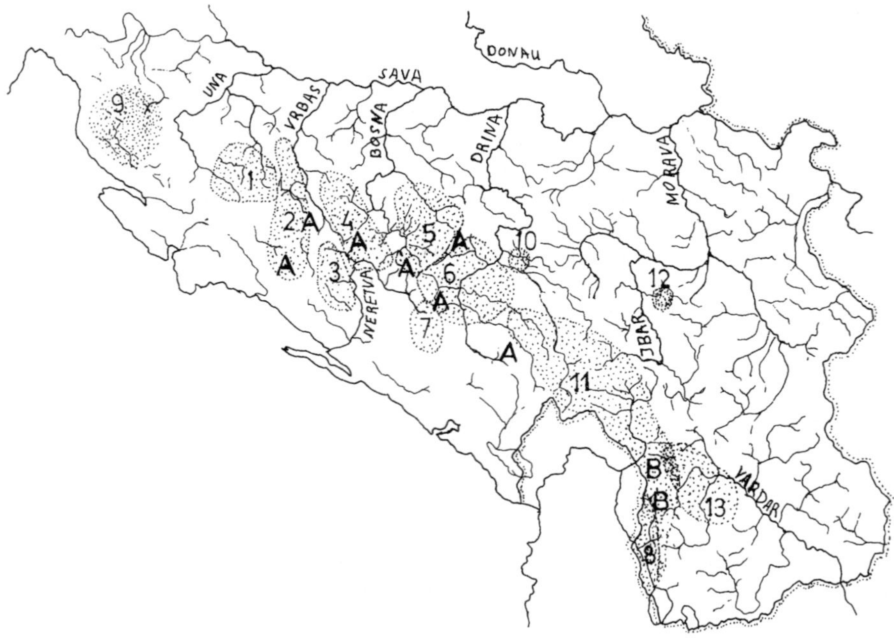

Fig. 1. Distribution of the endemic species of *Drusus* in Yugoslavia: 1 — *D. vespertinus*, 2 — *D. radovanovići septentrionis*, 3 — *D. ramae*, 4 — *D. medianus*, 5 — *D. bosnicus*, 6 — *D. klapaleki*, 7 — *D. radovanovići radovanovići*, 8 — *D. plicatus*, 9 — *D. croaticus*, 10 — *D. serbicus*, 11 — *D. discophorus*, 12 — *D. botosaneanui*, 13 — *D. macedonicus*, A — *D. schmidi*, B — *D. tenellus*.

PODNETIĆ, 1976) that all these species are closely related, descending probably from a single ancestor, which might have covered a larger area of distribution in the Dinarides.

The origin of the greater number of related species may also be noticed in the genus *Hydropsyche.* It has been shown by the work of BOTOSANEANU & MARINKOVIĆ-GOSPODNETIĆ (1966) on *Hydropsyche.* of the group *fulvipes-instabilis* and by the study of KUMANSKI & BOTOSANEANU (1974) on the species of *Hydropsyche* of the group *guttata.* The endemic species of these two groups inhabit mostly the central and eastern parts of the Balkan peninsula; their centres of differentiation have probably been there. The Dinarides are a centre of differentiation of other *Hydropsyche* species, those of the group *pellucidula.* The recently discovered species *Hydropsyche botosaneanui* MARINKOVIĆ-GOSPODNETIĆ 1966 belongs here as well as two related species, *H. smiljae* n. sp. and *H. dinarica* n. sp. (Their description will be published in God. Biol. inst. Sarejevo, 30, 1977.) The widest area of distribution is that of *H. dinarica* which has been found in the river system of all right tributaries of the river Sava as well as in the river system of Morača, a tributary of Scutary Lake. *H. smiljae* occupies only the lower course of the river Neretva and its tributaries (Fig. 2), while *H. botosaneanui* lives in the river Rama, a tributary of the middle course of the river Neretva, and in the river systems of Bosna, Drina and Ibar.

One of the zoogeographical characteristics of the Balkan fauna of Trichoptera is that it includes many related species, whose areas are vicarious. Such displacements have been already manifested in the range of the Dinarides. In this regard, a very interesting example is furnished by two species of the genus *Annitella.* The areas of *A. apflebecki* and *A. triloba* meet in the middle of the range of the Dinarides and are slightly overlapping (Fig. 3). The range of *A. apfelbecki* is to the west from their common boundary, while that of *A. triloba* is to the east, extending to Bulgaria. A similar case is that of two closely related species of *Chaetopteryx* of the group *villosa*, where the area of *C. bosniaca* from the Dinarides meets that of *C. cissylvanica* from Roumania and Bulgaria on the banks of the river Drina (Fig. 4). It seems that this river also forms a boundary between the distribution of *Psylopteryx bosniaca* and *P. montanus.* A pronounced example of the vicarious ranges of related species is found in the genus *Rhyacophila,* in the group *tristis* (Fig. 5). The area of *R. vranitzensis*, until now known only from the Dinarides, meets in the east with the range of the related species *R. obtusa*, so far regarded as the endemic species of Bulgaria. *Rhyacophila orghidani*, the endemic species of the Roumanian Carpathians, is related to these two species. *Rhyacophila cibinensis* inhabits the Carpathian mountains also, being closely related to *R. bosniaca* from Yugoslavia, mostly from the Dinarides. Another example of the vicarious areas of *Rhyacophila* species of the group *tristis* is that of *R. trescavicense* from the Dinarides and western Macedonia, and *R. pendayica* from western Greece.

All these, are examples of the affinity of the entire Balkan fauna of Trichoptera, and of its ties with the fauna of the Carpathians. At the same time these findings

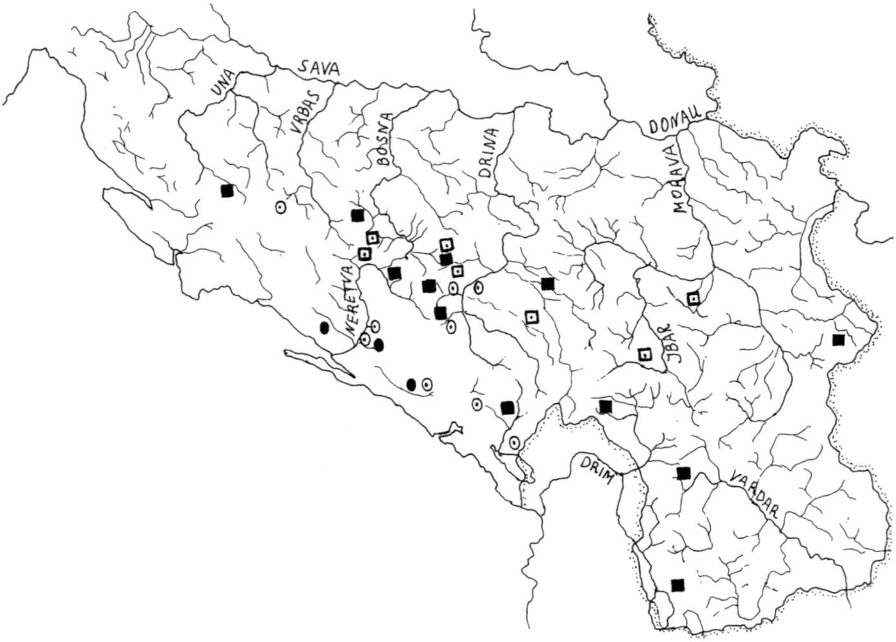

Fig. 2. Distribution of *Hydropsyche* species of the group *pellucidula* in Yugoslavia: ⊡ *H. botosaneanui*, ■ *H. tabacarui*, ⊙ *H. dinarica*, ● *H. smiljae*.

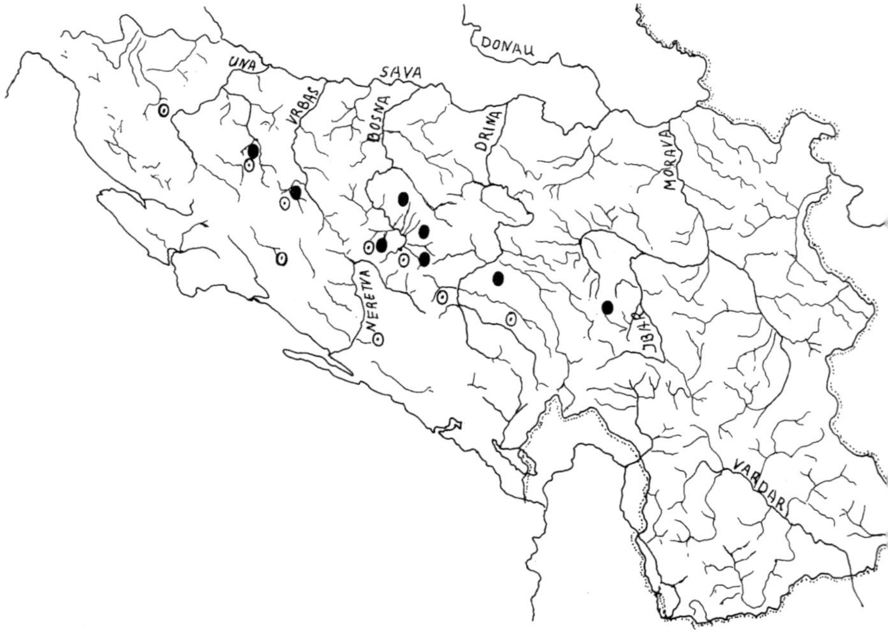

Fig. 3. Distribution of *Annitella* species in Yugoslavia: ⊙ *A. apfelbecki*, ● *A. triloba*.

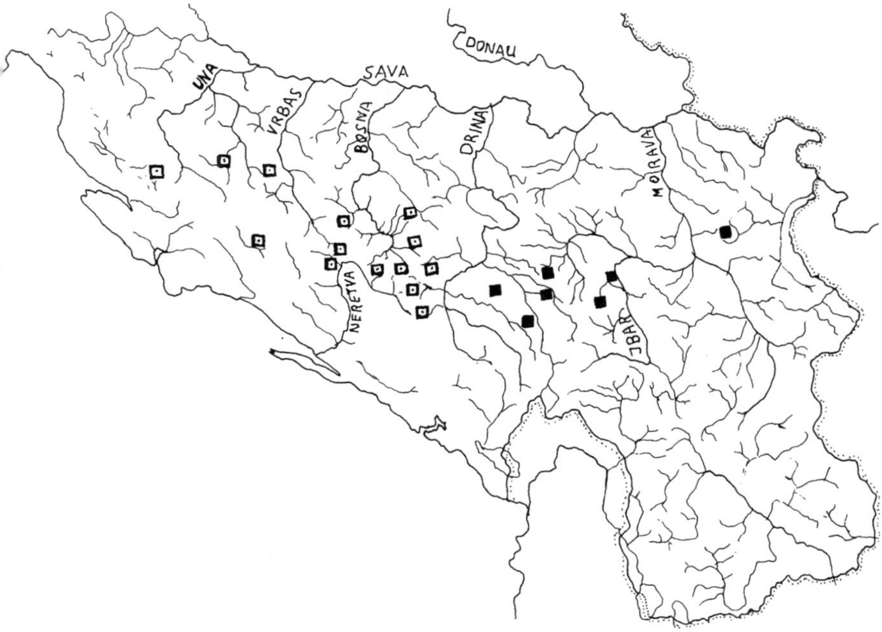

Fig. 4. Distribution of *Chaetopteryx bosniaca* / ⊡ / and *C. cissylvanica* / ■ / in Yugoslavia.

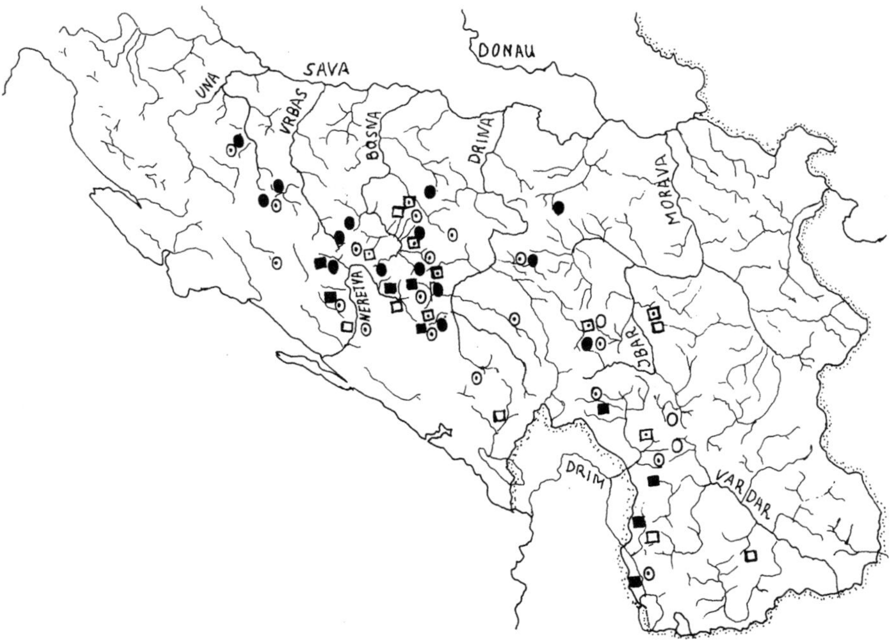

Fig. 5. Distribution of the endemic species of *Rhyacophila* in Yugoslavia: ⊙ *R. balcanica*, ● *R. vranitzensis*, ○ *R. obtusa*, □ *R. bosniaca*, ■ *R. trescavizense*, □ *R. loxias*.

suggest that the Balkan peninsula is an important centre of differentiation of species.

References

BOTOSANEANU, L. 1960. Trichoptères de Yougoslavie recueillis par le Dr. F. Schmid. Deutch. ent. Ztschr. 7: 261-293.

—. & MARINKOVIĆ-GOSPODNETIĆ, M. 1966. Contribution à la connaissance des *Hydropsyche* du groupe *fulvipes-instabilis* étude des genitalia males (Trichoptera). Annls. Limnol. 2: 503-525.

KLAPALEK, F. 1898. Fünf neue Trichopteren-Arten aus Ungarn. Termes. Füz. 21: 488-490.

—. 1900. Beiträge zur Kenntniss der Trichopteren- und Neuropterenfauna von Bosnien und Hercegovina. Wiss. Mitt. Bosn. Herzeg. 7: 671-682.

—. 1902. Zur Kenntnis der Neuropteroiden von Ungarn, Bosnien und der Hercegovina. Termes. Füz. 25: 161-180.

KUMANSKI, K. 1968. Beitrag zur Erforschung der Trichopteren Bulgariens (I). Faun. Abh. 2: 109-115.

—. 1971. Beitrag zur Erforschung der Köcherfliegen (Trichoptera) Bulgariens. III. Bull. Inst. Zool. u. Musee, Sofia, 33: 99-109.

KUMANSKI, K. & BOTOSANEANU, L. 1974. Les *Hydropsyche* (Trichoptera) du groupe de *guttata* en Bulgarie et en Roumanie. Mus. Maced. Sci. Nat. 14: 25-41.

MALICKY, H. 1974. Die Köcherfliegen (Trichoptera) Griechenlands. Übersicht und Neubeschreibungen. Ann. Mus. Goulandris 2: 105-135.

MARINKOVIĆ-GOSPODNETIĆ, M. 1966. New species of Trichoptera from Yugoslavia. Bull. sci. Cons. Acad. RSF Yougoslavie. Sec. A. 11: 110-112.

— 1973. Die Trichopteren-Fauna der Gebirgen Magliċ, Volujak und Zelengora. Wiss. Mitt. Bosn. Herzeg. 3: 131-144.

— 1975. Fauna of Trichoptera of Serbia. Recueil trv. fauna ins. Serb. 1: 221-236.

— 1976. The differentiation of *Drusus* species of the group *bosnicus*. Proc. 1st int. Symp. Trich., 77-85.

OBR, S. 1969. Ergebnisse der Albanien-Expedition 1961 des Deutschen Entomologischen Institutes: 80. Beitrag. Trichoptera. Beitr. Ent. 19: 937-960.

RADOVANOVIĆ, M. 1935. Trichoptere Jugoslavije. Gl. Zem. Muz. Sarajevo, 47: 73-84.

Proc. of the 2nd Int. Symp. on Trichoptera, 1977, Junk, The Hague

On the Trichoptera fauna of the Bükk Mountains, N. Hungary

O. KISS

Abstract

The trichopterous larvae of two stream systems in the Bükk Mountains were sampled monthly for one year, and bi-monthly during the following year.

The quantitative and qualitative composition of species seems to be closely connected with the presence of the substratum of plant origin, moss and algae, the changing water velocity, the variety of food and the increasing temperature of the water downstream in summer.

Introduction

The Bükk Mountains, consisting mainly of limestone, lie in the northern mountain range of medium height, separated from the central range of the Carpathians.

This study of the Trichoptera is part of a faunistic survey of Hungary, and the ecological work is connected with the hydroecological investigations which are being carried out by the Department of Zoology and Anthropology of the University of Debrecen.

Material and methods of collecting

The Bükk Mountains seem to be the coolest part of the country. The climatic zone is cool temperate, with maximum rainfall in summer. The mean annual temperature is 6-8°C. In summer the weather is temperate, in winter it is severe. In the highest parts of the mountains the annual rainfall is 700-800 mm.

The central part of the northern mountain range is the Magas Bükk. It is a 20 km long, 6-7 km wide limestone plateau, at an average height of 800-900 m, with denudational steps at 500-700 m, and then gentle lower slopes. The Magas Bükk consists mainly of Middle Triassic limestone; carrenfelds, dolinas and sink-holes give variety to its surface.

The Szalajka Valley (A) is situated in the north-western part, the rill Disznóskut (B) and stream Sebesviz (C) at the northern edge of the Magas Bükk (Fig. 1).

Sample collecting, observations and measurements were made monthly at 20 sampling stations from June 1974 to July 1975 and then bi-monthly from July 1975 to June 1976. Because of the different substrata, samples were taken by the method of KAMLER & RIEDEL (1960). In the case of the stony substratum the method of MACAN (1958) was also used.

Identification of species was done mainly from the works of HICKIN (1967), LEPNEVA (1966), McLACHLAN (1868), SCHMID (1951) and STEINMANN (1970). The species were mostly identified as larvae, and some as adults, either reared or caught in light traps.

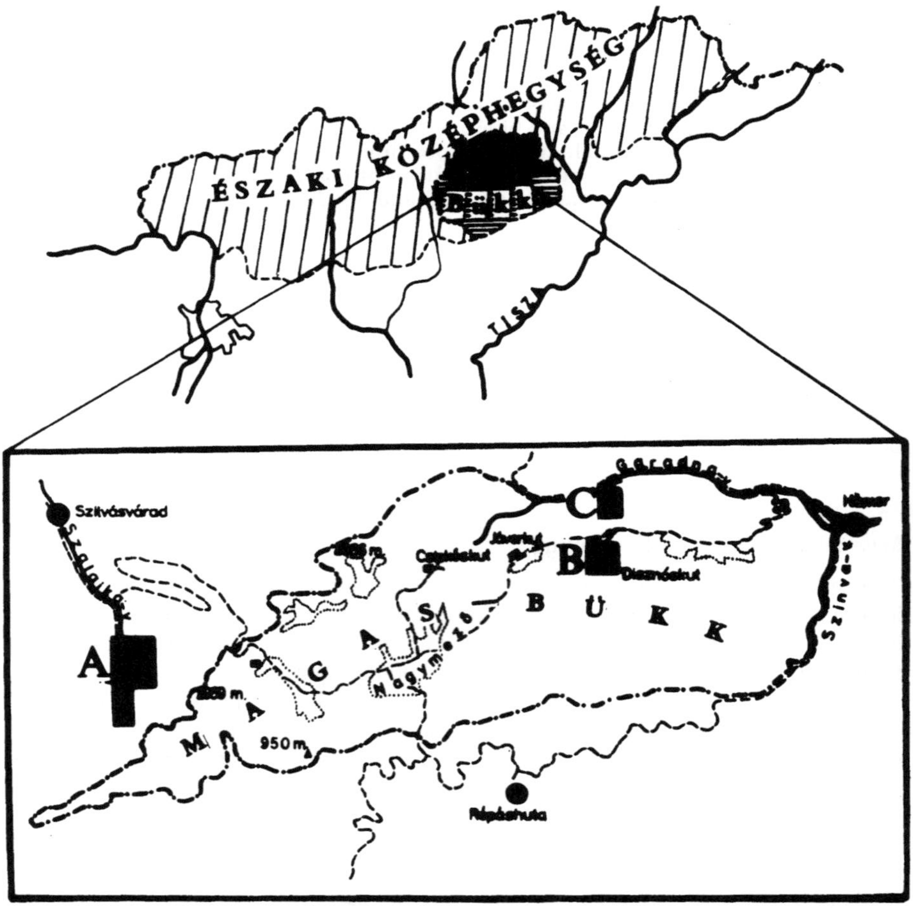

Fig. 1. The situation of the Bükk Mountains. The areas under investigation are A — stream Szalajka, B and C — stream system Disznóskut — Sebesviz.

Some typical habitats, mosaic patterns and Trichoptera communities

For the description of the stations, the mosaic pattern association theory of MARLIER (1951), KENDEIGH (1961) and OLÁH (1967) was used. The main limnological zones were the karst and detrital springs, flowing ponds, lotic and lenitic reaches of streams. For the ecological description of the stations the paper by SZABÓ *et al.* (1971) was also used.

From the fluctuations of water velocity it emerges that the characteristic substratum mosaics also show repetitive arrangements. In the characteristic springs and reaches of streams of the Bükk Mountains the following substratum mosaics can be differentiated: large stones, small stones and gravel, sand, slime, detritus, moss and algae.

First, the most typical habitats of the stream Szalajka with their mosaics and the relative dominance of the Trichoptera communities inhabiting them will be described.

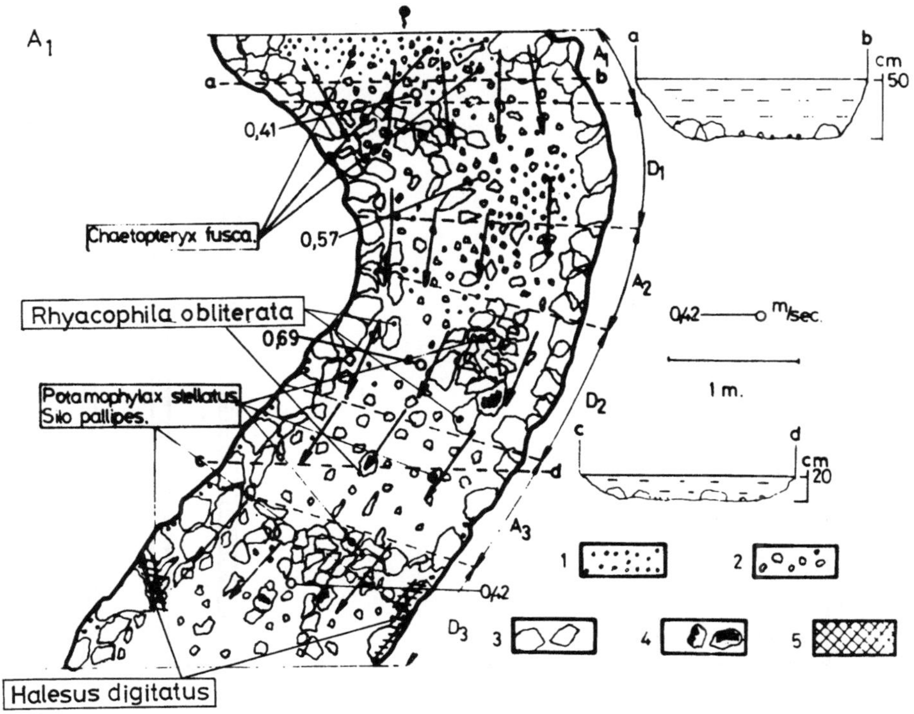

Fig. 2. The spring Upper Szalajka (A_1). Substratum mosaics: 1. sand, 2. small stones and gravel, 3. large stones, 4. moss, 5. detritus.

1. The spring Upper Szalajka (A_1) It is a karst spring, 460 m above sea level. The water originates from Triassic limestone. The water temperature range is 7.8-8.2°C, and the average water output is 1800-9000 l/sec. It is a typical rheocrene spring (Fig. 2).

Chaetopteryx fusca and *Potamophylax stellatus* were dominant among the five species collected (Fig. 3).

2. The stream Szalajka below the rapids (A_{11}) The stream bed is widening out, with a rapid water flow over projecting stones. The depth is 10-20 cm, and the temperature of the water is 7.2-12.8°C. *Petasitetum hybridi* DOST. and *Alnus glutinosa* L. are found on the banks. (Fig. 4).

Of the 11 species collected the dominant ones were *Odontocerum albicorne*, *Silo pallipes*, *Hydropsyche angustipennis* and species of *Rhyacophila* (Fig. 5).

In the area of the rill Disznóskut and stream Sebesviz the three following typical habitats were found:

(i) The spring Disznóskut (B_1) It is a detrital spring, emerging from debris of clay shale. Its height is 630 m above sea level. The temperature range is 6.1-8.0°C,

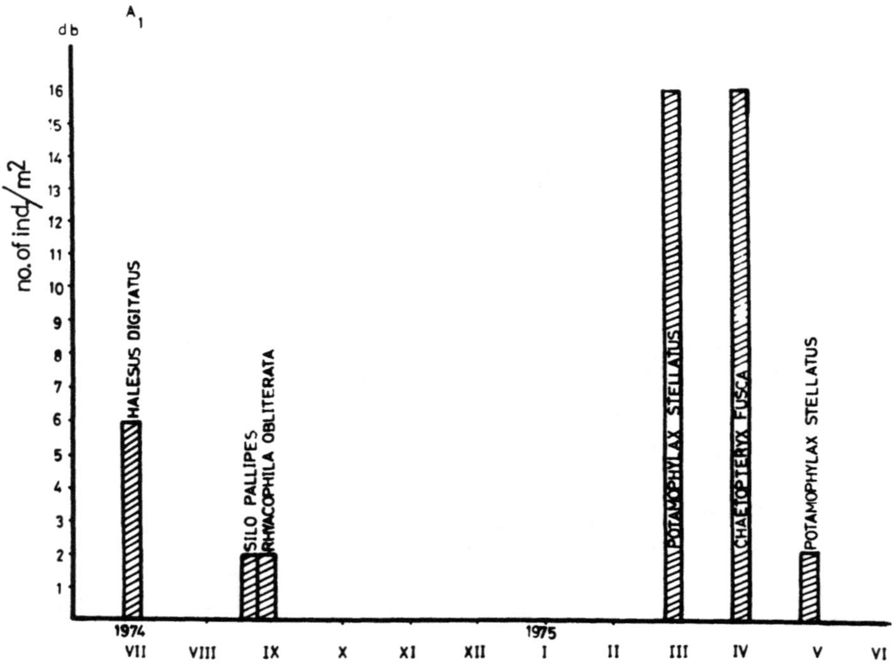

Fig. 3. The species found monthly and the number of individuals per m^2 at the station A_1.

A11

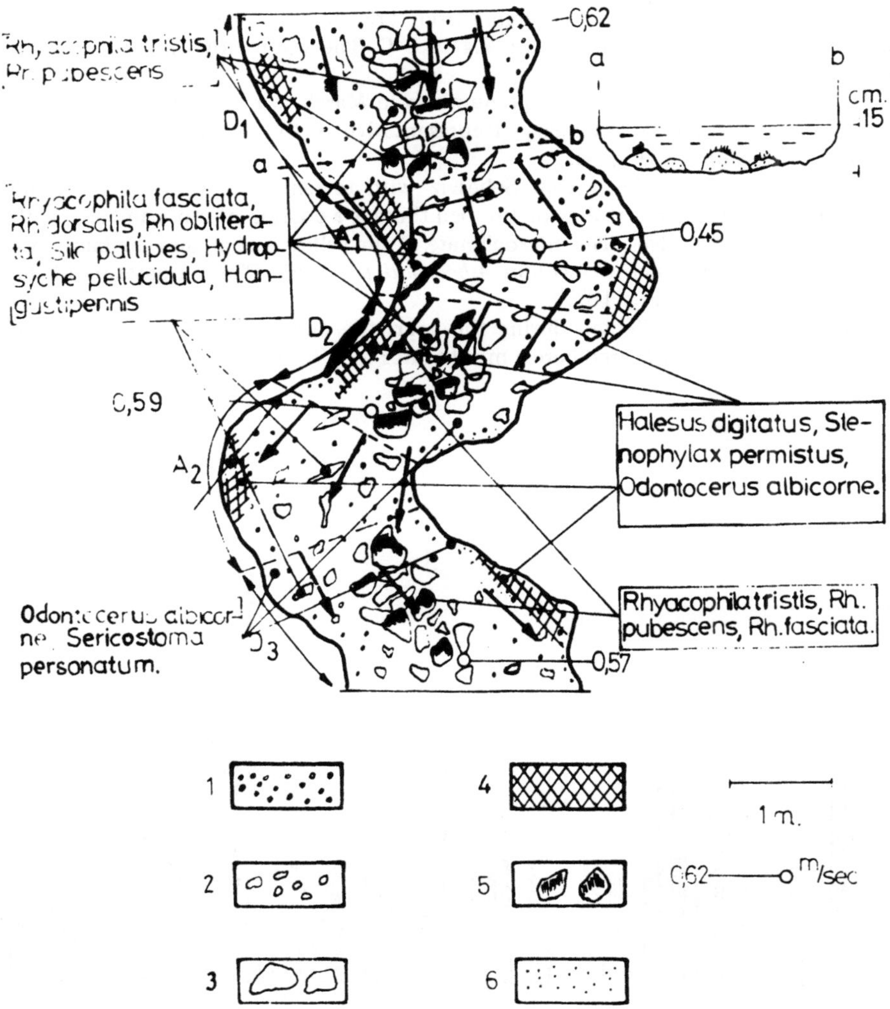

Fig. 4. The stream Szalajka below the rapids (A_{11}). Substratum mosaics: 1. sand, 2. small stones and gravel, 3. large stones, 4. detritus, 5. moss, 6. slime.

and average water output 0.3-0.6 l/sec. It is a limnocrene spring with *Petasitetum hybridi* DOST. growing along its banks (Fig. 6).

Among the species found here the dominant one was *Potamophylax nigricornis*, which was collected almost every month, except in September (Fig. 7).

(ii) The spring Huba (C_1) It is a karst spring of rheocrene type with abundant water, 560 m above sea level, emerging from Triassic limestone. Its average water output is 4.6 l/sec, and temperature 7.4-10.8°C. The water flows into the stream Sebesviz (Fig. 8).

Melampophylax nepos, with large numbers in March, was the dominant species; it is new to the fauna of Hungary. Representative of the petricolous fauna were species of *Rhyacophila* (Fig. 9).

(iii) The reach of the stream Sebesviz with torrents 400 m away from the spring (C_2). On the large stones of the bed there is a coating of moss and alga with algal brushes on the edges. The microclimate of the valley is cool, and the banks are shady and lined with *Melitti fagetum subcarpaticum* SOO. The temperature of the water was 5.4-10.8°C (Fig. 10).

Six species were collected during the year; the dominant one was *Melampophylax nepos*, which appeared here a month later than in the spring (Fig. 11).

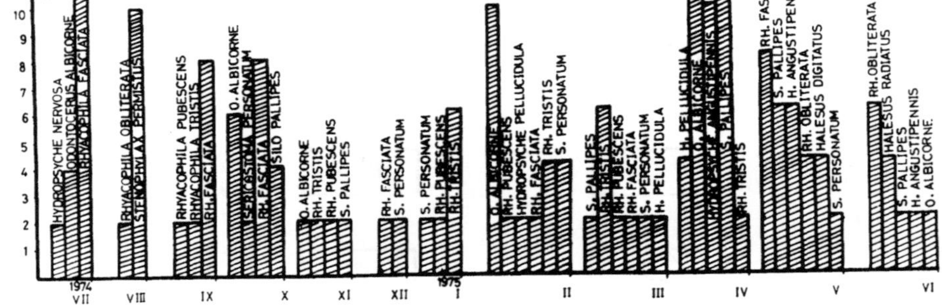

Fig. 5. The species found monthly and the number of individuals per m^2 at the station A_{11}.

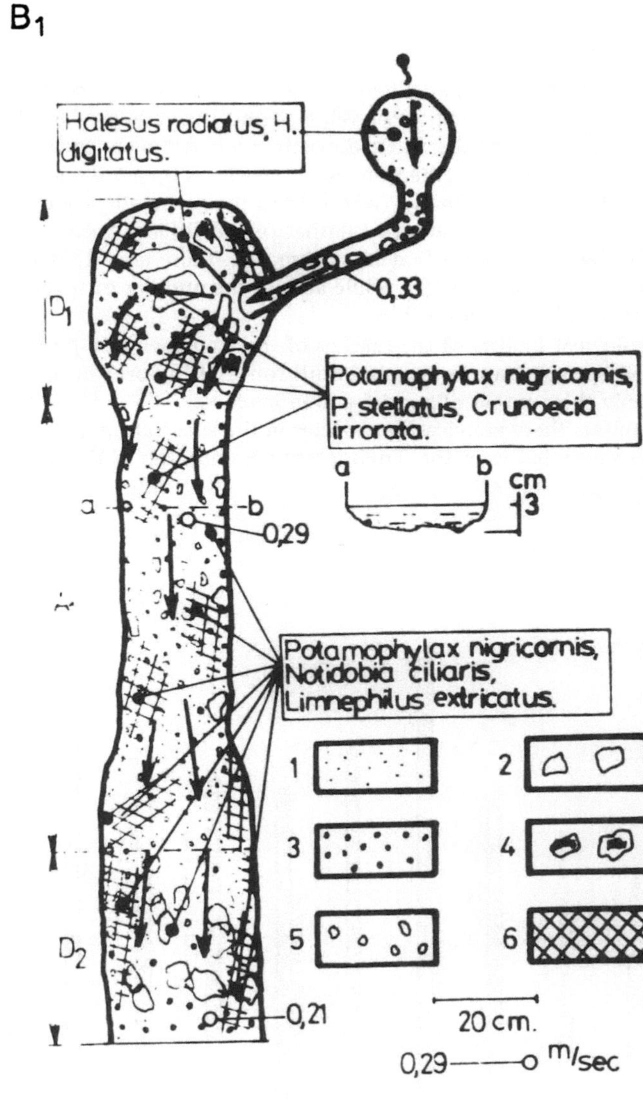

Fig. 6. The spring Disznóskut (B_1). Substratum mosaics: 1. slime, 2. large stones, 3. sand, 4. moss, 5. small stones and gravel, 6. detritus.

Conclusions

The increasing number of species and individuals of trichopterous larvae downstream from the spring can be explained by such ecological factors as the substratum of plant origin, the changing water velocity and the variety of food, as well as the increasing temperature of the water downstream in summer. At the station A_1, where there is no moss or slime, the number of species is 5, and the annual number of individuals is 107. At station B_1, where both moss and slime are present, the number of species is 7, and the annual number of individuals is 492. In the lower reaches of the stream it is found that at station A_{10} there are 11 species, and the annual number of individuals in 197, while at A_{12} the number of species is 17, and the annual number of individuals is 206.

Thus the ecological factors of the reaches of the stream change positively downstream from the spring towards the metarhithron and hyporhithron and they influence favourably the number of species and individuals.

In similar habitats there are vicarious species in the substratum mosaic. There are substantial differences between the Trichoptera communities in the similar mosaics

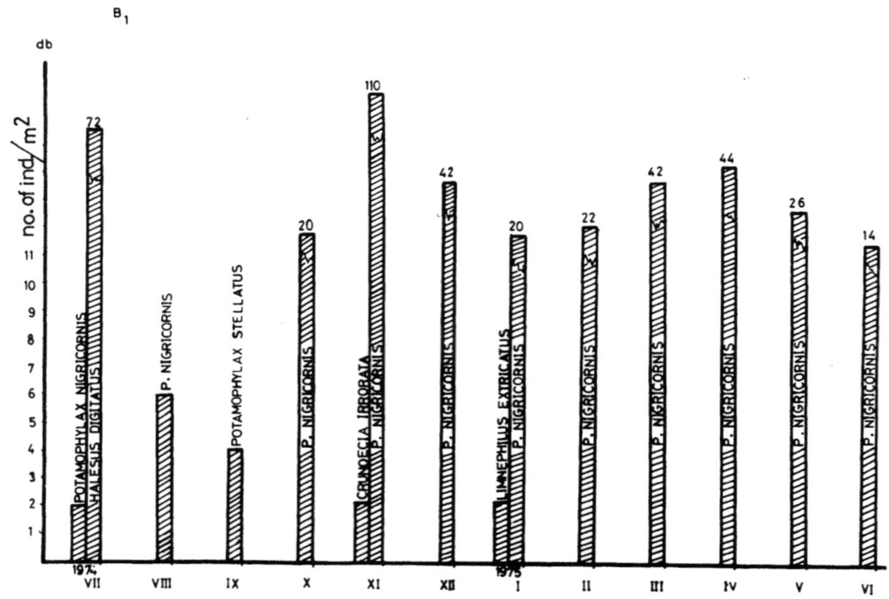

Fig. 7. The species found monthly and the number of individuals per m^2 at the station B_1.

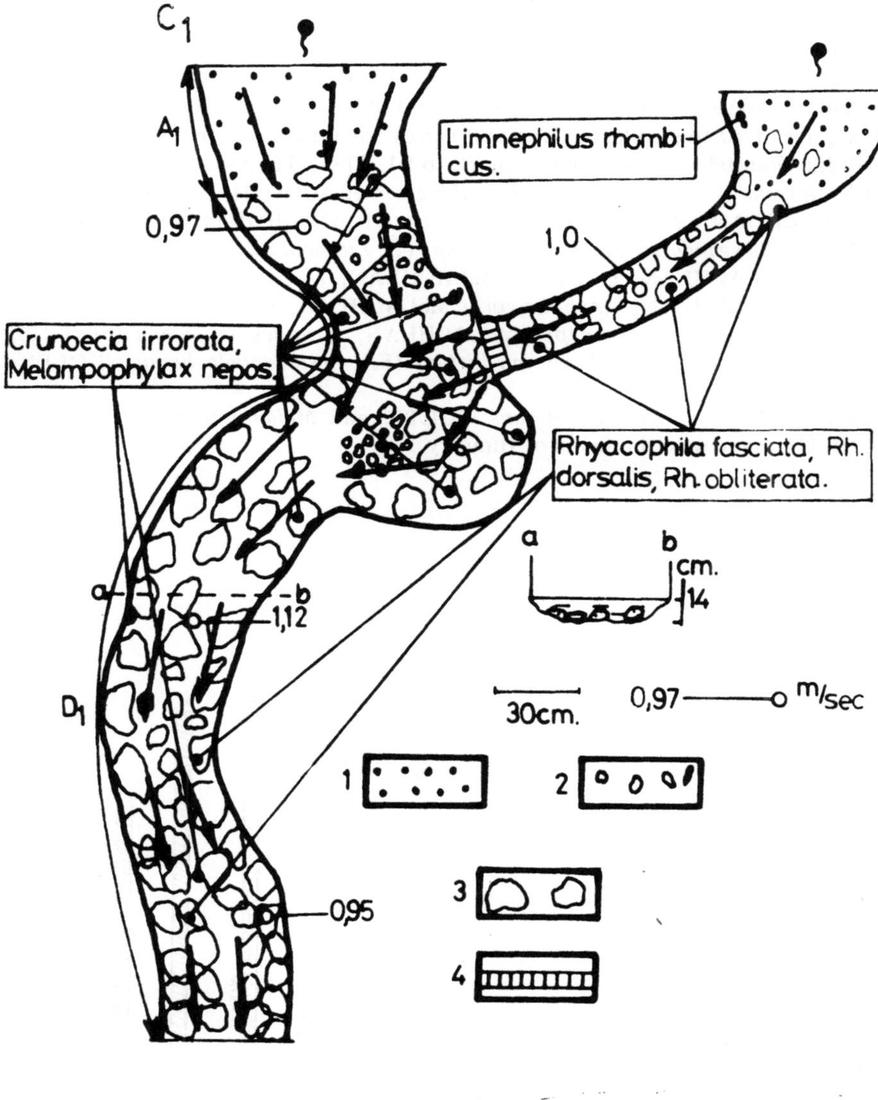

Fig. 8. The spring Huba (C_1). Substratum mosaics: 1. sand, 2. small stones and gravel, 3. large stones, 4. rapids.

of the stations A_1, A_2, A_3, A_4, A_{11}, B_2, B_3. Thus on the large stone mosaics of stations A_1, A_2, A_4, A_7 no *Rhyacophila fasciata* were found, but instead there was *Silo pallipes*. In the sand mosaic of station B_1 *Crunoecia irrorata* occurred; in the similar mosaics of station B_2 it was replaced by *Lepidostoma hirtum*.

As a result of this collecting the number of trichopterous species known from the Bükk Mountains is increased from 72 to 91. These 19 new species are: *Rhyacophila dorsalis* (CURTIS), *Wormaldia subnigra* McLACHLAN, *Plectrocnemia brevis* McLACHLAN, *Tinodes pallidulus* McLACHLAN, *Hydropsyche pellucidula* (CURTIS), *Phryganea striata* L., *Drusus* sp. (?), *Limnephilus elegans* CURTIS, *L. extricatus* McLACHLAN, *L. ignavus* McLACHLAN, *L. nigriceps* ZETTERSTEDT, *L. politus* McLACHLAN, *Anabolia soror* McLACHLAN, *Potamophylax latipennis* (CURTIS), *Melampophylax nepos* McLACHLAN, *Ernodes articularis* (PICTET), *Mystacides azurea* (L.), *Lepidostoma hirtum* (F.), *Lasiocephala basalis* (KOLENATI).

Two species, *Limnephilus elegans* CURTIS and *Melampophylax nepos* McLACHLAN, are new to the fauna of Hungary.

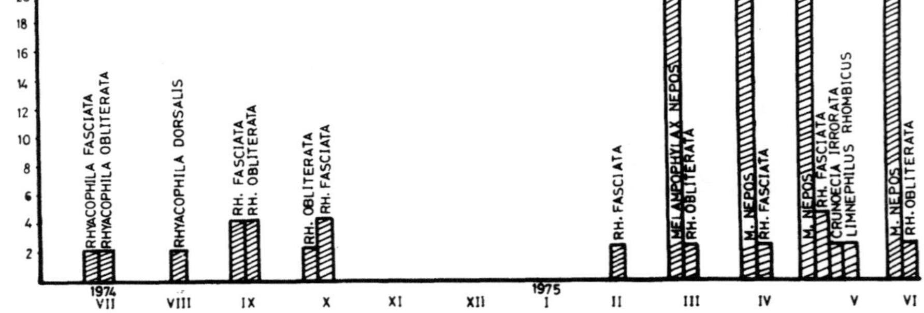

Fig. 9. The species found monthly and the number of individuals per m^2 at the station C_1.

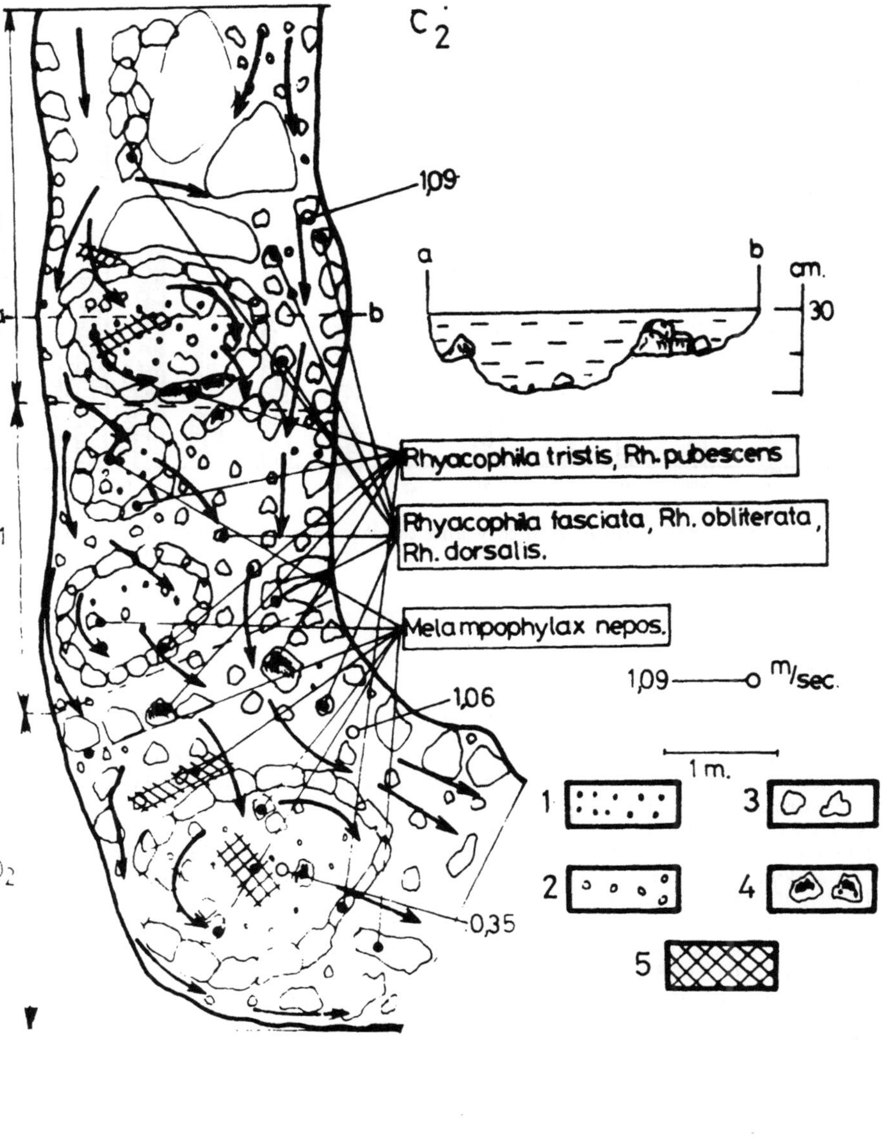

Fig. 10. The reach of the stream Szalajka with torrents (C_2). Substratum mosaics: 1. sand, 2. small stones and gravel, 3. large stones, 4. moss, 5. detritus.

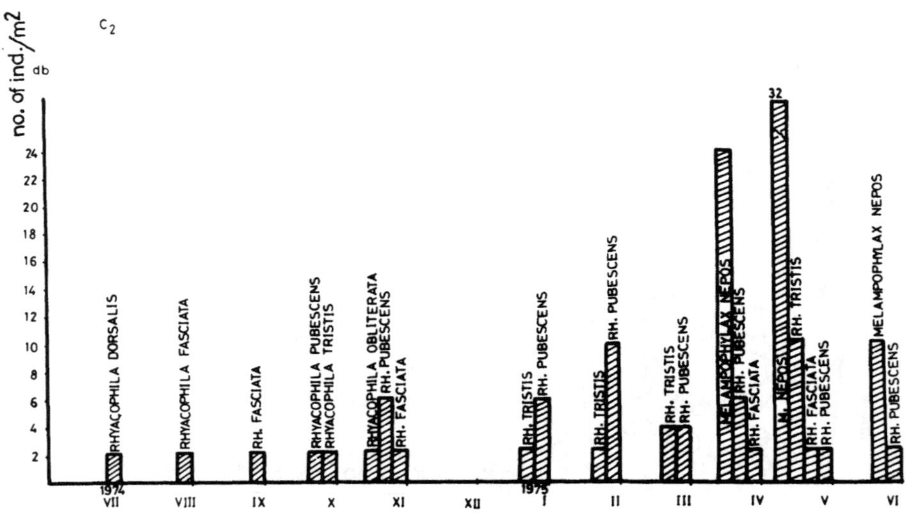

Fig. 11. The species found monthly and the number of individuals per m² at the station C_2.

References

HICKIN, N.E. 1967. Caddis larvae; larvae of the British Trichoptera. London: Hutchinson.

KAMLER, E. & RIEDEL, W. 1960. The effect of drought on the fauna of Ephemeroptera, Plecoptera and Trichoptera of a mountain stream. Polskie Archwm Hydrobiol. 8: 87-94.

KENDEIGH, C. 1961. Animal Ecology. London.

KISS, O. 1977. Trichoptera ökológiai vizsgálatok jellegzetes Bükk hegységi forrás-és patakvizekben (Szalajka-, Disznóskut-, Sebesviz). Thesis, Debrecen.

LEPNEVA, S.G. 1966. Fauna SSR/Akad. Nauk. SSSR, Moszkva, Tom I-II.

McLACHLAN, R. 1874-80. A monographic revision and synopsis of the Trichoptera of the European fauna. (Reprint) Hampton: Classey.

MACAN, T.T. 1958. Methods of sampling the bottom fauna in stony streams. International Association of Theoretical and Applied Limnology, Comm. 8.

MARLIER, G. 1951. La biologie d'un ruisseau de Plaine. Mem. Inst. Sc. nat. Belg. 114: 1-98.

OLAH, J. 1967. Untersuchungen über die Trichopteren eines Bachsystems der Karpaten. Acta biol. Debrecina, 5: 71-91.

SCHMID, F. 1951. Monographie du genere *Halesus* (Trich.). Trab. Mus. Cienc. Barcelona, I: 63-65.

STEINMANN, H. 1970. Tegzesek — Trichopterák. Fauna Hungariae. Budapest, XV.

SZABO, J. es tarsai. 1971. Hidroökologiai vizsgálatok a Bükk es a Zempléni-hegység vizeiben. II. rész Acta biol. Debrecina, 9: 187-195.

Discussion

STATZNER: What is the influence of the pools on the number of species downstream?

KISS: The pools may have a little influence on the number of species. In the stream Szalajka, which is connected with four pools, it is mainly the rheophile species characteristic of flowing water, which show an increasing number of species. This increase, both in the stream and the pools, seems to be very slow. At sampling station A_1 there were 5 species, with 2 also in the pool; at A_5 (250 m from the outflow of the pool) it was 12, with 4 species also in the pool; at A_{12} (500 m from A_5) it was 17 species, with 4 also in the pool.

WINTERBOURN: Do the larvae of *Potamophylax nigricornis* feed on beech leaves, and are leaves present throughout the year?

KISS: The larvae of *P. nigricornis* feed on the leaves which fall into the bed of the rill (i.e. on the förna). The leaves can be found there throughout the year.

Proc. of the 2nd Int. Symp. on Trichoptera, 1977, Junk, The Hague

A progress report on studies and some characteristics of the Bulgarian caddis fauna

K. KUMANSKI

Abstract

This report gives the present state of knowledge of Trichoptera in the north-eastern part of the Balkan peninsula.

Because of the absence of any large inland lakes, Bulgaria is one of the countries with lowest freshwater resources in Europe. On the other hand, despite the small size of the territory (c. 111,000 km^2) the altitude ranges from 0 to nearly 3000 m above sea-level. Thus there is a great variety of ecological conditions and a rich freshwater fauna, at least of Trichoptera, has now been confirmed.

There are now 212 species of Trichoptera known in Bulgaria, with some 15 others, which have been wrongly reported, or whose presence has yet to be confirmed. If the total number of species is 240 to 250, the recorded species represent 85-90% of the total fauna. Naturally, there are still problems about the distribution of some of the species, but the present knowledge of the order has allowed the completion of the appropriate volume of the series *Fauna Bulgarica*.

Almost all the families of Trichoptera known in Europe have been found in Bulgaria. The only ones absent are the Arctopsychidae and the Molannidae. Species of Molannidae may perhaps be found, but it seems almost certain that the family Arctopsychidae is absent.

The largest family is the Limnephilidae, with 66 species in Bulgaria, about one quarter of approximately 250 species known in Europe, which represents about the same proportion as the 212 species of Bulgarian caddis flies do to the 800 species found in Europe. The representation of other large families is given in Table 1.

The lowest representation is seen in the Rhyacophilidae, at about 18%, while the Bulgarian Hydropsychidae and Leptoceridae include about 45% of the European representatives of these families. This higher proportion of thermophilic families may be considered as a characteristic of the Bulgarian caddis fauna.

Some of the species described in Bulgaria illustrate another characteristic of its fauna, since most have a more or less limited distribution in the Balkan peninsula. The best defined endemics can be found among the mountain species. In Bulgaria (and especially in zone 7 after ILLIES (1967)) these are, for example, both *Chae-*

Table 1. Quantitative correlations between some caddisfly families in Europe and Bulgaria

Family	SPP. IN EUROPE (after Illies, '67)	SPP. IN BULGARIA	% OF THE BULGARIAN TO THE EUROPEAN SPP.
RHYACOPHILIDAE	72	13	18
LIMNEPHILIDAE	250	66	26
HYDROPTILIDAE	80	21	26
PHRYGANEIDAE	19	5	26
PSYCHOMYIIDAE	40	11	28
GLOSSOSOMATIDAE	35	11	31
LEPTOCERIDAE	70	31	44
HYDROPSYCHIDAE	31	14	45

topteryx maximus and *C. bulgaricus*, *Psilopteryx schmidi*, *Chionophylax monteryla*, *Potamophylax borislavi* (all of them in the Rila-Rhodops mountain system), *Drusus bureschi* and *D. discophorus balcanicus* from the Staraplanina mountain range. Much more numerous is another group of species, whose distribution includes more than one mountain region in the Balkan peninsula. These species occur in most cases in zones 5, 6 and 7 (ILLIES, 1967). The following caddisflies are examples of this group: *Chaetopteryx stankovici*, *C. bosniaca*, *Annitella triloba*, *Drusus discophorus discophorus*, as well as *Rhyacophila loxias* and *Odontocerum hellenicum*.

Another group of species, more or less widespread in the Balkan peninsula, is found also in the Carpathians. Such are *Rhyacophila fischeri*, *R. furcifera*, *Glossosoma discophorum*, *Agapetus belareca*, *Chionophylax mindszentyi* (with the nominate form in the North-Carpathians and the subspecies *bulgaricus* in the Staraplanina Mountains), *Drusus romanicus* (also with two subspecies), *Chaetopteryx cissylvanica* and *Chaetopterygopsis sisestii*.

All the examples mentioned above illustrate the important role of the Balkan and the Carpathian mountains as a genetic centre, which sometimes looks rather diffuse. We find a good example of this in the genus *Drusus*, where a series of species, endemics from separate mountain massives, have been recently discovered. On the other hand, all the species mentioned above have their related forms in other parts of Europe. Sometimes the related species coexist in the Balkan peninsula. The species of *Chaetopterygopsis* are an example of this fact; thus *C. sisestii*, which is abundant in the Carpathians and the high mountains of Bulgaria, may coexist with *C. maclachlani*. The density of the latter, however, is very low and there is no doubt that its southern boundary of distribution passes through our country. In many other cases a total replacement of species or even genera takes place. Thus *Chaetopteryx villosa*, the most widespread representative of the genus as a whole, is absent from the Balkans. Instead, *C. stankovici* (from the same group) is the commonest Chaetopterygini species in the central and the eastern part of the Balkan peninsula. The single Balkan *Annitella* species is also very common in our mountains. Much the same position is found in *Psilopteryx* and *Chionophylax*. It is worth mentioning that *Asynarchus lapponicus* is the only Bulgarian caddisfly with an arcticalpine type of distribution.

One must remember that the greatest part of the Bulgarian species not mentioned so far have a wide distribution in Europe and even in the Palaearctic and Holarctic.

The southern elements in the Bulgarian fauna are not so numerous and yet they leave an original mark on it. This includes a group of species more or less widespread in the Mediterranean area whose origin is not homogeneous. On the one hand it includes some of the most southern species of the Limnephilidae, the largest family in the northern hemisphere, those from the complex of *Stenophylax-Micropterna*. Thus, for instance, *Micropterna caesareica* and *M. malaspina* have their most northern localities in Bulgaria. The same may perhaps be said of *Limnephilus tauricus*. All of them have been described from Asia Minor. On the other hand, there is *Calamoceras illiesi*, of the family Calamoceratidae, which takes the place of the Iberian *C. marsupus*, whose Bulgarian localities represent the northernmost extension of that subtropical and tropical family.

Thanks mainly to the intensive investigations by MALICKY a series of *Tinodes* species in the Aegean region has been discovered during the past years. Most of them are closely related island endemics. The influence of this genetic centre upon the Bulgarian fauna, though limited, can be demonstrated by the endemic *Tinodes popovi* from the east Balkan mountains. The inhabitant of Continental Greece, *T. janssensi*, has also been found in the southern parts of Bulgaria. *Tinodes raina* is also probably a southern species whose *terra typica*, the most southern branches of the Carpathians, could be considered as a northern refuge for it.

Some *Wormaldia*-forms also show the relation between the caddisflies of the Balkans and Asia. This is a group recently established, with a basic species *W. khourmai* from N-Iran (KUMANSKI, in press). A separate subspecies occurs both on the Balkan peninsula and in Asia Minor and another probable subspecies inhabits eastern Anatolia. A second species belonging to the group of *W. khourmai* has also been found in Bulgaria.

Species common both to the Balkan peninsula and Asia Minor can also be found in some other families, e.g. *Lithax musaca* from the goerids, *Hydropsyche valkanovi* from the hydropsychids, *Hydroptila vichtaspa* from the hydroptilids, as well as *Setodes viridis* from the leptocerids. The last of these exists in four subspecies and has a distribution from Central Europe to both Israel and Iran to the south-east. *Setodes alexanderi*, found in Bulgaria and Portugal, is another example of a Mediterranean leptocerid.

In general, the southern elements are less than 10% of the Bulgarian caddis flies. In spite of this they can be considered as one of its most significant characteristics.

Table 2. Zoogeographical analysis of the Bulgarian fauna of Trichoptera

ZOOGEOGRAPHICAL CATEGORIES	NUMBER OF SPP.	% OF TOTAL
HOLARCTIC	8	3.5
HOLOPALAEARCTIC	14	6.5
W-PALAEARCTIC	54	26.0
HOLOEUROPEAN	33	16.0
CENTRAL AND S-EUROPEAN	36	17.5
MEDITERRANEAN	17	8.0
CARPATHIAN-BALKAN	15	7.0
BALKAN	19	8.5
PROBABLE ENDEMICS FROM NE-BALKANS (BULGARIA)	16	7.0
TOTAL	212	100.0

Table 2 shows the absolute and the specific rates of the zoogeographical categories forming the caddis fauna of Bulgaria.

No Trichoptera species overwinter as adults in Bulgaria. The earliest date for finding adults is at the beginning of April, so it can be supposed that the earliest emergences take place in March. Living adults have been collected up to the middle of December. Some different groups of species can be distinguished when analysing the phenology of the adults. The best formed are the two antipodean groups — one of autumnal and the other of vernal species, with about 15 species in each group. The autumnal species are very homogeneous, being mainly limnephilids, with all the Chaetopterygini and two *Halesus* species, and a single rhyacophilid, *Rhyacophila obliterata*. Emergence begins not earlier than September. The group of vernal species includes representatives from various families: Glossosomatidae, Polycentropodidae, Psychomyiidae, Limnephilidae, Goeridae etc., none of them dominating. Emergence usually finishes before the end of June. Some of these species in Bulgaria are *Agapetus belareca*, *Holocentropus stagnalis*, *Polycentropus ierapetra*, *Tinodes raina*, as well as both *Chionophylax monteryla* and *C. mindszentyi bulgaricus*. The two following phenological groups include the majority of our Trichoptera: that of vernal-aestival species (c. 70), emerging between April and September, and that of the aestival species (c. 60), emerging between June and September. The species variety is very high and almost all families are represented. Another small group is that of aestival-autumnal species, whose adults emerge from June and July up to November. *Rhyacophila loxias*, *Potamophylax latipennis*, *P. pallidus* and *Allogamus uncatus* are examples of this group. Naturally, some corrections for isolated species may be made in the future. In this sense, the last phenological grouping is the most stable one; it includes species with very extended periods of emergence, from the spring till the late autumn; 38 species have such a period in

Bulgaria and it seems probable that the real number is larger.

The varied relief of the territory explains the existence of some different altitudinal groups in the Bulgarian Trichoptera, and so far about three-quarters of the species can be classified in this way.

The small group of about 10 high-mountain species contains mainly limnephilids. The lowest line of distribution is 1300-1400 m above sea-level. All members are cold-requiring stenotope species, inhabiting flowing and lake waters in the coniferous and the subalpine mountain zones. Examples are *Asynarchus lapponicus*, *Drusus romanicus* and *D. discophorus discophorus*, *Chionophylax monteryla*, *C. mindszentyi bulgaricus* and *Limnephilus coenosus*.

The most numerous is the group of about 65 mountain species, with a wider altitudinal range. The important feature here is the lowest line of distribution, usually between 600 and 1000 m. As an exception, some mountain species have been found at still lower levels in the Strandzha Mountains. The Limnephilidae, with 27 species, is dominant here. The Rhyacophilidae are present with 9 species which is 70% of the Bulgarian species. Other families are also represented in this group with isolated species, e.g. some *Hydropsyche* (*fulvipes*, *instabilis* and *tabacarui*), *Philopotamus*, *Silo*, *Beraea*, *Lasiocephala basalis*, *Odontocerum hellenicum* and *Adicella filicornis*.

As a contrast to the two mountain groups mentioned above, the species of the low altitude group are clearly missing from the mountains. They are found from sea-level up to 600-700 m. The group is relatively large, with about 45 species. Dominant families here are the Leptoceridae (both absolutely, 22 spp., and in percentage, as c. 70% of the species), and Hydroptilidae and Hydropsychidae with half of their Bulgarian species. Single representatives of other families are, for example, *Helicopsyche bacescui*, *Calamoceras illiesi*, *Neureclipsis bimaculata* and *Lype reducta*. Naturally, this group typifies the most warm-requiring elements of the Bulgarian Trichoptera.

Finally, for 23 other species, the limiting role of the altitudinal factor is, generally, unimportant. Their distribution therefore extends from sea-level up to more than 1500 m; in some extreme instances, e.g. *Wormaldia triangulifera asterusia*, *Plectrocnemia conspersa*, *Hydropsyche pellucidula*, some *Limnephilus* (*affinis*, *flavicornis*, *griseus*, *vittatus*) and *Micropterna sequax*, the altitudinal range reaches to 2200-2300 m.

Together with faunistics, the taxonomy of adults is the other most advanced aspect of our knowledge of Trichoptera in Bulgaria. In some isolated cases only (*Wormaldia*, *Tinodes*, *Polycentropus*, part of *Hydropsyche* and some other genera mainly from Hydropsychoidea) the females remain indistinguishable. Our knowledge of preimaginal taxonomy is very different. Together, with the immature stages of most of the local Balkan species, those of many of the widespread caddis flies are wholly or, at least, practically unknown. Sometimes larval determination cannot even be brought to generic level. Such, for instance, are the larvae of the *Stenophylax-Micropterna*-complex, as well as those of the genera *Potamophylax*, *Alloga-*

mus, Rhadicoleptus, Drusus, Ecclisopteryx, Chaetopteryx, Annitella, Chaetopterygopsis, Psilopteryx, Oecismus and *Sericostoma.*

I am not sure if the situation is much different in other European countries, but a work comparable to that of WIGGINS (1977) on the larvae of the N-American genera, is much needed in Europe.

References

ILLIES, J. 1967. Limnofauna Europaea. Stuttgart: Gustav Fischer.

KUMANSKI, K. (in press) To the knowledge of the genus *Wormaldia* (Philopotamidae, Trichoptera) in the Balkans and Asia Minor.

WIGGINS, G.B. 1977. Larvae of the North American caddisfly genera (Trichoptera). Toronto: University of Toronto Press.

Discussion

MORETTI: Est-ce-que vous avez trouvée des formes d'eau saumâtre en Bulgarie?

KUMANSKI: Yes, the hydroptilid *Agraylea sexmaculata* Curt. is the only species found in brackish waters, at the Black Sea coast near Varna.

Proc. of the 2nd Int. Symp. on Trichoptera, 1977, Junk, The Hague

Larval and imaginal diapauses in Limnephilidae

C. DENIS

Abstract

The features of larval diapause differ according to the species. In females undergoing imaginal diapause, the state of immaturity varies from one species to another.
Diapauses are conditioned by the action of the photoperiod during the entire larval life. In the case of larval diapause, it is the third instar which is particularly sensitive to this action and in the case of imaginal diapause, it is the fifth.

Several species of Limnephilidae enter diapause as a result of the long summer days, either at the end of their larval period of growth or during their adult life. NOVAK (1959) and then NOVAK & SEHNAL (1963 and 1965) were the first to observe these phenomena in Trichoptera, which result in the egg-laying periods being synchronised in the autumn. Thus, a larval diapause delays metamorphosis, the adults only appearing in September or October. At this time the eggs are almost fully developed and are laid a few days after the females emerge. An imaginal diapause inhibits ovarian development during summer. In this case the adults appear as early as spring but when they emerge the females are immature and their eggs will only ripen in the autumn.

For several years now, I have been engaged in the study of these diapauses from comparative and deterministic points of view.

1. Comparative study

a. *Larval diapause*

In the region surrounding Rennes, species which clearly manifest a larval diapause are: *Anabolia nervosa*, *Halesus radiatus*, *H. digitatus*, *Chaetopteryx villosa*. These insects enter diapause at the end of May or in June, metamorphosing at the end of August or in September. Although these species' cycles appear to be synchronous, differences are manifested during the last period of the larval life. The larvae of *A. nervosa* plug the two openings of their case in autumn, just before metamorphosis, whereas those of the other three species can build the anterior sieve membrane of their case as early as June, i.e. before or during the diapause.
A comparative study of *A. nervosa* and *H. radiatus* gives the following results. From

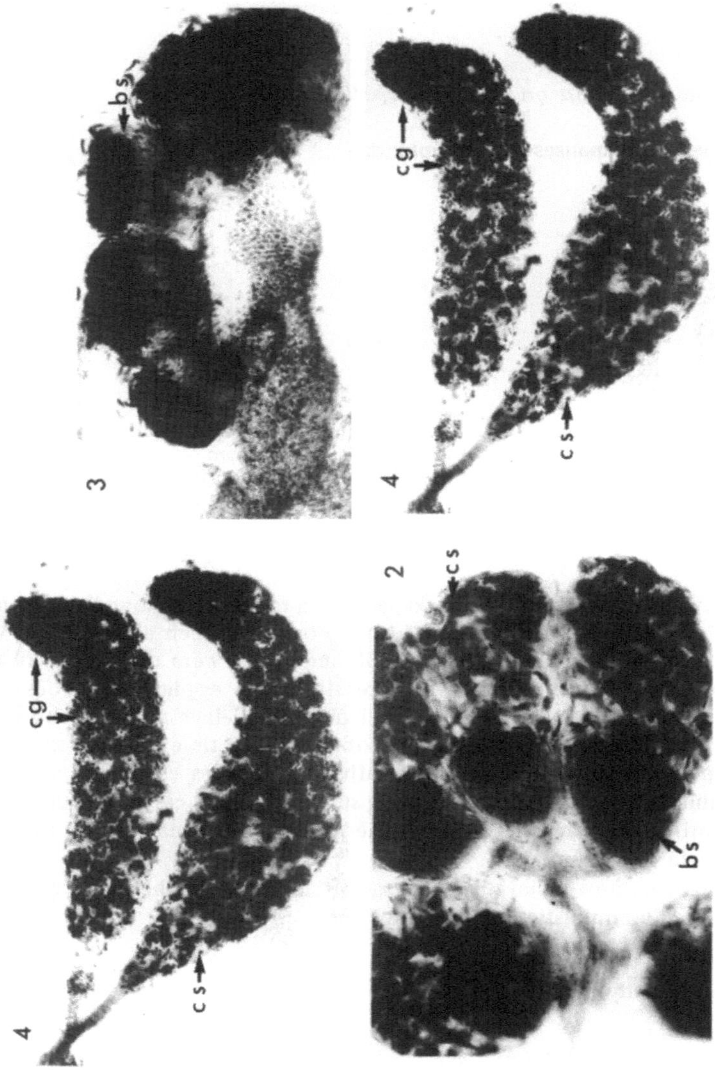

Plate I. Testicular development of *Anabolia nervosa*, 1. Typical lacunose aspect of testicular lobes during diapause, the central area being empty, 2. At the beginning of the prepupal period, the cellular development has resumed and the central area of the lobes is full of cysts in spermatogenesis, 3. At the imaginal moult all the bundles of spermatozoïds are formed and arranged in parallel lines at the base of the lobes, 4. The spermatogenesis is delayed in males without diapause. Aspect of testicular lobes at the beginning of the prepupal period: spermatogenesis is only beginning (to be compared with photo 2), b s: bundles of spermatozoïds arranged in parallel lines at the base of the lobes, c e: central area empty, c g: young cysts and germarium, c s: cysts in spermatogenesis.

a chronological point of view, the final stages of larval life differ in both species. At a temperature of 15°C., *A nervosa* builds the anterior and posterior membranes of its pupal case ten and eight days respectively before the pupal moulting. Although the posterior sieve membrane of the case of *H. radiatus* is likewise built eight days before the pupal moulting, the anterior one is built from 20 days to 3 months before this moulting. From the internal development point of view, the larvae of *A. nervosa* undergo no change during the diapause and the imaginal disks remain latent. Males whose spermatogenesis began before the end of larval growth have testicles with a characteristically lacunose appearance: the bundles of the formed spermatozoïds are arranged in parallel lines at the base of the lobes whose apex is occupied by the young cysts and the germarium, the central area appearing empty (Plate I, fig. 1). Spermatogenesis will only recommence after the diapause. Similarly certain larvae of *H. radiatus* in diapause which have latent imaginal disks and lacunose testicles have not begun to plug their cases. In the others, however, pupal organs have begun to grow and in the testicles the central area of the lobes is full of cysts undergoing spermatogenesis. In previous studies it has been shown that the larvae of *A. nervosa* enter diapause with either a summer or a winter photoperiod, whilst the larvae of *H. radiatus* only do this with a summer photoperiod (DENIS, 1972 and 1973).

b. *Imaginal diapause*

Although all females which have an imaginal diapause are immature after emerging, the state of development of their ovaries differs according to the species (LE LANNIC, 1976). *Glyphotaelius pellucidus* shows the greatest immaturity: the first follicles to be formed are only distinguished with difficulty, their ovocyte and trophocytes stay very small. *Limnephilus lunatus* and *Grammotaulius atomarius* have more distinct follicles; ovocyte and trophocytes having increased in size. The follicles of *L. rhombicus* and especially of *L. centralis* are even greater in volume.

Such differences may explain the gaps observed between the egg-laying periods. The egg-masses laid by *L. centralis* may be seen at the beginning of September whilst those of *L. lunatus* and *G. pellucidus* can only be observed at the end of September or beginning of October.

The special localisation of the egg-masses laid by *G. pellucidus* must be mentioned here. Along forest streams, LE LANNIC and I have observed hundreds of them on the vegetation along the banks and especially on the foliage of the trees up to a height of three meters above the bed of the stream. For the species I have studied, a diapause only occurs with a summer photoperiod.

2. Necessary conditions for diapause and for obtaining cycles without diapause

In the laboratory it has been possible to obtain cycles without a diapause by rearing Limnephilid larvae under a long photoperiod. This led me to study the necessary conditions to obtain cycles with diapause and without diapause.

Table 1. Experimental study of diapause-conditioning in *Anabolia nervosa.*

PHOTOPERIOD DURATION					RESULTS
1 st instar	2 nd instar	3 rd instar	4 th instar	5 th instar	
12	12	12	12	12	in each experimental case:
18	12	12	12	12	DIAPAUSE
12	*18*	12	12	12	most of the animals died during diapause; however a few of them went into metamorphosis 10 or 11 months after the last larval moulting
12	12	12	*18*	12	
12	12	12	12	*18*	DIAPAUSE... thereafter, metamorphosis from 5 to 6 months after the last larval moulting
12	12	12	*18*	*18*	DIAPAUSE... thereafter, metamorphosis from 5 to 6 months after the the last larval moulting
12	12	*18*	*12*	*12*	in each experimental case:
12	12	*18*	*18*	12	- Several animals had a
12	*18*	*18*	12	12	DIRECT DEVELOPMENT
18	*18*	12	12	12	... and for them, metamorphosis began from 1 tot 2 months after the last larval moulting
18	*18*	*18*	12	12	
12	*18*	*18*	*18*	12	- But the other animals went into DIAPAUSE
18	*18*	*18*	*18*	12	... and all these animals died during diapause; some of them did 10 or 11 months after the last larval moulting
18	*18*	*18*	*18*	*18*	in each experimental case, the pupal instar began from 2 to 4 months after the last larval moulting:
12	*18*	*18*	*18*	*18*	- The animals with a 5 th instar shorter than 3 months had a DIRECT DEVELOPMENT
12	12	*18*	*18*	*18*	- The animals with a 5 th instar longer than 3 months had a DIAPAUSE

a. *Larval diapause*. Research carried out on *Anabolia nervosa*

Batches of 20 to 30 larvae were subjected to either a short photoperiod (12 hours) or a long photoperiod (18 hours) during each instar of their development. These experiments and their results are recorded in Table I.

Note: The presence or absence of a diapause was determined both by the existence or non-existence of a long gap between the end of feeding and metamorphosis and by the state of the testicles. At the beginning of the prepupal period the lobes of the testicles are lacunose if the insect has undergone a diapause. If there has been no diapause, the whole area is occupied by cysts which are more or less developed.

A diapause necessarily occurs if one only of the instars, with the exception of the third, or if the two final instars take place during a long photoperiod. It was noticed that larvae which undergo a diapause during a long photoperiod all metamorphose from three and a half months to six months after the last larval moulting. On the other hand, almost all larvae undergoing a diapause during a short photoperiod have their development definitely stopped.

Non-stop development can be obtained by the action of a long photoperiod during the third instar or during several consecutive instars. In these cases, if the fifth instar takes place with a short photoperiod, the insects stop feeding one or two months after the last larval moulting, and some of them begin to pupate. The others will start feeding again and then definitely enter diapause. The proportion of the ones to the others seems to be a result of their feeding habits. Thus, those fed on grass will practically all follow a line of non-stop development, whilst the majority of those fed on dead leaves will enter diapause. If the fifth instar takes place with a long photoperiod, all the insects start to pupate between two and four months after the last larval moulting. An examination of the males' testicles, when the front opening of the case has just been plugged, reveals that the animals having started to pupate less than three months after the last larval moulting had non-stop development (full testicular lobes), whilst the others underwent a diapause (lacunose testicular lobes).

Note: Compared with the case of a normal development with a diapause, non-stop development tends to provoke a delay in the chronology of the development of the testicle (Plate I).

b. *Imaginal diapause.* Research carried out on *Limnephilus rhombicus* and *L. lunatus*

Insects were exposed to either short or long photoperiods during the successive stages of their development. The results of these experiments are recorded in table II.

Table II shows that it is during the fifth instar that conditioning necessary for the diapause is produced. If the fifth instar takes place with a short photoperiod

Table 2. Experimental study of diapause-conditioning in *Limnephilus rhombicus* and *L. lunatus*.

PHOTOPERIOD DURATION				
1-4 th instars	5 th instar	pupal instar	adult instar	RESULTS
12	12	12	*18*	DIAPAUSE
12	12	*18*	*18*	DIAPAUSE
18	*18*	12	*18*	*DIRECT OVARIAN MATURATION*
18	*18*	*18*	*18*	*DIRECT OVARIAN MATURATION*
18	12	12	*18*	DIAPAUSE
18	12	*18*	*18*	DIAPAUSE
12	*18*	12	*18*	*DIRECT OVARIAN MATURATION*
12	*18*	*18*	*18*	*DIRECT OVARIAN MATURATION*

Table 3. Effect of a photoperiod variation during the 5th larval instar in *Limnephilus lunatus*. In all cases, pupal and imaginal instars have had a photoperiod of 18 hours.

PHOTOPERIOD DURATION				
1-4 th		5 th instar		RESULTS
instars	0-10 days	10-20 days	20-30 days	
12	*18*	12	12	DIAPAUSE
12	12	*18*	12	DIAPAUSE
12	12	12	*18*	DIAPAUSE
12	12	*18*	*18*	DIAPAUSE
12	*18*	*18*	12	DIAPAUSE
18	*18*	*18*	12	*DIRECT OVARIAN MATURATION*
18	*18*	12	12	*DIRECT OVARIAN MATURATION*

then the adults undergo a diapause, but with a long photoperiod, the females show direct ovarian development.

After other experiments during the course of which the photoperiod was modifield during the fifth instar (Table III), it appeared that the action of a long photoperiod during only a fraction of the instar was not sufficient to bypass the diapause unless the four previous instars had also taken place under the long photoperiod.

References

DENIS, C. 1972. Etude du cycle biologique de *Limnephilus lunatus* (Trichoptera, Limnephilidae). Obtention de deux générations annuelles. Bull. Soc. Sci. Bretagne, 47: 33-38.

—. 1972. Etude, au laboratoire, du cycle biologique *d'Anabolia nervosa* Curt. (Trichoptera Limnephilidae). Bull. Soc. Sci. Bretagne, 47: 43-48.

—. 1973. Obtention d'un cycle biologique sans diapause chez *Limnephilus rhombicus* et *Anabolia nervosa* (Trichoptera, Limnephilidae). Bull. Soc. Sci. Bretagne, 48: 197-207.

—. 1973. Influence de la photopériode sur le cycle biologique de *Halesus radiatus* (Limnephilidae). Bull. Soc. Sci. Bretagne, 48: 193-196.

LE LANNIC J. 1976. Développement de l'appareil reproducteur de quelques Trichoptères Limnéphilides et premières données expérimentales sur son fonctionnement. Thèse 3ème cycle, Rennes.

NOVAK, K. 1959. Entwicklung und Diapause der Köcherfliegenlarven *Anabolia furcata* Br. (Trichoptera). Čas. čsl. Spol. ent., 57: 207-212.

—. & SEHNAL, F. 1963. The development cycle of some species of the genus *Limnephilus* (Trichoptera). Čas. čsl. Spol. ent., 60: 68-80.

—. & —. 1965. Imaginaldiapause bei den in periodischen gewässern lebenden Trichopteren. Proc. XII Int. Congr. Ent. London, 434.

Discussion

BOUVET: Comment déterminez-vous l'age des larves de cinquième stade?

DENIS: Chez *Limnephilus lunatus*, utilisé dans la seconde série d'expériences (tableau 3) la 5ème stade dure environ 30 jours à 15°C ou plus, quelle que soit la photopériode. Pour d'autres espèces telles que *L. rhombicus* le problème serait plus délicat car le 5ème stade dure environ 1.5 a 2 mois sous photopériode longue et jusqu'à 2.5 mois sous photopériode courte.

MORETTI: Quelle est votre opinion sur le mécanisme physiologique de la diapause déclenchée par la durée de la photopériode et surtout sur les différentes sensibilités des divers stades?

DENIS: Je ne puis apporter actuellement de réponse à ces questions. Je compte bien étudier ces problemes prochainement.

WALLACE: From June onwards I have taken in Britain larvae of *Halesus radiatus* and *digitatus* which have plugged the anterior part of the case and are attached to decayed wood ready for pupation. They become active if disturbed, but otherwise I presume that they remain there until pupating. (I have however not taken *Chaetopteryx villosa* in this condition.) I am pleased that you have proved that such insects are in diapause.

Proc. of the 2nd. Int. Symp. on Trichoptera, 1977, Junk, The Hague

Adaptations physiologiques et comportementales des *Stenophylax* (Limnephilidae) aux eaux temporaires.

YVETTE BOUVET

Summary

Physiological and behavioural adaptations of *Stenophylax* (Limnephilidae) to temporary waters.

The caddisflies *Stenophylax*, which live in temporary streams as larvae and pupae, show two kinds of adaptations to the drying up of their environment.
1. Physiological adaptations: embryology may take place during emersion, larval quiescence, respiratory metabolism is adapted to the irregular mode of the streams, pupation may take place in an aerial environment, imaginal ovarian diapause.
2. Behavioural adaptations: oviposition may take place outside water, first instar larvae remain in the mucilaginous mass of the eggs during emersion, larvae hide in the ground when their environment dries up, emerging in the adult form, *Stenophylax* migrates into the subterranean environment.

Lors du premier symposium sur les Trichoptères, à Lunz (1974), j'ai présenté la position systématique, le cycle biologique et le milieu de vie des Trichoptères du groupe de *Stenophylax*.

Les derniers résultats de mes travaux sur ce groupe (BOUVET, 1977) mettent en évidence l'importance des caractéristiques des cours d'eau qui hébergent la phase larvaire des espèces considérées, essentiellement la nature temporaire de ces ruisseaux.

Pendant le développement embryonnaire, larvaire et nymphal (durée = 8 à 10 mois) le niveau de l'eau varie, les Trichoptères subissent des crues violentes ou un assèchement total séparés par des périodes en régime hydrologique moyen. Pour résister à ces variations du régime du cours d'eau, les *Stenophylax* ont développé deux types d'adaptations: 1. les adaptations physiologiques, 2. les adaptations comportementales.

1. Adaptations physiologiques

1.1. Développement embryonnaire en milieu aérien

Les oeufs sont déposés très souvent avant la remise en eau du ruisseau. Ils résistent à la desiccation grâce à la masse mucilagineuse qui les enveloppe et entretient autour d'eux un taux d'humidité compatible avec le développement embryonnaire. L'éclosion des oeufs peut se produire avant l'immersion de la ponte.

1.2. Quiescence larvaire

Lorsque le ruisseau s'assèche au cours de développement larvaire, les larves entrent en quiescence; les processus physiologiques de croissance larvaire reprennent dès la remise en eau lors d'une crue.

1.3. Adaptation du métabolisme respiratoire aux variations du taux d'oxygène dissous liées au régime irrégulier des cours d'eau

Les larves de *Stenophylax* consomment plus d'oxygène lorsqu'elles vivent en eau agitée qu'en eau calme. Ceci correspond à la possibilité d'adapter leur métabolisme respiratoire aux périodes de crue et d'à-sec. La consommation d'oxygène importante en eau courante est corrélative d'une vie en milieu bien oxygéné qui 'excite' l'animal par son agitation, tandis que la consommation plus basse en eau stagnante correspond à un réchauffement du milieu naturel qui fait baisser le taux d'oxygène dissous.

1.4. Nymphose en milieu aérien

Dès la mue nymphale, les processus de métamorphose conduisant au stade imaginal ne peuvent plus être arrêtés par une exondation. La métamorphose nymphale se déroule normalement lorsque le taux d'humidité de l'atmosphère est compatible avec la survie de la nymphe (proche de 100%).

1.5. Diapause imaginale ovarienne

A l'émergence, les ovaires de *Stenophylax* sont au stade I ou II (BOUVET, 1971). Les femelles subissent une diapause de 8 à 10 semaines, correspondant à l'assèchement estival des cours d'eau qui hébergent les stades larvaires.

2. Adaptations comportementales

2.1. Oviposition en milieu aérien

Les femelles gravides peuvent déposer leurs oeufs enrobés d'une masse mucilagineuse dans le lit du ruisseau encore à sec ou sous les pierres de la berge, au dessus du niveau de l'eau.

2.2. Séjour dans la ponte des larves de premier stade

A l'éclosion, les larves de premier stade restent dans la masse mucilagineuse de la ponte si le ruisseau est encore à sec. Elles peuvent ainsi résister plusieurs jours (7 à 10 j.).

2.3. Enfoncement des larves dans le substrat

Dès que le ruisseau s'assèche, les larves s'enfoncent entre les blocs rocheux et s'enfouissent dans le substrat où elles entrent en quiescence.

2.4. Migration des imagos

A l'émergence, les imagos en diapause migrent vers des biotopes souterrains (grottes et gouffres, fissures, tunnels artificiels, etc...), qui leur fournissent des conditions d'estivation favorables (température, humidité, stabilité des paramètres physiques). Les adultes peuvent vivre jusqu'à la remise en eau des ruisseaux qui hébergent leurs larves, en consommant les réserves accumulées dans leurs tissus adipeux.

On retrouve les différentes adaptations aux eaux temporaires chez d'autres espèces de Trichoptères (WIGGINS, 1973), mais le groupe de *Stenophylax* présente deux originalités:

— la migration des imagos vers des biotopes souterrains où ils estivent.

— le regroupement de tous les types d'adaptations aux eaux temporaires chez les mêmes espèces.

Références bibliographiques

BOUVET, Y. 1971. La diapause des Trichoptères cavernicoles. Bull. Soc. zool. Fr. 96: 375-384.

——. 1977. Conditions de vie des Trichoptères subtroglophiles (Insectes, Limnephilidae); leurs réactions aux variations des factuers du milieu. Thèse Doctorat d'Etat et Sciences Naturelles, Lyon.

WIGGINS, G.B. 1973. A contribution to the biology of caddisflies (Trichoptera) in temporary pools. Life Sc. Contrs. R. Ont. Mus. 88: 1-28.

Discussion

MORETTI: Vous avez très bien travaillé sur ce champ de recherches. Lorsque le cours d'eau s'assèche, avez-vous pu découvrir des larves ayant pénétré en épaisseur dans le fond du cours d'eau, en devenant de cette façon phréaticoles?

BOUVET: Oui, les larves du groupe de *Stenophylax* s'enfoncent dans le substrat, lorsque la granulométrie leur convient, et peuvent poursuivre leur développement dans le sous-écoulement du ruisseau.

DENIS: Lorsque les larves s'enfoncent sous terre peut-on distinguer une sortie de galerie? Comment parviennent-elles a respirer?

BOUVET: Non car l'eau dépose, avant de disparaître complètement, du limon; ce limon colmate également le fourreau des larves. La respiration est très ralentie puisque les larves sont en quiescence.

Proc. of the 2nd Int. Symp. on Trichoptera, 1977, Junk, The Hague

The effects of flight behaviour on the larval abundance of Trichoptera in the Schierenseebrooks (North Germany)

B. STATZNER

Abstract

Imaginal ecology is usually neglected in studies on the factors limiting the distribution of caddisflies. Comparing the distribution of imagines and larvae in two lake-fed brooks in North Germany it becomes evident that imaginal migration and the presence or absence of specific swarming places influence the larval distribution of several species. Flight behaviour is expected to be affected mainly by channel pattern, riparian and emergent aquatic vegetation, width of the water surface, and also by the flight of other species.

Flight behaviour of the following species is described: *Neureclipsis bimaculata, Mystacides azurea, M. longicornis, M. nigra, Athripsodes aterrimus, A. cinereus, Ceraclea dissimilis.*

1. Introduction

Classical studies on aquatic insects of running waters try to explain the occurrence, absence, or the abundance of a species in a particular habitat by reference to factors which influence the aquatic phases of their lives, i.e. temperature of water, current speed, substratum, food, and others. Although these factors doubtless have considerable effects on the running water communities, this approach disregarded factors influencing the species during their aerial life.

Only in Diptera (because of their role in human diseases) and Odonata (by reasons of the easy observation) has the aspect of the dependence of larval distribution on the imaginal ecology been worked out in some detail.

In the case of Trichoptera one finds several short references to flight behaviour in the literature and some publications deal more comprehensively with the swarming behaviour of caddis (MORI & MATUTANI, 1958; GRUHL, 1960; SCHUMACHER, 1969, BENZ, 1975); all describe fixed swarming places for the species investigated. Another topic in studies on flight behaviour of Trichoptera is upstream migration, especially of females (ROOS, 1957; NISHIMURA, 1967; LEHMANN, 1970; SVENSSON, 1974). This upstream migration, it is suggested may compensate for the downstream drift of larvae ('colonization cycle', MÜLLER, 1954a), but it is evidently not a generally occurring phenomenon in running water Trichoptera (ELLIOTT, 1967; SCHUMACHER, 1970).

This background and my own observations of flight behaviour of caddis during a comprehensive study of a running water in North Germany led me to investigate the relation between the flight behaviour and the distribution of larval caddisflies.

2. Methods and the study area

During 1974/75 nearly 14,000 larvae and pupae of Trichoptera were collected in quantitative benthos, drift, and artificial substrate samples. About 3,000 imagines were sampled by net captures from swarms, for the purpose of localizing the specific swarming places.

The running waters investigated, the 'Upper' (= Obere = OSB) and the 'Lower' (= Untere = SB) Schierenseebrook are situated between three eutrophic lakes in the 'Naturpark Westensee' near Kiel.

The two brooks differ especially in channel pattern and riparian vegetation. While the OSB is meandering and almost entirely lined by a well developed belt of *Alnus*, the USB is quite straight without a comparable dense growth of alders (Fig. 1). Thus, the bank of the OSB is very uneven with many of the alders standing in the water, while that of the USB is almost straight. In the USB there was a dense stand of *Phragmites* in the brook in K7/K8, and, mainly in summer, well developed *Ranunculus circinatus* in K9. In the other parts of the brooks aquatic macrophytes were rare, because of the type of riparian vegetation which kept irradiation low.

A detailed description of the methods used and the area investigated will be given elsewhere.

3. Results

The results are presented under the headings of single species or species groups.

3.1. *Neureclipsis bimaculata* (L.)

The highest number of flying imagines of *N. bimaculata* were caught above the outflows just below the lakes (Fig. 2). In the OSB, accumulations of flying imagines were regularly observed downstream of alders which reached far to the middle of the brook. In the USB, such accumulations (at such developed alders) were rare and occurred mainly downstream of the bridge at m 305. There the brook is narrowed by big stones, and on cold evenings the highly turbulent water increased the air temperature by as much as 2°C.

Eggs were found in the USB only in the outflow, but in the OSB they occurred further downstream. This was the case on natural substrata (eggs were nearly always laid on submerged branches) as well as on vertical bamboo sticks provided as oviposition sites (Fig. 3). In addition, it must be mentioned that suitable natural substrata were relatively rare in the upper run of the USB, due to the lack of alders.

In summer samples, first instar larvae were most abundant on the bamboo sticks

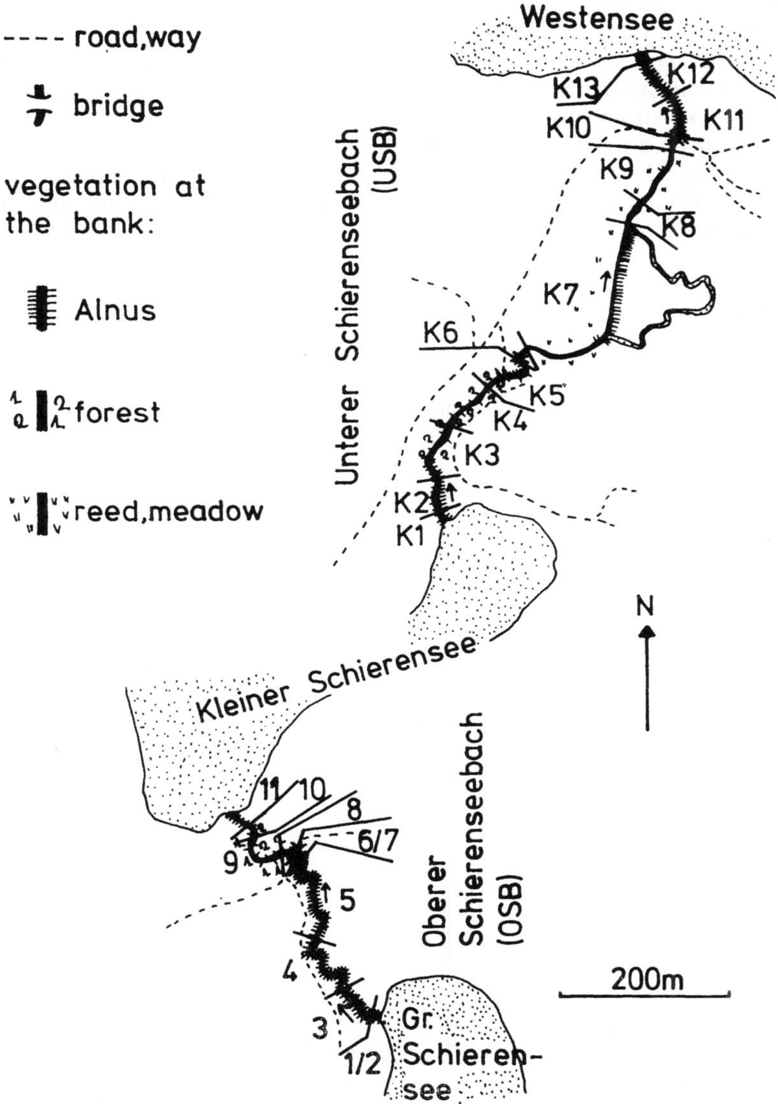

Fig. 1. The area studied in the 'Naturpark Westensee' near Kiel.

below the lakes (Fig. 3), except at the first sampling point in the OSB which had a very low current and represented lake conditions.

Later, small larvae were found further downstream. In the USB the distribution pattern of first instar larvae changed drastically from summer to autumn, parallel with a decrease in current speed at the outflow station.

The pattern of distribution in benthos samples was almost the same in both brooks with a maximum density occurring more than 100 m downstream of the lakes (Fig. 4). These results are interpreted as follows: After emergence, the imagines of *N. bimaculata* fly upstream. Straightness of the bank, determined by chan-

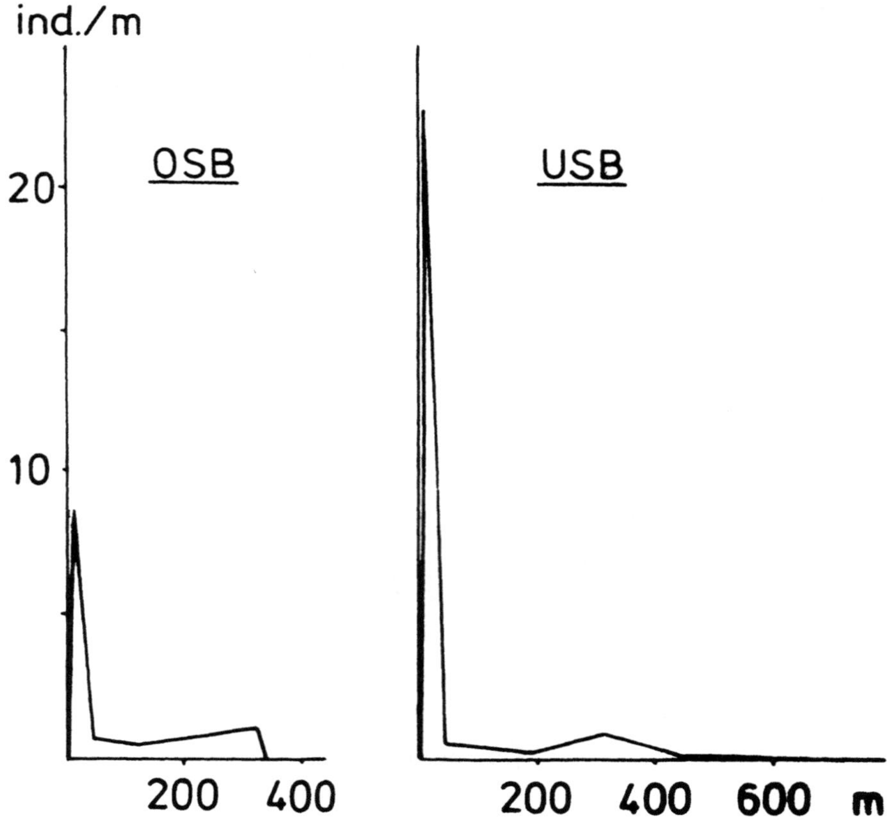

Fig. 2. *Neureclipsis bimaculata*: abundance of imagines in captured individuals (flight period 1975) per m run in sections of both brooks at different distances from the lakes.

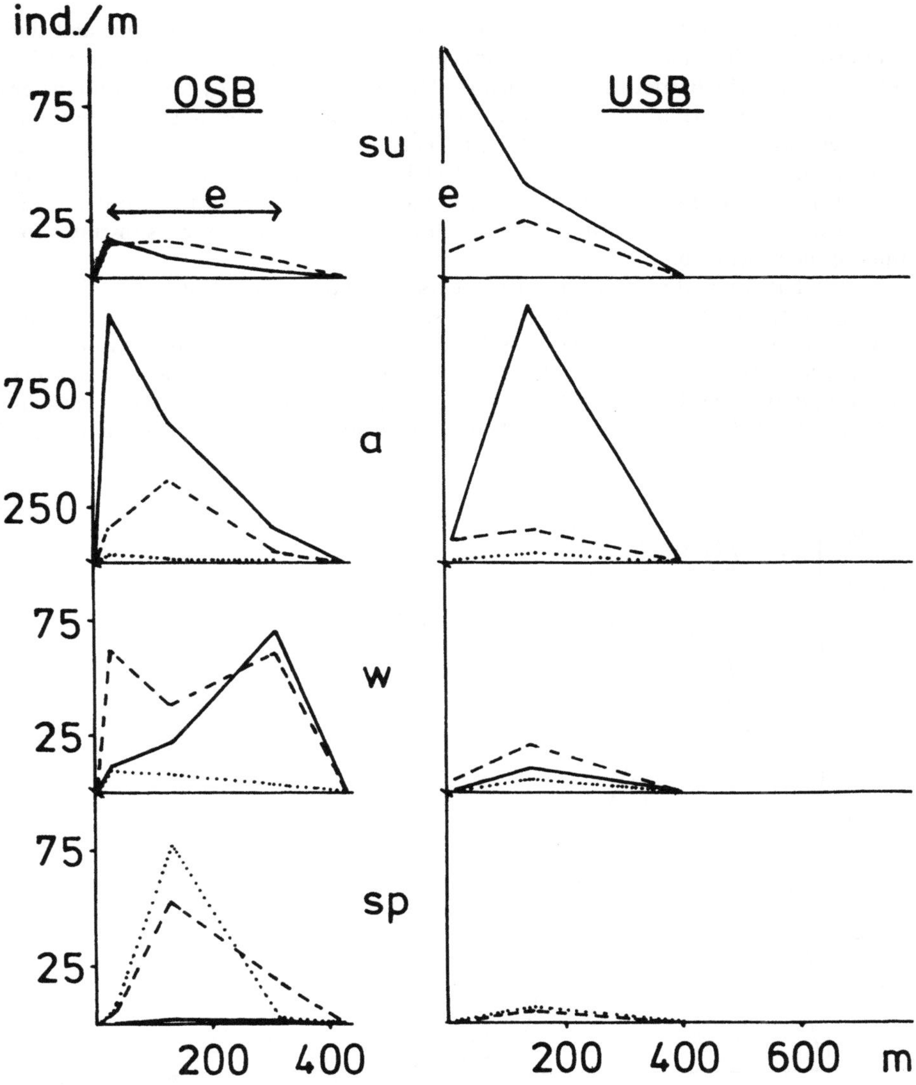

Fig. 3. *Neureclipsis bimaculata*: abundance of larval instars per m of vertical exposed bamboo sticks at different distances of the lakes in the different seasons. ——: first, – – –: second, . . .: third instar. The occurence of eggs is indicated by e.

nel pattern and riparian vegetation, influence the upstream flight distance and the proportion of imagines which reach the place just below the outflow. The lake represents a flight barrier for the imagines. This may be caused by optic stimulus or by the change in current (flying imagines touch the water surface at short intervals), while positive anemotaxis (cf. ROOS, 1957) cannot lead to the observed accumulations of imagines. After copulation (contact between the sexes occurs on riparian vegetation) it is expected that oviposition takes place in a part of the brook not far away from the point of mating. Oviposition is dependent on suitable substrates for the eggs being available.

After hatching, downstream drift occurs, although this is not yet proved by making drift captures with suitable mesh size during this time. This drift may be caused by low current speed (cf. MISHALL & WINGER, 1968) and/or overcrowding, because often I observed intraspecific aggression between larvae in a running water aquarium (cf. GLASS & BOVBJERG, 1969; SCHUMACHER, 1970). Particularly in the USB, the difference between the distribution of imagines and eggs,

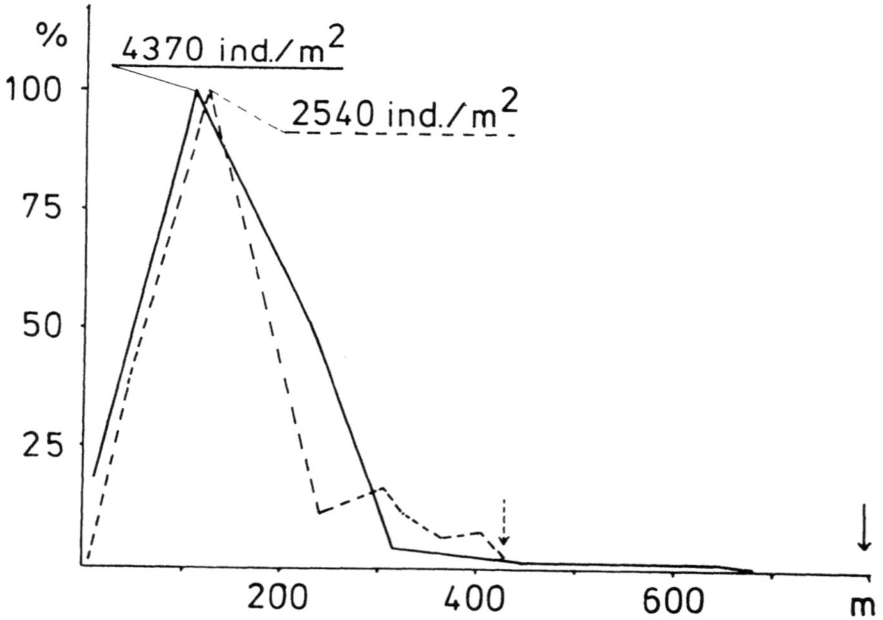

Fig. 4. *Neureclipsis bimaculata*: Distribution of larvae (annual average) in the OSB (— — —) and USB (. . .) at different distances from the lakes, expressed as percentage of density at the places of maximal abundance. The vertical arrows indicate the inflow of each brook into the next lake.

compared with that of larvae, indicates a clear colonization cycle, which seemed to be related to larval population density as well as to environmental factors.

3.2. *Mystacides* spp.

In the area investigated three species of this genus occur: *M. azurea* (L.), *M. longicornis* (L.), and *M. nigra* (L.). Swarming males of these three species were found in a narrow area, where they occupied well-distinguished different habitats. This could

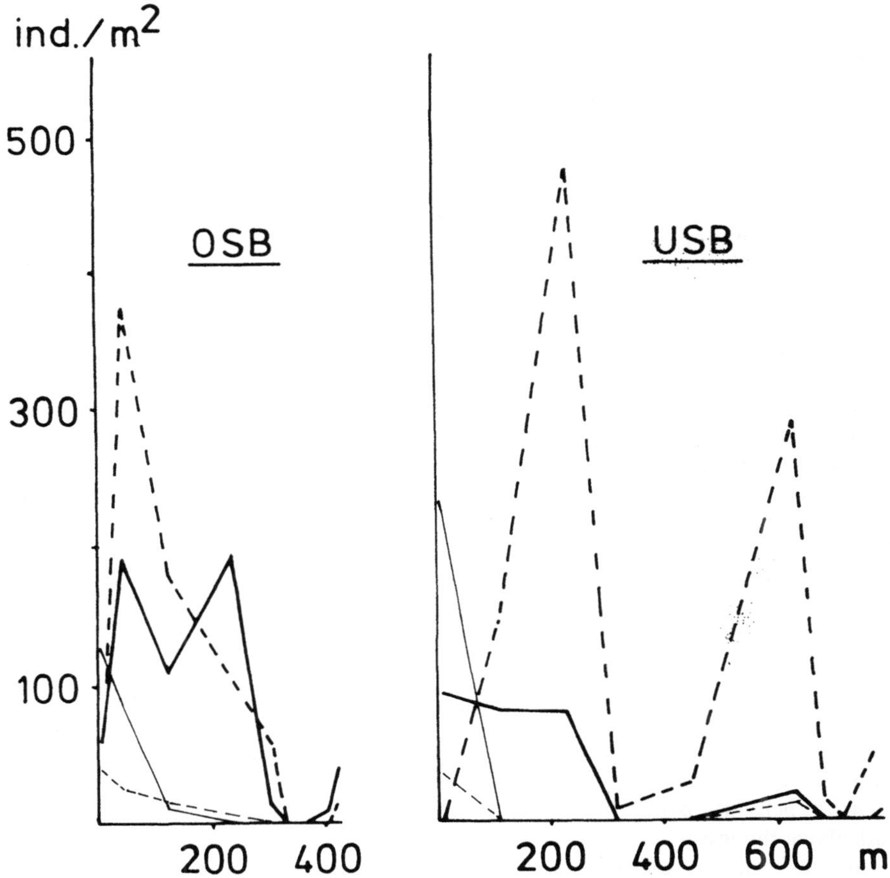

Fig. 5. *Mystacides* spp.: abundance (annual average) of larvae at different distances from the lakes. *M. azurea*: thick line; *M. longicornis*: thin line; ——: at the place of maximal current; — — —: near the bank.

be observed at the outflows and inflows of the lakes as well as at the banks of the lakes, where a well-developed belt of *Phragmites* existed.

M. azurea swarmed just above the water surface and avoided approaching closer than 40 cm to the reed-belt. *M. nigra* flew at the same height over the water surface as *M. azurea*, but in the reeds. The third species, *M. longicornis*, swarmed just above the reed-belt. The swarming place of one species was never used by males of the other two species.

Thus, K7/K8 in the USB, where well-developed *Phragmites* occurred, seemed to be the only suitable parts of both brooks for swarms of *M. nigra* and *M. longicornis*. In fact, only a few small swarms of *M. longicornis* were found in those parts of the USB.

Except for K7/K8, swarms of *M. azurea* were found in nearly all parts of the brooks. They never swarmed near the bank. Where a big stone or a trunk stood out from the water, a hole in the swarm existed, because in this case also the males avoided flying nearer than 40 cm to any object. In some places the growth of vegetation from the bank narrowed the open surface of the water to less than 80 cm during the flight period. Then the swarms of *M. azurea* disappeared from those places, but swarming continued upstream and downstream of them.

The distribution of larvae of *Mystacides* in the brooks is shown in Fig. 5. *M. nigra* did not occur there. *M. longicornis* was mainly found in the outflows, but some larvae were captured in K9. *M. azurea* had a more general distribution pattern, but did not occur where the current speed and substrata were not suitable. In the *Phragmites*-area of the USB the situation was different. There ideal substrata for *M. azurea* (leaves and litter with detritus) and low current were found near the bank, but only few larvae were captured compared with equivalent places upstream (K4) and downstream (K9).

The occurrence of larvae of *M. longicornis* mainly in the outflows, the only parts of the brooks where flying females of this species were observed, not far away from the swarming places, as well as the low abundance of larvae of *M. azurea* in the *Phragmites*-area of the USB, in spite of suitable substrata, suggests that oviposition occurred close to the swarming places. Thus the swarming behaviour of the males of *M. azurea* and *M. longicornis* is expected to influence larval distribution.

3.3. *Athripsodes* spp. and *Ceraclea dissimilis* (STEPH.)

The swarming places of the males of the two *Athripsodes* species, *A. aterrimus* (STEPH.) and *A. cinereus* (CURT.) were well separated. While *A. aterrimus* flew in a narrow belt just beside the bank, *A. cinereus* swarmed some distance from the bank above the water surface.

At the beginning of the flight in the early evening, males of *A. cinereus* were well dispersed over the water surface. With progressive decrease of the light, the males assembled at places where a sharp gradient of light intensity existed, e.g. near bridges, near particular trees, or in the curves of a meander. In the OSB (section

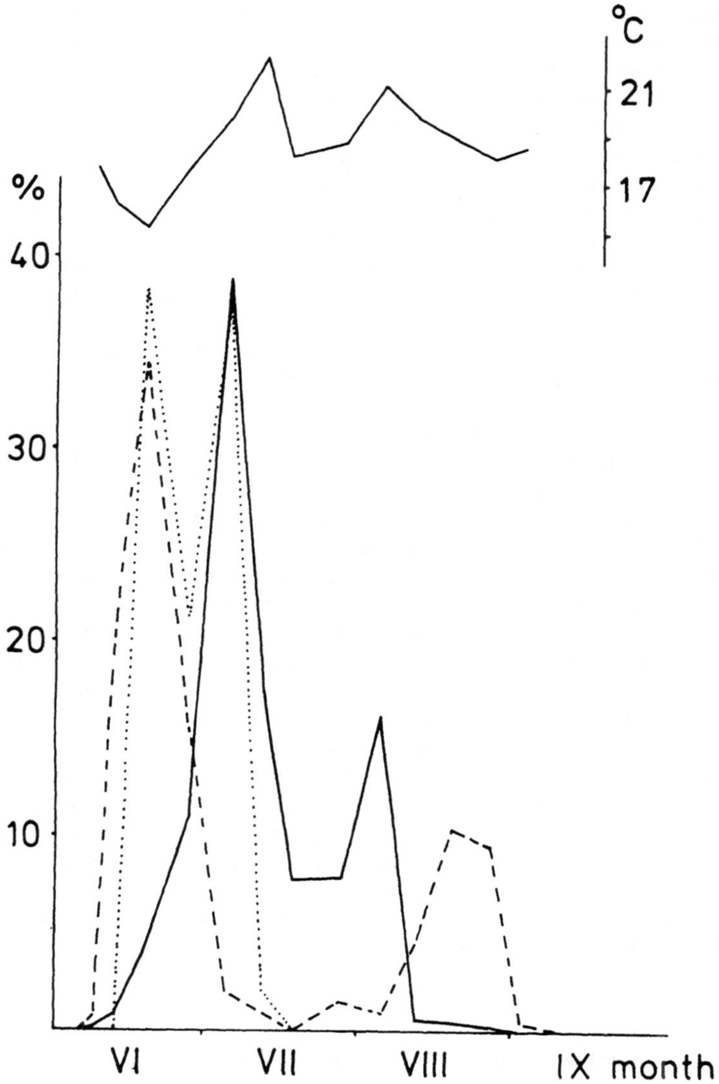

Fig. 6. *Athripsodes* spp. and *C. dissimilis*: Flight intensity (expressed as percentage of the total of captured imagines) in 1975. ——: *A. cinereus*; . . .: *A. aterrimus*; — — —: *C. dissimilis*. Above: mean air temperature two hours before sunset.

3-5) the males formed such dense clouds in the eastern-lying curves close to trees on the eastern bank, that the noise of the beating wings could be clearly heard. Such accumulations were also observed in K2/K3 and near the bridges in K6 and K10. In K7-K9 and downstream of K10 they never occurred, presumably because sharp light gradients did not exist on the meadow or in the dense growth of alders before the inflow into the Westensee. Also, the stand of *Phragmites* in K7/K8 is expected to suppress swarming of *A. cinereus* (c.f. *M. azurea*).

Another common leptocerid, *Ceraclea dissimilis* used the same swarming places and showed the same behaviour as *A. cinereus.* In *C. dissimilis*, I observed the contacting of mates on several occasions. When the dense clouds were established, the females flew into a cloud, grasped a male and disappeared with it to a nearby tree.

If two species use the same swarming places and if the swarms facilitate the contact of a mate, then the swarms are expected to be separated in time. In fact, *C. dissimilis* started its swarming period before *A. cinereus*, and stopped after the latter. But, during the flight period of *A. cinereus*, flying *C. dissimilis* became rare (Fig. 6). Two flight periods in the year were reported by RESH (1976) for *Athripsodes augustus* (now known as *Ceraclea transversa*) and explained by the different speed of larval development in relation to the availability of its food of freshwater sponge. Although *C. dissimilis* belongs to the group of non-sponge feeders of this genus (RESH et al., 1976), this splitting into cohorts was also observed in its larvae in the Schierenseebrooks, but was not very clear. This presumably was caused by the two swarming periods, which are expected to be the result of the suppression of swarms of *C. dissimilis* by swarming *A. cinereus*, i.e. of interspecific competition for the same swarming places. If both species occurred together in the same light trap (CRICHTON, 1960), this may be the result of the different sampling method.

A. aterrimus, which swarmed apart from *A. cinereus* and *C. dissimilis*, overlapped in time with these two species (Fig. 6). Comparing the abundance of swarms with the distribution of larvae (Fig. 7) in *A. cinereus*, it is evident that suitable swarming places locally increased larval abundance. In *C. dissimilis* this relation is not very clear, and the reasons for the different larval distribution pattern of *A. cinereus* and *C. dissimilis* are not yet known.

The third species, *A. aterrimus*, was relatively rare in the brooks, i.e. larval numbers in the samples were too low to work out a reliable distribution.

4. Discussion

Although the number of species investigated is relatively low and the method of direct observation of the flight behaviour cannot lead to a final interpretation of the phenomena, some aspects of the influence of imaginal behaviour on the larval distribution become evident.

The upstream flight of *N. bimaculata*, which causes the concentration of larvae in the outflows, has an enormous effect for the whole benthic community. In the

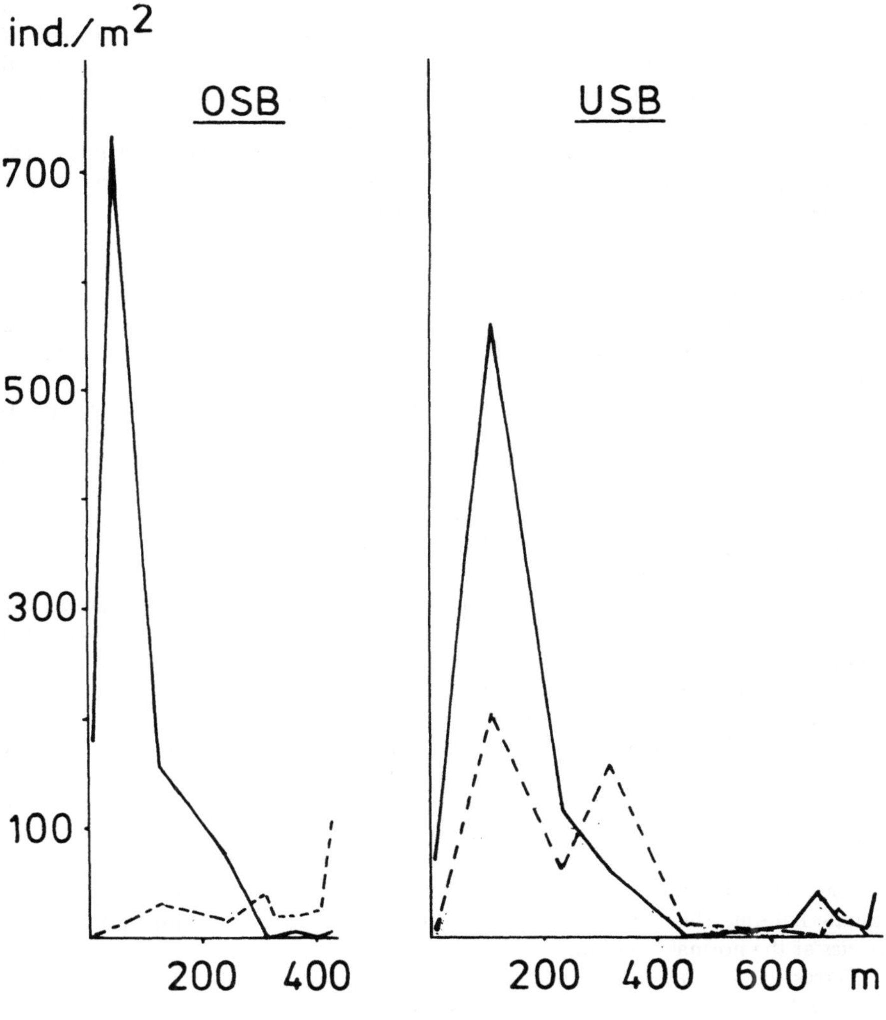

Fig. 7. *A. cinereus* (——) and *C. dissimilis* (– – –): abundance of larvae (annual average) at different distances from the lakes.

Schierenseebrooks the character and the production of the outflow community is based on that migration to a great extent (STATZNER, in press). The distance of upstream flight is likely to be modified according to environmental factors, and different distribution patterns might result. But usually, *N. bimaculata* is reported from lake outflows (MÜLLER, 1954 b, 1955; BRICKENSTEIN, 1955; ILLIES, 1956; TOBIAS & TOBIAS, 1971, and others). The usual explanation for this distribution refers to the filter feeding of its larvae and the high planktonic drift in lake outflows but this seems to be of minor importance in this connection.

While the upstream flight in general may be influenced by environmental factors (cf. ROOS, 1957; ELLIOTT, 1967; NISHIMURA, 1967; LEHMANN, 1970; SCHUMACHER, 1970; and above) and, therefore, would be expected to differ in different habitats, the constant swarming places in the Schierenseebrooks indicates that this kind of flight is less variable in these species. However, there exist similar and different observations in the literature.

According to GRUHL (1960), *M. azurea* swarmed above the water surface as well as above the reeds. *M. longicornis* was found above the reeds (GRUHL, 1960) and, in Japan, above trees (MORI & MATUTANI, 1958). WESENBERG-LUND (1913) observed swarms of *M. nigra* above the reeds. *A. cinereus* swarmed above the water surface (BERG, 1938; GRUHL, 1960; BENZ, 1975) and below trees up to 50 m from the water (BENZ, 1975). MORGAN (1956) observed *A. aterrimus* in a small band just by the bank.

It is not yet clear whether these differences are restricted to local areas, as a response to particular habitat topography, whether they are due to particular microclimate conditions (cf. SOLEM, 1976), or whether they indicate that euryök species are able to use less than optimum swarming places in certain parts of their range. However, the results of this study demonstrate that in some species a clear relationship exists between larval abundance and suitable swarming places for the adults, despite the fact that the area investigated is relative small, i.e. the contrasts are expected to be decreased by the female's flight before oviposition and larval drift.

This leads to the conclusion that a lack of optimum swarming places may decrease the abundance or prevent the occurrence of a species in a particular aquatic habitat. This may also happen, if the swarming place is occupied by another species at the normal swarming time.

Terrestrial vegetation and the area of uninterrupted water surface seem to be the most important factors in this connection. Emergent aquatic vegetation as well as the width of the open water is expected to affect those species which swarm above the water surface. Terrestrial vegetation may influence swarming behaviour by providing optical stimuli, either directly (treetops, etc.) or indirectly (light gradients). Further, it provides shelter from wind, which above 3m/sec suppressed swarming in most of the species of the Schierenseebrooks.

The few examples above, which will be complemented elsewhere by observations on other insect groups, demonstrate, that a knowledge of imaginal ecology will lead to a better comprehension of the distribution of aquatic insects and reduce the chance of misinterpretation.

5. Acknowledgements

Thanks are due to Prof. Dr. K. BÖTTGER (Kiel) for equipment and much helpful advice and to Dr. N.V. JONES (Hull) who corrected the English manuscript. The research was supported by a GraFöG studentship which is gratefully acknowledged.

References

BENZ, G. 1975. Über die Tanzschwärme der Köcherfliege *Hydropsyche pellucidula* Curtis (Trichoptera: Hydropsychidae). Mitt. schweiz. ent. Ges. 48: 147-157.

BERG, K. 1938. Studies on the bottom animals of Esrom Lake. K. danske vidensk. Selsk. Skr. 8: 1-255.

BRICKENSTEIN, C. 1955. Über den Netzbau der Larve von *Neureclipsis bimaculata* L. (Trichopt., Polycentropidae). Abh. Bayer. Akad. Wiss. 69: 1-44.

CRICHTON, M.I. 1960. A study of captures of Trichoptera in a light trap near Reading, Berkshire. Trans. R. ent. Soc. Lond. 112: 319-344.

ELLIOTT, J.M. 1967. Invertebrate drift in a Dartmoor stream. Arch. Hydrobiol. 63: 202-237.

GLASS, L.W. & BOVBJERG, R.V. 1969. Density and dispersion in laboratory populations of caddisfly larvae (*Cheumatopsyche*, Hydropsychidae). Ecology 50: 1082-1084.

GRUHL, K. 1960. Die Tanzgesellschaft der *Hydropsyche saxonica* McLach. (Trichoptera). Mitt. dt. ent. Ges. 19: 76-83.

ILLIES, J. 1956. Seeausfluß-Biozönosen lappländischer Waldbäche. Ent. Tidskr. 77: 138-156.

LEHMANN, U. 1970. Stromaufwärts gerichteter Flug von *Philopotamus montanus* (Trichoptera). Oecologia 4: 163-175.

MINSHALL, G.W. & WINGER, P.V. 1968. The effect of reduction in stream flow on invertebrate drift. Ecology 49: 580-582.

MORGAN, N.C. 1956. The biology of *Leptocerus aterrimus* Steph. with reference to its availability as a food for trout. J. Anim. Ecol. 25: 349-365.

MORI, S. & MATUTANI, K. 1958. Daily swarmings of some caddis fly adults and their habitat segregation. Zool. Mag. Tokyo 62: 191-198.

MÜLLER, K. 1954a. Investigations on the organic drift in north Swedish streams. Rept. Inst. Freshwat. Research Drottingholm 35: 133-148.

— 1954b. Faunistisch-ökologische Untersuchungen in nordschwedischen Waldbächen. Oikos 5: 77-93.

— 1955. Produktionsbiologische Untersuchungen in nordschwedischen Fließgewässern. 3. Die Bedeutung der Seen und Stillwasserzonen für die Produktion in Fließgewässern. Rept. Inst. Freshwat. Research Drottingholm 36: 148-162.

NISHIMURA, N. 1967. Ecological study on net-spinning caddis-fly, *Stenopsyche griseipennis* Mc L. II. Upstream-migration and determination of flight distance. Mushi 40: 39-46.

RESH, V.H. 1976. Changes in the caddis-fly fauna of Lake Erie, Ohio, and of the Rock River, Illinois, over a fifty year period of environmental deterioration. Proc. 1st int. Symp. Trich., 167-170.

—, MORSE, J.C. & WALLACE, I.D. 1976. The evolution of the sponge feeding

habit in the caddisfly genus *Ceraclea* (Trichoptera: Leptoceridae). Ann. ent. Soc. Am. 69: 937-941.
ROOS, T. 1957. Studies on upstream migration in adult stream-dwelling insects I. Rept. Inst. Freshwat. Research Drottingholm 38: 167-193.
SCHUMACHER, H. 1969. Das Schwärmverhalten von *Hydropsyche borealis* Martynov (Insecta, Trichoptera). Zool. Anz., Suppl. 33: 555-558.
——. 1970. Untersuchungen zur Taxonomie, Biologie und Ökologie einiger Köcherfliegenarten der Gattung *Hydropsyche* Pictet (Insecta, Trichoptera). Int. Rev. ges. Hydrobiol. 55: 511-557.
SOLEM, J.O. 1976. Studies on the behaviour of adults of *Phryganea bipunctata* and *Agrypnia obsoleta* (Trichoptera). Norw. J. Ent. 23: 23-28.
STATZNER B. (in press) Factors that determine the benthic secondary production in two lake outflows — a cybernetic model. Verh. Int. Ver. Limnol.
SVENSSON, W. 1974. Population movements of adult Trichoptera at a South Swedish stream. Oikos 25: 157-175.
TOBIAS, W. & TOBIAS, D. 1971. Köcherfliegen und Steinfliegen einiger Gewässer in Sör Varanger (Nord-Norwegen) (Trichoptera, Plecoptera). Senckenbergiana biol. 52: 227-245.
WESENBERG-LUND, C. 1913. Fortpflanzungsverhältnisse: Paarung und Eiablage der Süßwasserinsekten. Fortschr. Naturw. Forsch. 8: 161-268.

Discussion

JONES: Did you examine the distribution of oviposition sites?

STATZNER: As mentioned above, I examined the distribution of oviposition sites of *N. bimaculata.* This was easy because the females attach the eggs to the substratum in the water. If the egg-mass is dropped above the water surface, as I observed once in *M. azurea*, the site of the eggs may not be identical with the site of oviposition. I observed no further ovipositions of Leptoceridae, which presumably may occur during the night thus I have no information about oviposition sites in this family. Indeed information about places of mating, migration of females after mating, and oviposition sites will make possible a more comprehensive interpretation of the phenomena I observed. Despite the lack of such additional information I want to point out that besides current speed, substratum, and others, i.e. factors that influence the aquatic phases of aquatic insects, terrestrial factors that control the imagines may influence the distribution in the aquatic habitat.

BADCOCK: Current speed may not only influence larval distribution but also be a factor affecting oviposition if the female imago enters flowing water to deposit her eggs. I watched *Hydropsyche angustipennis* dive into the water, swim to the lower surface of an inclined stone and start egg-laying. I wondered if she would have succeeded if the current had been much faster.

STATZNER: I agree that species which enter the water may need suitable physical conditions for succesful oviposition according to the specific oviposition behaviour. I have observed in several species that the males and females touched the water surface at short intervals during their flight. Were these adult caddis measuring current speed?

Finally I want to emphasize that swarming as well as oviposition, hatching, larval moulting, pupation, emergence, and all the other phases in a life cycle of a caddis fly as well as of other aquatic insects need suitable environmental conditions. Aquatic life is one period, aerial life another one, but both must be passed by the same individual.

Proc. of the 2nd Int. Symp. on Trichoptera, 1977, Junk, The Hague

Flight periods and ovarian maturation in Trichoptera in Iceland

G.M.GÍSLASON

Introduction

This paper examines the flight periods of Icelandic Trichoptera and ovarian maturation of the females.

The life histories of adult Trichoptera have been studied by many workers (BOTOSANEANU, 1957, BRINDLE, 1956, 1958, 1965, CORBET et al., 1966, CORBET & TJÖNNELAND, 1956, CRICHTON, 1960, 1965, 1971, 1976, DECAMPS, 1967, GOWER, 1967, GÖTHBERG, 1970, HIRVENOJA, 1960, NOVAK & SEHNAL, 1963, PHILIPSON, 1957, SVENSSON, 1972, TOBIAS, 1968, 1969, ULFSTRAND, 1969, 1970). These workers have usually employed light traps and analysed the collections obtained in terms of flight periods, sex ratios and, sometimes, diel patterns of activity.

NOVAK & SEHNAL (1963) examined the ovarian maturation in several *Limnephilus* species and SVENSSON (1972) several Limnephilidae and non-Limnephilidae.

Material and methods

Adult Trichoptera were collected in Iceland during the years 1955 to 1976. Most of the material was collected in 1974, when the study was most intensive, but most of the earlier specimens and some collected in 1974 and 1975 came from the collection of Mr. HÁLFDÁN BJÖRNSSON, a naturalist living at Kvisker, Öraefi. They were mainly swept from vegetation or caught on the wing by a net. In Thjorsarver in the central highlands in 1972 they were also caught in pit-fall traps, and in 1972 and 1974 traps catching flying insects were used. Mr. ERLING ÓLAFSSON, a zoologist at Lund University, collected the Trichoptera in Thjorsarver, central highlands in 1972 and 1973 (ÓLAFSSON, in press).

In order to study the stage of maturation of the female ovaries they were dissected under a binocular microscope. From Iceland 714 females of the eleven species were examined. The females were stored in 70% alcohol before dissection. The different developmental stages were defined according to the classification proposed by NOVAK & SEHNAL (1963).

Flight periods

The flight periods are expressed on a weekly basis, assuming four weeks per month. Each month has been arbitrarily allotted weeks of seven and eight days alternately, i.e. week 1 includes the period 1st to 7th of each month, week 2 includes the period 8th to 15th, week 3 covers the dates between and including 16th to 22nd, whilst week 4 the 23rd to the 30th (or 31st).

In addition to records from present work, records from the literature, where exact dates of capture of Trichoptera are given (LINDROTH, 1931, LINDROTH et al. 1973, ÓLAFSSON, in press, SIGURJÓNSDÓTTIR, 1974, TJEDER, 1964), provided the information on flight periods presented in Figs. 1-11.

FRISTRUP (1942) expresses the flight period of the Icelandic Trichoptera to the nearest month, the period extending from the first month to the last month of capture of each species. He does not give any dates of capture so his data could not be included in the diagrams presented here. He found that the following species had longer flight periods than is shown in the figures presented here: *Apatania zonella* and *Grammotaulius nigropunctatus* were found in September, and *L. elegans* and *Agrypnia picta* in August. The present study gives longer flight periods for all other species, except *L. decipiens*, *L. fenestratus* and *L. picturatus*.

Details of the flight periods in the lowlands and the highlands are given in Figs. 1-11. It is possible to divide the insects into two classes according to the time of emergence: those that first appear in the spring (earlier than early June) and those that appear in mid-summer (late June or July). The first group includes *A. zonella*, *L. affinis*, *L. griseus* and *L. sparsus*. These species had also long flight periods of more than three months which extend into the autumn (late August or September). The larvae of these species were usually found in streams and lakes, which are covered with ice for much shorter periods than other water bodies, and with warmer temperatures during the winter. In the second group are *L. decipiens*, *L. fenestratus*, *L. picturatus*, *G. nigropunctatus*, *P. cingulatus* and *A. picta*. The larvae of all these species, except *P. cingulatus* are found nearly exclusively in pools and swamps, which are covered with ice and snow for longer periods in winter than lakes and streams, and presumably freeze solid during the winter. These species have their main flight periods in July and August. *P. cingulatus* larvae are commonest in streams. Pupae of this species were found from May to August and empty pupal cases were found throughout the whole summer. Probably *P. cingulatus* has a continuous flight period from late June to September. The data obtained, however, show absence of adults in July and late August. This is most likely because collecting in *P. cingulatus* areas was not intensive.

In those species which occur in both the lowlands and the central highlands, emergence of adults starts earlier in the lowlands with the exception of *L. griseus* and *P. cingulatus* (Figs. 6 and 10). From the occurrence of pupae, however, it is probable that *P. cingulatus* may emerge in late June in the lowlands, adult records being missing because of lack of collecting at that time.

P. cingulatus presumably starts emerging in late June in the lowlands. Some pupae and empty pupal cases are found at that time, and probably *L. griseus* might appear earlier in the lowlands from streams than records indicate. The adult *L. griseus* found in late May in the highlands (Fig. 6) probably emerged from spring-fed streams there, which are the only icefree water bodies present at that time, and pupae do not appear in pools and swamps in the area until June.

The later appearance of adults in the highlands is presumably due to the fact that climate is colder there, and the region becomes snow-free about one or two months later than in the lowlands.

Comparison of flight periods in Iceland and in other regions

Clearly flight periods are often shorter in Iceland than in more southerly countries. In Britain *L. affinis* is found from late April to November (CRICHTON, 1971, 1976), *L. decipiens* from mid-July to November, *L. elegans* from May to July, *L. griseus* from early May to mid-October and *P. cingulatus* from early June to mid-October (BRINDLE, 1965; CRICHTON, 1971);

SVENSSON (1972) and ULFSTRAND (1969) studied *L. griseus*, *L. sparsus*, *G. nigropunctatus* and *P. cingulatus* in South-Sweden and found that they emerge at a similar time as in Iceland but their flight periods were about a month longer. HIRVENOJA (1960) found that *L. griseus* and *A. picta* had similar flight periods in South-Finland as in Iceland. The flight periods in Lapland in Northern Sweden (64°-69°N) (similar latitude to Iceland) were found to be later in the year than in Iceland (GÖTHBERG, 1970; TOBIAS, 1968, 1969; ULFSTRAND, 1970).

Ovarian development

NOVAK & SEHNAL (1963, 1965) established that all *Limnephilus* species of Czechoslovakia they studied have an ovarian quiescence during the summer, and this has been confirmed by GOWER (1967), LE LANNIC (1975) and SVENSSON (1972). NOVAK & SEHNAL (1963) state that these species do not undergo ovarian quiescence in the highlands of Czechoslovakia. They suggested the following classification of development stages which has been adopted in the present study.

Stage A. The immature female.
The reproductive organs are very small, the ovaries are thin and no separate ova can be seen. The colleterial glands are small. The fat body is distended.

Stage B. The maturing female.
The ovaries are thicker, and separate ova are visible. The colleterial glands are rather small. Fat body is still distended.

Stage C. The mature female before oviposition.

The ovaries are big, ovulation has taken place and the lateral oviducts are distended with thickly packed fully mature eggs. The ovarioles contain only small egg rudiments. The colleterial glands are very big and often nearly as long as the whole abdomen.

Stage D. The female after oviposition.
The ovaries are very much reduced and corpora lutea may be seen in the ovarioles. The colleterial glands have collapsed, the fat body disappeared, and the abdomen consists only of the sclerotized skin.

It is found that different species emerge in either stage A or B (Fig. 1-11).

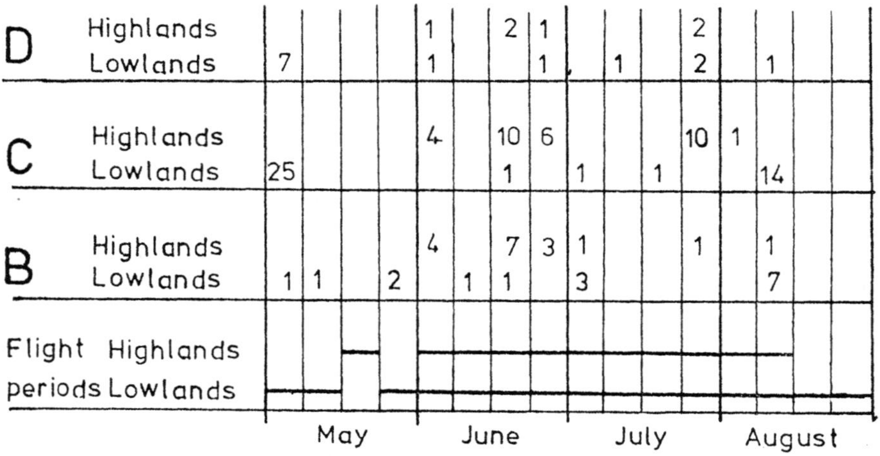

Fig. 1. Apatania zonella (Zett.).

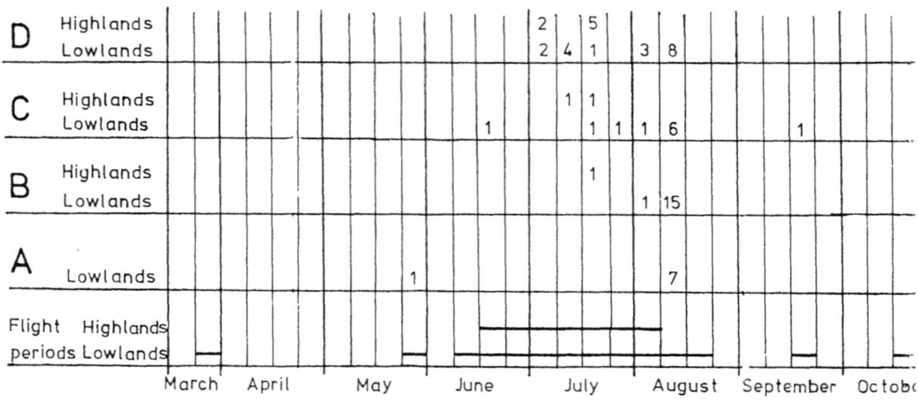

Fig. 2. Limnephilus affinis Curt.

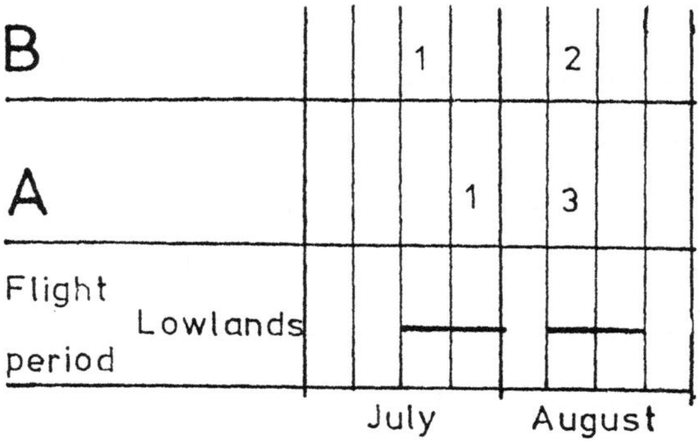

Fig. 3. *L. decipiens* (Kol.).

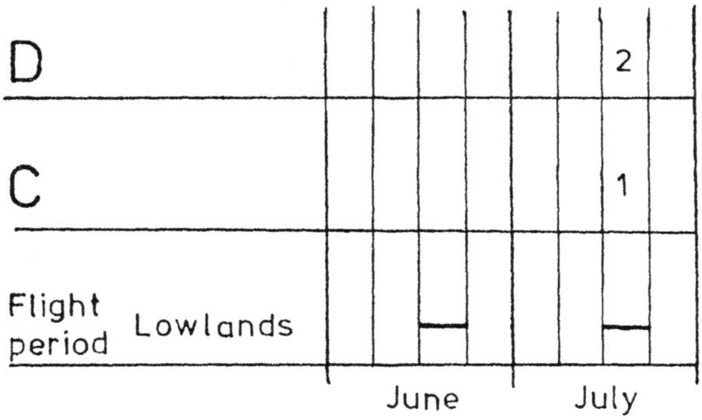

Fig. 4. *L. elegans* Curt.

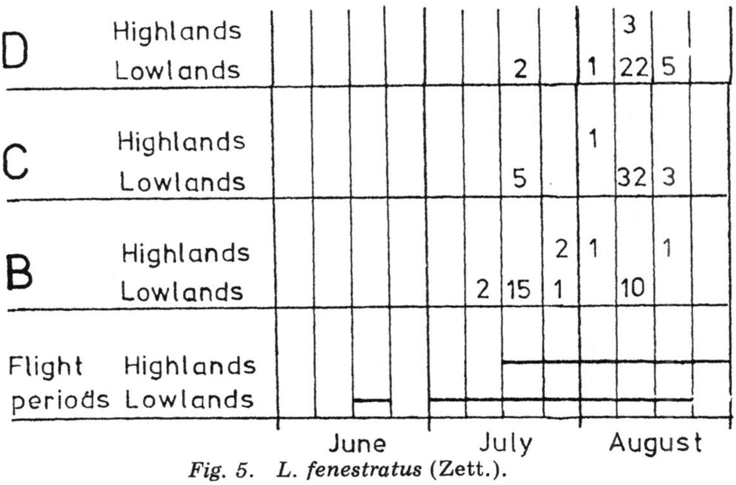

Fig. 5. *L. fenestratus* (Zett.).

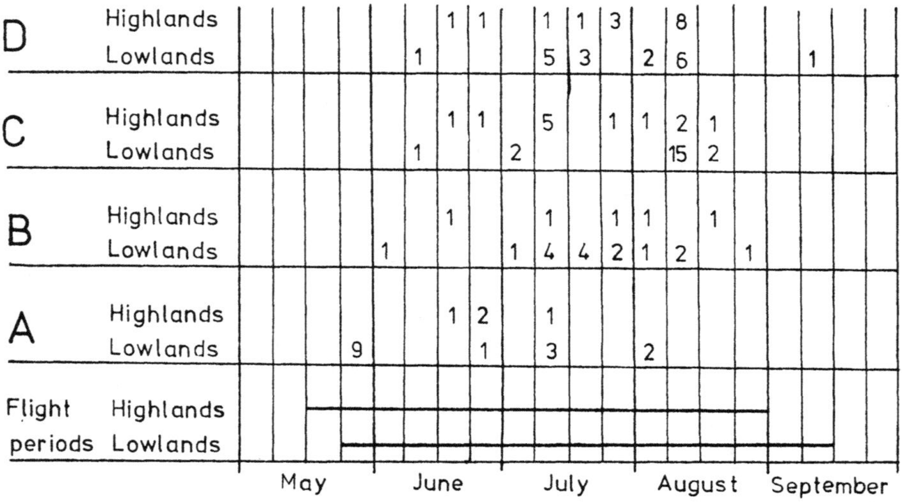

Fig. 6. ***L. griseus*** (L.).

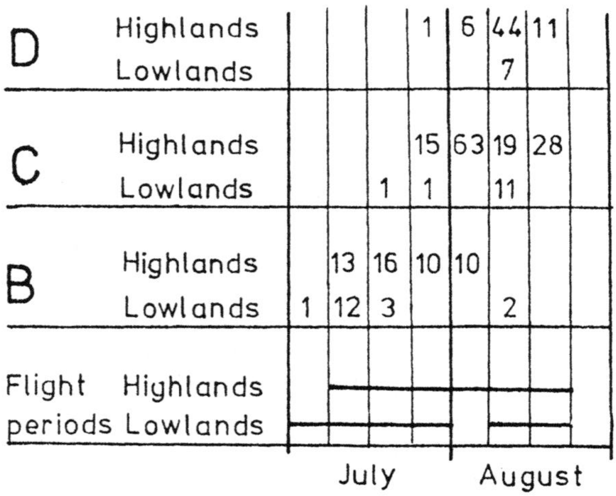

Fig. 7. ***L. picturatus*** McL.

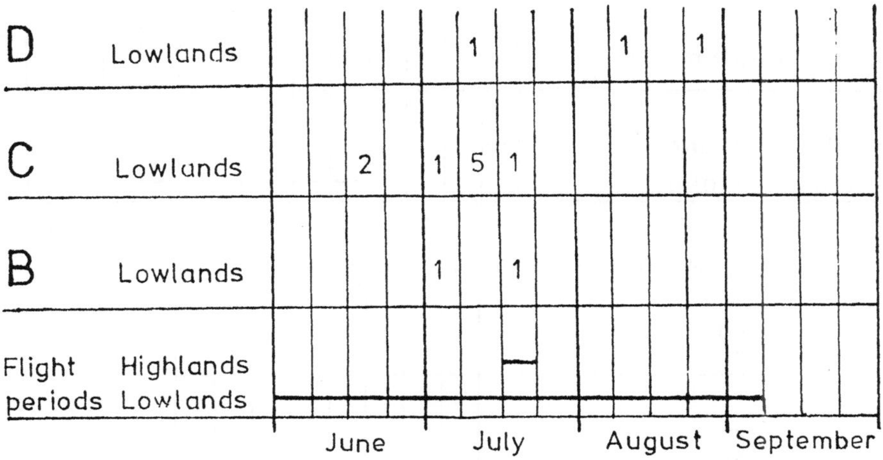

Fig. 8. *L. sparsus* Curt.

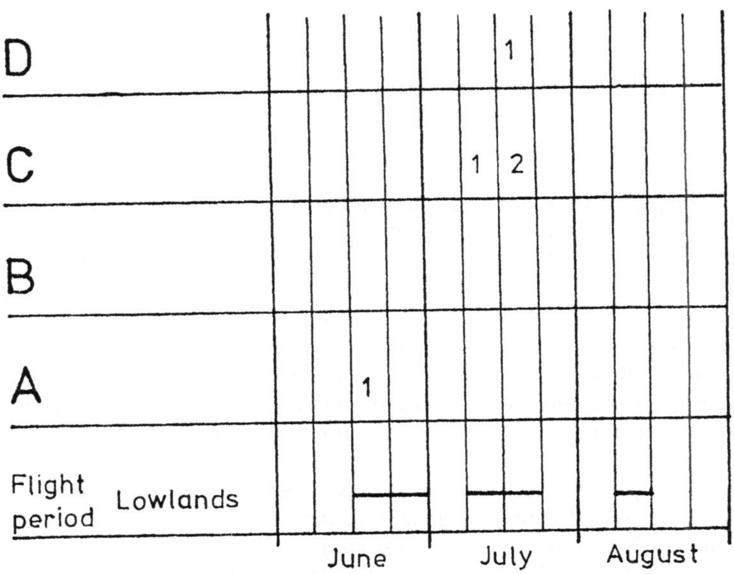

Fig. 9. *Grammotaulius nigropunctatus* (Retz.).

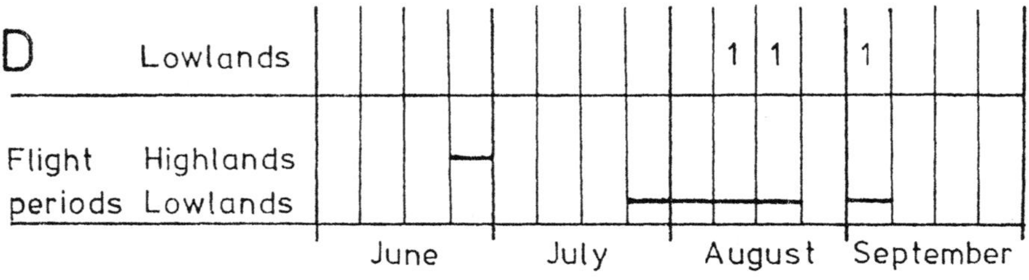

Fig. 10. *Potamophylax cingulatus* (Steph.).

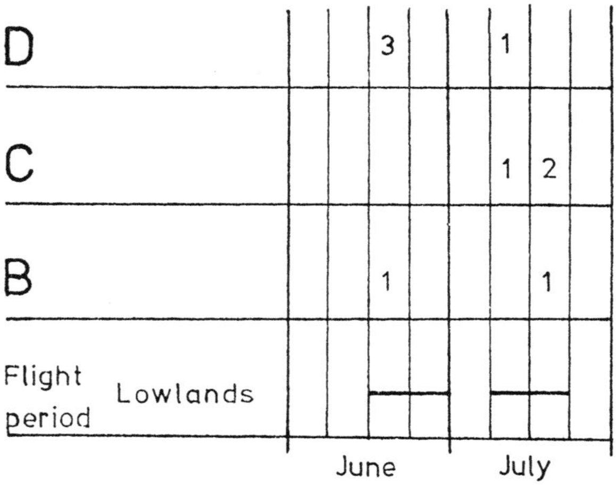

Fig. 11. *Agrypnia picta* Kol.

Figs. 1-11. Flight periods and reproductive cycles in different Trichoptera species in the lowlands and highlands of Iceland. Horizontal lines: flight periods expressed as occurrence per week. A, B, C and D different ovarian maturation stages explained in the text with the number of females examined found in each stage each week.

The species emerging in stage A are *L. affinis*, *L. decipiens*, *L. griseus* and *G. nigropunctatus* and probably *L. elegans*. *L. sparsus* emerges in stage A, based on observations in NE-England, where it has been taken in stage A in early summer.

The species emerging in stage B are *A. zonella*, *L. fenestratus*, *L. picturatus* and *A. picta*.

P. cingulatus probably emerges in stages B and C according to SVENSSON (1972), but it was only found in stage D in Iceland.

No Icelandic Trichoptera undergo ovarian quiescence (Figs. 1-11), neither in the lowlands nor the highlands. They are found in all stages during the summer but with higher proportions of females in stages A and B early in their flight period. It is clear from their annual life history (GÍSLASON, 1977) that adults are emerging throughout the summer. Pupae of *A. zonella* are found from May to September. *L. affinis* pupae are found from December until September, *L. decipiens* pupae are found in August and *L. elegans* pupae in July. *L. griseus* pupae occur from early May until late August, and adults were observed emerging on 12 August 1973. *L. picturatus* is found in pupal stages from early June to mid-July, *L. sparsus* pupae from June to August, and *P. cingulatus* pupae from May to August. It would thus appear that *A. zonella, L. affinis, L. griseus* and *L. sparsus* emerge during the whole summer, and lay eggs shortly after emergence. Egg masses of *A. zonella* have been found in early May and of *L. affinis* from early June to September. The other species are more confined in their emergence periods. They usually emerge in mid-summer, most of the population emerging at the same time (e.g. *L. fenestratus* and *L. picturatus*) and the pattern of development is gradual from stage B to D over a short period of time.

L. affinis, L. griseus and *L. sparsus* outside Iceland (NOVAK & SEHNAL, 1963, 1965; SVENSSON, 1972, and own observations on *L. affinis* and *L. griseus* in Northumberland) emerge over a short period in spring or early summer, stay in stage A until in late summer, when days get shorter, then start to develop their ovaries. Outside Iceland these species emerge from temporary water bodies which become dry in spring and fill up again in autumn, but in Iceland the larvae occupy permanent water bodies.

Discussion

Icelandic Trichoptera have shorter flight periods than the same species further south. In Iceland their flight periods are shorter in the highlands than in the lowlands. This may be associated with temperature, since Iceland has much lower summer temperatures than e.g. Britain and Southern Sweden where studies on Trichoptera flight periods have been carried out (BRINDLE, 1965; CRICHTON, 1965, 1971, 1976; SVENSSON, 1972). In Swedish Lapland at similar latitudes to Iceland the flight periods were similar to the highlands of Iceland (GÖTHBERG, 1970).

Trichoptera occupying streams and springs, *A. zonella, L. affinis, L. griseus, L. sparsus*, emerge earlier than species occupying pools and swamps only, *L. fenestratus, L. decipiens, L. elegans, L. picturatus, A. picta*. This may again be associated with temperature, pools and swamps being frozen solid during winter, whereas springs and streams always have water and do not freeze solid.

According to NOVAK & SEHNAL (1963) and SVENSSON (1972) species emerging in stage A have imaginal quiescence, but those emerging in stage B or C have passed stage A in larval or pupal life and do not have a later quiescence period, which is passed in stage A.

The Icelandic Trichoptera do not have any ovarian quiescence as observed among the Limnephilidae in other countries by NOVAK & SEHNAL (1963, 1965),

GOWER (1967) and SVENSSON (1972). Ovarian quiescence has not been observed at high altitudes in Europe (NOVAK & SEHNAL, 1963; MORETTI et al., 1976). In warmer countries *Limnephilus* species often occupy temporary water bodies, and the ovarian quiescence is an adaptation enabling species to occupy these habitats. For some species it is controlled entirely by light. The ovaries either do not develop during long days which coincide with the period when the pools and ponds are dry (NOVAK & SEHNAL, 1963) or may develop slowly as shown experimentally for *L. affinis* and *L. griseus* from Northumberland, when long day decreased the rate of development compared with short day (GÍSLASON, 1977).

In Iceland on the other hand these same species occupy permanent water bodies, and it would be disadvantageous to them to have an ovarian quiescence. Long photoperiod has apparently no effect on the ovarian development in Iceland, since nights in Iceland are light in mid-summer, and day is much longer there between spring and autumn equinox than further south.

A. zonella, *L. affinis*, *L. griseus* and *L. sparsus* emerge throughout the summer, laying eggs shortly after emergence. In this way they can utilize the summer for larval growth (GÍSLASON, 1977), larvae from eggs laid in early summer being nearly fully grown at the onset of winter and emerging early the following summer. Larvae from eggs laid in late summer spend the winter in earlier instars, continue growing in the spring and emerge in late summer. *L. picturatus* uses the first half of the summer for larval growth, and probably *L. fenestratus* too, but these species emerge in mid-summer in stage B, stage A being passed in the larva or pupa. If the species in Iceland had ovarian quiescence they would probably not be able to complete their life cycles in one year.

Acknowledgements

I am indebted to the staff of the Institute of Biology, University of Iceland for helping me when I was collecting caddis flies, and especially to Mr. ERLING ÓLAFSSON and Mr. HÁLFDÁN BJÖRNSSON for allowing me to use their collections of Trichoptera. I am very grateful to Prof. ARNTHOR GARDARSSON and Dr. G.N. PHILIPSON for reading over the manuscript and to Mrs. DÓRA JAKOBSDÓTTIR for typing it. I also must thank Dr. PHILIPSON for supervising this work, whilst I was working on my Ph.D. thesis at Newcastle University. This work was supported by grants from the Icelandic Governments Loans Fund (Lanasjodur islenskra namsmanna), and the Icelandic Science Fund (Visindasjodur).

References

BOTOSANEANU, L. 1957. Recherches sur les Trichoptères (imagos) de Romanie. Bull. ent. Pologne 26: 383-433.

BRINDLE, A. 1956. The ecology of north-east Lancashire Trichoptera with special reference to the use of light as a collecting method. Entomologist's Gaz. 7: 179-184.

—. 1958. Night activity of Trichoptera. Entomologist's mon. Mag. 94: 38-42.
—. 1965. The flight period of the Trichoptera (caddisflies) in Northern England. Entomoligist's Rec. 77: 148-159.
CORBET, P.S. & TJÖNNELAND, A., 1956. The flight of twelve species of East African Trichoptera. Univ. Bergen Årb. Naturv. Rekke 1955, 9: 1-49.
—. et al, 1966. The Trichoptera of St. Helen's Island, Montreal. I. The species present and their relative abundance at light. Can. Ent. 98: 1284-1298.
CRICHTON, M.I. 1960. A study of captures of Trichoptera in a light trap near Reading, Berkshire. Trans. R. ent. Soc. Lond. 112: 319-344.
—. 1965. Observations on captures of Trichoptera in suction- and light-trap near Reading, Berkshire. Proc. R. ent. Lond. (A), 40: 101-108.
—. 1971. A study of caddisflies (Trichoptera) of the family Limnephilidae, based on the Rothamsted Insect Survey, 1964-68. J. Zool. Lond. 163: 533-563.
—. 1976. The interpretation of light trap catches of Trichoptera from the Rothamsted insect survey. Proc. 1st int. Symp. Trich.: 49-57.
DÉCAMPS, H. 1967. Écologie des Trichoptères de la vallée d'Aure (Hautes-Pyrénées). Annls. Limnol. 3: 399-577.
FRISTRUP, B. 1942. Neuroptera and Trichoptera. Zool. Icel. 3: 1-23.
GISLASON, G.M. 1977. Aspects of the biology of Icelandic Trichoptera, with comparative studies on selected species from Northumberland, England. Ph. D. Thesis, University of Newcastle upon Tyne.
GOTHBERG, A. 1970. Die Jahresperiodik der Trichopteren-Imagines in zwei lappländischen Bächen. Öst. Fisch. 23: 118-127.
GOWER, A.M. 1967. A study of *Limnephilus lunatus* Curtis (Trichoptera: Limnephilidae) with references to its life cycle in watercress beds. Trans. R. ent. Soc. Lond. 119: 283-302.
HIRVENOJA, M. 1960. Ökologische Studien über die Wasserinsekten in Rükimäke (Südfinnland) III. Trichoptera. Ann. Ent. Fenn. 26: 201-221.
LE LANNIC, J. 1975. Contribution a l'etude du développement et de la maturation de l'appareil reproducteur de *Limnephilus rhombicus* L. Bull. Soc. Zool. Fr. 100: 539-550.
LINDROTH, C.H. 1931. Die Insektenfauna Islands, und ihre Probleme. Zool. Bidr. Uppsala 13: 105-600.
—. et al. 1973. Surtsey, Iceland. The development of a fauna, 1963-1970. Terrestrial invertebrates. Ent. scand. Suppl. 5: 1-280.
MORETTI, G.P. et al, 1976. The Trichoptera population of a temporary ecosystem of the Ubrian Apennines (Perugia, Italy). Proc. 1st int. Symp. Trich., 111-115.
NOVAK, K. & SEHNAL, F. 1963. The development cycle of some species of the genus *Limnephilus* (Trichoptera). Čas. čsl Spol. ent. 60: 68-80.
—. & —. 1965. Imaginaldiapause bei den in periodischen Gewässern lebenden Trichopteren. Proc. XIIth Int. Congr. Ent. 1964: 434.
ÓLAFSSON, E. (in press). Thjorsarver. Rannsoknir a skordyrum og attfaettlum 1971-1974. Orkustofnun, Reykjavik.
PHILIPSON, G.N. 1957. Records of Caddis Flies (Trichoptera) in Northumberland, with notes on their seasonal distribution in Plessey Woods. Trans. nat. Hist. Soc. Northumb. 12: 77-92.
SIGURJONSDOTTIR, H. 1974. Könnun a utbreidslu liddyra a snidi upp Esju. Univ. Icel. 33 pp. manuscript.
SVENSSON, B.W. 1972. Flight periods, ovarian maturation, and mating in Trichoptera at a South Swedisch stream. Oikos 23: 370-383.
TJEDER, B. 1964. Neuroptera, Trichoptera and Diptera-Tibulidae from Iceland with a redescription of *Rhabelomastrix parva* Siebke. Opusc. Ent. 29: 143-151.

TOBIAS, W. 1968. Die Trichopteren der Lule Lappmark (Sweden). I. UV-Lichtfänge am Stora Lulevattn. Ent. Z. 78: 12-16.
——. 1969. Die Trichopteren der Lule Lappmark (Sweden): II. Verzeichnis der Arten, Fundorte und Flugzeiten. Ent. Z. 79: 77-92.
ULFSTRAND, S., 1969. Nattsländorna (Trichoptera) vid en skånsk bäck. Fauna Flora 64: 122-130.
——. 1970. Trichoptera from River Vindelälven in Swedish Lapland. Ent. Tidskr. 91: 46-63.

Discussion

MALICKY: Are there any detailed laboratory studies on the presumed parthenogenesis of *Apatania zonella*? The different ratios of females, found by several authors, suggest a similar situation to that of *Solenobia triquetella*. This moth was studied by Seiler in Switzerland, where it has three races, two of which are parthenogenetic with differing sex ratios.

GISLASON: I do not know of any such studies on *A. zonella*. I did not find any spermatozoa in the females of *A. zonella*, which indicated that they had not copulated.

MACKAY: Could the absence of diapause be caused by the unusually long photoperiod typical of Iceland in summer?

GISLASON: For most species in Iceland the larvae grow in autumn, spring and summer, and the pupal stage extends from early spring into summer. The larvae and pupae experience a long photoperiod. As shown by Denis at this Symposium when limnephilid larvae and pupae experience a long photoperiod the adult females do not have an ovarian quiescence. Therefore it is quite possible that this is the reason for the absence of diapause.

WALLACE: How old is the Icelandic caddis fauna?

GISLASON: The last ice age ended in Iceland about 10,000 years ago. Trichoptera have colonized Iceland since then, with the possible exception of *A. zonella*. This species has been found on nunataks on one of the large glaciers, which suggests that it could have survived the last ice age.

NIELSEN: Males of *Apatania zonella* appear to be common at two high mountain lakes in Jotunheimen in Norway, under extreme conditions.

Proc. of the 2nd Int. Symp. on Trichoptera, 1977, Junk, The Hague

Diel flight activity of some Trichoptera in Northern Sweden

A. GÖTHBERG

Abstract

The investigation was carried out in 1970-73 at Rickleå Field Station, Umeå (64°N) and at Messaure Ecological Station, Jokkmokk (66°N), using suction traps provided with a device for changing jars automatically every two hours.

The catches represented the diel flight activity of the following 19 species: *Rhyacophila nubila* (ZETT.), *Agapetus ochripes* CURT., *Hydroptila forcipata* (EAT.), *H. simulans* MOS., *H. tineoides* DAL., *Stactobiella risi* (FELB.), *Oxyethira frici* KLAP., *Hydropsyche saxonica* McL., *Plectrocnemia conspersa* (CURT.), *Polycentropus flavomaculatus* (PICT.), *Wormaldia subnigra* McL, *Philopotamus montanus* (DON.), *Psychomyia pusilla* (F.), *Micrasema gelidum* McL., *Athripsodes cinereus* (CURT.), *A. commutatus* (ROST.), *Halesus tesselatus* (RAMB.), *Potamophylax cingulatus* (STEPH.) and *P. latipennis* (CURT.).

There was a range in the time of activity of these species, from *P. montanus*, which was active between 0600 and 2200, to *P. pusilla* flying between 2200 and 0800 hours.

There was no special time niche segregation between species belonging to the same genus group, e.g. *P. cingulatus-latipennis-nigricornis*, *A. cinereus-commutatus*, *H. forcipata-simulans-tineoides*, *H. tessellatus-digitatus* and *P. flavomaculatus-irroratus*.

The sex ratio changed through the 24-hour period in *W. subnigra* and *Apatania stigmatella* (ZETT.), the males being dominant during the period of greatest activity. In *P. montanus* and *S. risi* more males flew in the morning, and more females in the evening. In many species the proportion of males decreased during the summer. In *O. frici* the bimodal flight period was clearly accompanied by a bimodal curve in the sex ratio.

In *R. nubila*, which flew throughout the summer, the activity peak changed from 2200-2400 at the beginning of the flight period to 1800-2000 in September. Later in the year it appeared that the low night temperatures caused the species to become day-active.

(This paper will be published elsewhere.)

Proc. of the 2nd Int. Symp. on Trichoptera, 1977, Junk, The Hague

An application of the linear discriminant function and a multivariate analysis to caddis larvae taxonomy

H.B. BUHOLZER

Abstract
Computer programs are used for the discrimination of two related species *R. dorsalis* and *R. vulgaris* (Rhyacophilidae, Trichoptera). One program is a discriminant analysis for two groups, including the calculation of the coefficients for a linear discriminant function with which an unknown individual can be allocated to one or other species. A second program represents a multiple discriminant analysis in a stepwise manner by entered and removed variables. Fourteen measurements originating in the head were used, from which 4 variables were needed for the first program and 5 other variables for the second program.

The inducement for the morphometric studies by computer programs is the taxonomic studies of caddis larvae of the family Rhyacophilidae (DÖHLER, 1950; DECAMPS, 1965; LEPNEVA, 1964). According to SCHMID (1970) the *vulgaris* group of the genus *Rhyacophila* is a much-branched unit which includes 10 species in Switzerland.

The subgroups of *torrentium*, *intermedia* and *praemorsa* can be separated as larvae from the subgroup *vulgaris*, in which the species *R. simulatrix* is distinct from *R. vulgaris*. In the subgroup *nubila* the *fasciata*-complex can be separated from the *dorsalis*-complex in which *R. aurata* is distinct from *dorsalis* which is often misidentified as *nubila*. Investigations in the last years have shown that two species *R. vulgaris* and *dorsalis* have a large variability in the formation of characters and in the degree of sclerotisation depending on the different sampling localities, altitude, pollution etc. of the rivulets and rivers.

The aim of these studies and the use of this statistical analysis are to find a discriminant function which hydrobiologists and inexperienced workers may use to facilitate definite determination of larvae of *R. vulgaris/dorsalis*. Computer programs from a permanent file from the Computer Centre of the Swiss Federal Institute of Technology were used, written as Biomedical Computer Program for Multivariate Analysis at the University of California, Los Angeles (DIXON, 1971; 1973).

General description of BMD 04 M. discriminant analysis for two groups

This program computes a linear function of variables measured on each individual of two groups. This function can serve as an index for discrimination between the groups. It is determined from the criterion of best in that the difference between the mean indices for the two groups divided by a pooled standard deviation of the indices is maximized. The variances and covariances between the characters are first calculated, yielding the three mXm matrices, these give a pooled within-groups variance-covariance matrix W. The multiplication of the inverse matrix by the vector δ_{VD} which consists of the mean difference of each character gives the discriminant function as a vector z. The discriminant scores were plotted in a table of z_i in order of algebraic size. Three reference scores for the centroid VULG, the centroid DORS and for the point midway between them, DS_V, DS_D and $DS_{0.5}$ are calculated.

$$DS_V = \text{mean } Z_V \qquad DS_{0.5} = 1/2\,(DS_V + DS_D)$$
$$DS_D = \text{mean } Z_D$$

The absolute difference between the scores DS_V and DS_D is equal to D^2. According to MAHALANOBIS (1936), D^2 can be calculated between any pair of points f and g. Then the square root of D^2 is simply the Euclidean distance in the D-space. For the test whether the centroids are significantly different the F-test for D^2 is used.

General description of BMD 07 M, stepwise discriminant analysis.
This program performs a multiple discriminant analysis in a stepwise manner. At each step one variable is entered into the set of discriminatory variables. The variable with the largest F-value is entered. The program also computes canonical correlations and coefficients for canonical variables. It plots the first two canonical variables to give an optimal two-dimensional picture of the dispersion of each case in the two groups. In addition the means of each character for each taxon are required representing the centroids of the taxa.

Samples and measurements
For an optimal procedure it should be necessary to work with larvae which are reared from eggs of each species. Since this was not possible larvae from such samples where pupae and adults of only one species have been found were used. Over a period of three months in 1975, 23 specimens of each species were collected for the calculations. Their distribution in the rivers and provinces (Kantone) in Switzerland was as follows:

Rhyacophila vulgaris	Wasserauen (Appenzell)	St. Gallen
	Schwendibach	Goldach
	15 ex.	8 ex.

Rhyacophila dorsalis	Sihlbrugg (Zug) Sihl 10 ex.	Studen (Schwyz) Sihl 3 ex.	Arth a/Rigi (Schwyz) Rigiaa 10 ex.

All measurements were made by a stereomicroscope WILD M 5 with an eyepiece micrometer that has been calibrated against a stage micrometer to the nearest 0.01 mm. Errors and biases in measurements may arise from personal experience, personal visual aberrations, measuring habits and inadequacies of the mensuration system. The reliability of measurements should be tested (KIM et alia, 1966). The following 14 characters were measured on each individual using the designations of LEPNEVA (1964):

1. head length
2. head width posterior
3. head width anterior
4. clypeus width posterior
5. clypeus width anterior
6. clypeus length
7. distance setae 9-14
8. distance setae 14-15
9. distance frontal setae-15
10. distance setae 15-16
11. hypocranial-suture length
12. distance setae 18-8
13. distance setae 8-8
14. distance setae 17-17

It seems to be necessary to group in classes by comparing the original values and their frequency distribution (SOKAL & ROHLF, 1973). The Chi-square distribution test method proves that the means are well distinct. However, one or two variables are not sufficient for good discrimination regarding the overlap of the distributions of both species. It is necessary to use more variables and if possible some with transformation.

Results from the first program BMD 04 M

A good discrimination between the two groups results by using all 14 characters. For the practical application a linear discriminant function must be found with as few characters as possible. For that reason the characters with the best discriminatory power were selected. The misidentification for each character was calculated by the distance between character means of both species, divided by the pooled standard deviation (KIM et alia, 1966)

$$z = \frac{X_1 - X_2}{2\,s}$$

This value refers to the table of normal probabilities to obtain the per cent overlap (SACHS, 1973). The characters showing low probability of misidentification $P < 0.30$ can be used as diagnostic characters. The original characters 3, 4, 5 and 13 with their transgenerated new characters were selected. The discriminant scores are as follows:

$DS_V = 8.227$ $DS_D = 7.906$ $DS_{0.5} = 8.067$

The linear discriminant function:

$$z = 19.503\, x_3 + 1.12\, x_4 - 9.666\, x_5 - 13.872\, x_{13} + 21.866\, x_{3'} + 0.092\, x_{4'} - 4.098\, x_{5'} - 8.423\, x_{13'}$$

where x_3, x_4 ... are the measurements of the original variables and $x_{3'}$, $x_{4'}$ ... are their inverses. Specimens giving a z-value higher than this $DS_{0.5}$ would be assigned to *vulgaris*, those giving a lower value to *dorsalis*. The probability of misidentification of a single individual in the linear function can be calculated on the assumption of a multivariate normal distribution, so that the unknown does belong to V or D. If an unknown DS_U lies upon the DS_V side of $DS_{0.5}$ one can ask how many standard errors it is from DS_D and consult tables of the normal distribution. It is calculated by the difference between DS_U and the DS_V divided by the standard deviation (SOKAL & SNEATH, 1970).

Results from the second program BMD 07 M

This program performs a multiple discriminant analysis in a stepwise manner. At each step one variable is entered into the set of discriminatory variables. Data of all 14 characters were inputed, but five steps were necessary only to obtain an optimal discrimination over all 46 individuals. The stepping criterion was that the character with the largest F-value was entered. All individuals of both groups were separated in the right taxon and plotted in the two dimensional graph, indicating by the first two canonical variables which are constructed by the five selected characters. The meancoordinates or centroids are -9.457/ 0.000 for *R. vulgaris* and +9.457/ 0.000 for *R. dorsalis*. For the calculation of the coordinates of an unknown individual the products of the coefficients for canonical variables $f_3 f_9$... and the corresponding measurements x_3, x_9, x_{12}, x_{13}, x_{14}, must be summarized with a constant of canonical Var A and B. The coordinate A must be calculated for the practical application only.

$$A = - 83.183\, x_3 + 172.740\, x_9 + 69.518\, x_{12} - 86.122\, x_{14} + 91.525.$$

If the coordiante A lies on the minus side the unknown individual must be affiliated to *vulgaris* and if on the plus side to *dorsalis*. In order to obtain a definite determination in both groups it is always necessary to use the fifth larva instar.

It seems reasonable to conclude that discriminant functions may increase the speed and accuracy of determinations by inexperienced workers. In addition, discriminant functions may be useful for detecting small morphological differences that are beyond the scope of the classical descriptive methods (BIGELOW & REIMER, 1954).

References

BIGELOW, R.S. & REIMER, C. 1954. An application of the linear discriminant function to insect taxonomy. Can. Ent. 86: 69-73.

DECAMPS, H. 1965. Larves pyrénéennes du genre *Rhyacophila*. Ann. Limn. 1: 51-72.

DIXON, W.J. 1971. Biomedical computer programs. University of California. Publications in Automatic Computation. University of California Press.

——. 1973. Biomedical Computer Programs. University of California Press.

DÖHLER, W. 1950. Zur Kenntnis der Gattung *Rhyacophila* im mitteleuropäischen Raum (Trichoptera). Arch. Hydrobiol. 44: 271-293.

KIM, K.C., BROWN, B.W. & COOK, E.F. 1966. A quantitative taxonomic study of the *Hoplopleura hesperomydes* complex (Anoplura, Hoplopleuridae) with notes on a posteriori taxonomic characters. Syst. Zool. 15: 24-45.

LEPNEVA, S.G. 1964. Larvae and pupae of Annulipalpia, Trichoptera. Fauna of the U.S.S.R. Zool. Inst. Akad. Nauk. S.S.S.R., 88: 1-560 (trans. Israel Program Sci. Trans. Inc. 1970).

MAHALANOBIS, P.C. 1936. On the generalized distance in statistics. Proc. Nation. Inst. Sci. India 12: 49-55.

SACHS, L. 1973. Angewandte Statistik. Berlin: Springer Verlag.

SCHMID, F. 1970. Le genre *Rhyacophila* et la famille des Rhyacophilidae. Mémoires de la Soc. Ent. du Canada No. 66.

SOKAL, R.R. & SNEATH, P.H.A. 1970. Numerical Taxonomy: the principles and practice of numerical classification. San Francisco: Freeman.

——. & ROHLF, F.J. 1973. Introduction to Biostatistics. San Francisco: W.H. Freeman & Co.

Proc. of the 2nd Int. Symp. on Trichoptera, 1977, Junk, The Hague

The taxonomic significance of eye proportions in adult caddisflies

H. MALICKY

Abstract

The eye proportions of adult caddisflies are shown to be valuable additional taxonomic characters, as found in *Hydropsyche* spp. and several Glossosomatidae.

Many authors have recorded that *Hydropsyche exocellata* has extremely big eyes, but it seems that nobody has attempted to note the exact size and distance apart of the eyes. When I tried to define these proportions, I found that not only could *exocellata* be clearly separated from other species by this method, but that within the European *Hydropsyche* species several well-defined groups could be separated.

My method was to make outline drawings of the dorsal aspect of the head and eyes using a stereo microscope and drawing mirror. From these drawings measurements were made and relations calculated. Usually 15-20 specimens from the whole distribution area of a species were measured. To avoid too much confusion in the figures, the values falling within the encircled areas in the figures are omitted. Each species showed a certain variation, partly explained by inaccuracies in drawing. Despite this, the following groupings could be distinguished (Fig. 1):

1. A group with extremely big eyes and a small distance between them, i.e. *H. exocellata* and *H. tobiasi*. The latter has been mistaken for *exocellata*, and therefore many literature records of *exocellata* are not reliable. The examination of the holotype of *exocellata* (MALICKY, 1977), however, has shown that really two species with these particular eye proportions exist in Europe.
2. Rather smaller eyes were found in a few species which are absent in Central and Northern Europe, but some of them, e.g. *H. demavenda* and *H. sattleri*, were sometimes misidentified as *exocellata* too.
3. A group of many species with medium sized eyes. Within this group several species could be distinguished according to the shape of the eyes, e.g. very rounded and conspicuous in *H. timha* and *H. resmineda*, and rather flat in *H. maderensis*.
4. A small group of *H. fulvipes*, *H. silfvenii* and *H. bulbifera* (and a few others), which have still smaller eyes more widely spaced from each other.
5. Very small eyes, widely spaced, i.e. *H. tibialis*, *H. tabacarui* and *H. ressli*.

Only the Central and Northern European species of the genus *Hydropsyche* are figured in Fig. 1. In MALICKY (1977) all European species of the *guttata* group

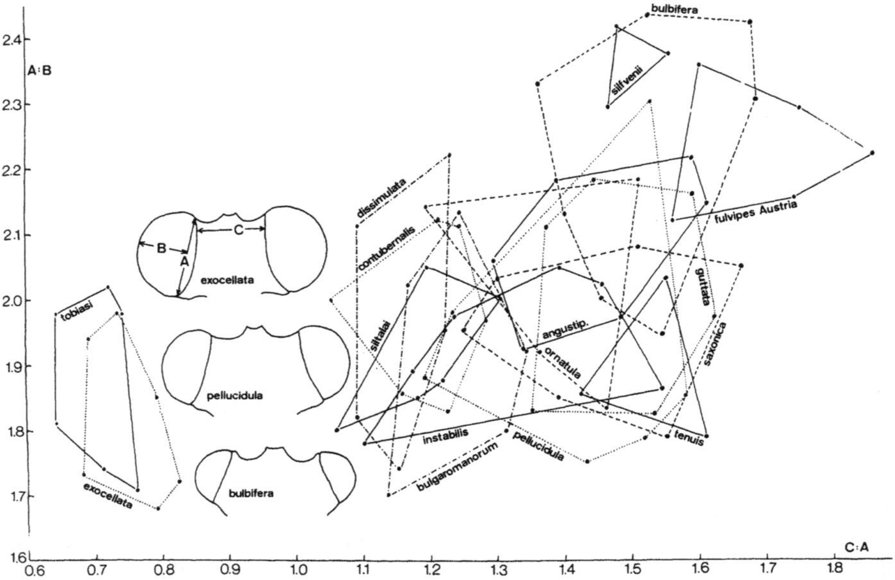

Fig. 1. Eye proportions in Central and Northern European species of *Hydropsyche*, with outlines of the heads of three species in dorsal aspect.

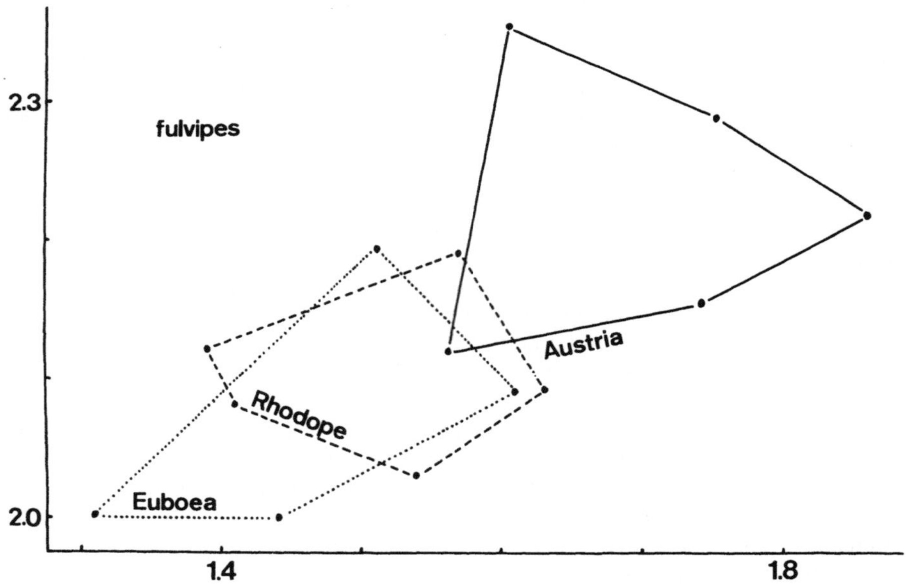

Fig. 2. Differences in the eye proportions in *H. 'fulvipes'* from different origin. Same scale as in fig. 1.

(i.e. those without digitiform appendices on the 10th segment of the males) are figured in the same presentation.

In general the eye proportions did not vary with distribution except in *H. fulvipes* where specimens from Austria differed clearly from those from Greece and Bulgaria (Fig. 2), but all should be *fulvipes* according to the genitalia. They are probably not conspecific. The *instabilis* group in the Mediterranean area has still to be revised on the basis of more material. The differences expressed in the eye proportions suggest that also in *H. 'fulvipes'* better distinguishing characters in the genitalia should be looked for.

Only males of *Hydropsyche* were compared. Differences may exist also in the females, but the identity of most females is still not clear.

Finally, I tried to discover similar differences in other caddisfly groups. They were evident in glossosomatids, e.g. in *Agapetus*, *Glossosoma* and *Catagapetus*. The differences in *Wormaldia* (Philopotamidae) and in several sericostomatids were not convincing.

In *Agapetus* a certain correlation was found between eye size and the color of the animals and their time of activity. Day-active species which were more or less dark colored (e.g. *A. nimbulus*, *A. fuscipes*, *A. caucasicus*) had smaller eyes than night-active which were light brownish or yellowish (e.g. *A. ochripes*, *A. delicatulus*). The differences existed in both males and females. It may also be noted that the *Hydropsyche* species with the smallest eyes (*H. tibialis*, *H. tabacarui*, *H. ressli*, *H. fuscipes*) have also very dark brown wing coloration.

Reference

MALICKY H. (1977); Ein Beitrag zur Kenntnis der *Hydropsyche guttata*-Gruppe.— Z. ArbGem. öst. Ent. 29: 1-28.

Proc. of the 2nd Int. Symp. on Trichoptera, 1977, Junk, The Hague

The genital segments of female Trichoptera

A. NIELSEN

Abstract

28 species, which represent all the subfamilies occurring in North Europe, have been studied. On a subfamily and even on a generic level the female genitalia are still more variable than those of the males. Some Annulipalpia have a 'Legeröhre' like most Mecoptera and some Lepidoptera. In other Annulipalpia and in the Integripalpia conditions might seem more plesiomorphic, but this is explained as secondary. The apparent ventral side of segment IX actually is the united gonopods VIII and IX, the latter again united with segment IX itself. The true ventral side of segment IX together with part of segment X form the dorsal wall of the genital chamber. The lateral walls of the latter are formed of gonopods IX, the ventral wall of gonopods VIII. In polycentropines and hydropsychids conditions are complicated by the formation of an atrium with an upper and a lower lip. In the Integripalpia the united external parts of segments IX and X have an exceedingly variable and often extremely complicated shape. The genital chamber of Trichoptera no doubt is unrivalled in complexity among insects. Most Trichoptera have a bursa copulatrix, probably homologous with that of Mecoptera and Lepidoptera.

The genital segments are VIII-X (XI), though segment VII generally is more or less, in the philopotamids highly, modified. In rhyacophilids, glossosomatids, philopotamids and hydroptilids these segments form a 'Legeröhre' as in most Mecoptera and some Lepidoptera, and segments VIII and IX are each provided with a pair of long apodemes (fig. 1, ap. VIII, ap. IX). Those of the latter segment are, though in a reduced state, rather persistent throughout the order. The 'Legeröhre' may be considered a plesiomorphic condition, though an apparently more plesiomorphic condition is found in the advanced families, segment VIII having a tergum and a sternum like the preceding segments. A closer examination, however, reveals that these sclerites are not serially homologous with those of the pregenital segments.

In *Rhyacophila* (fig. 1) the front end of segment VIII is synscleritous and produced backward into a pair of valves (valv.), which partly enclose the rest of the segment. In the glossosomatids segment VIII, though superficially very different, in principle is built as in *Rhyacophila.* Homologues of the valves are found as dorsal (d.pl.) and ventral plates (v.pl.) in *Hydropsyche* (fig. 12) and as ventral plates (v.pl.) in polycentropids (figs. 7, 9 and 10). They are found, though in a reduced state, in the hydroptilid *Orthotrichia* and even in *Agrypnia* among the Integripalpia (fig. 3, valv.).

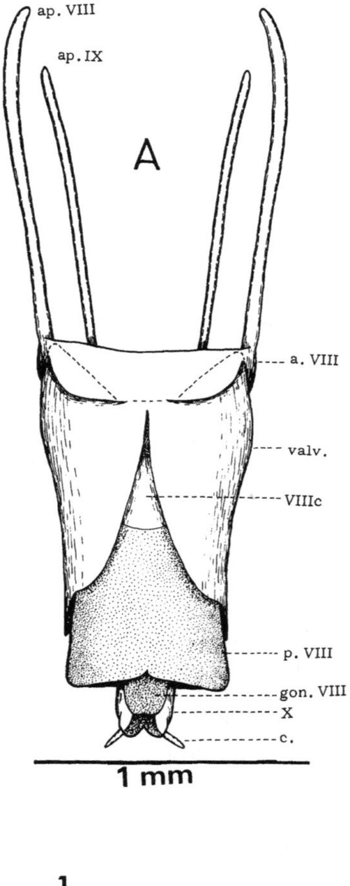

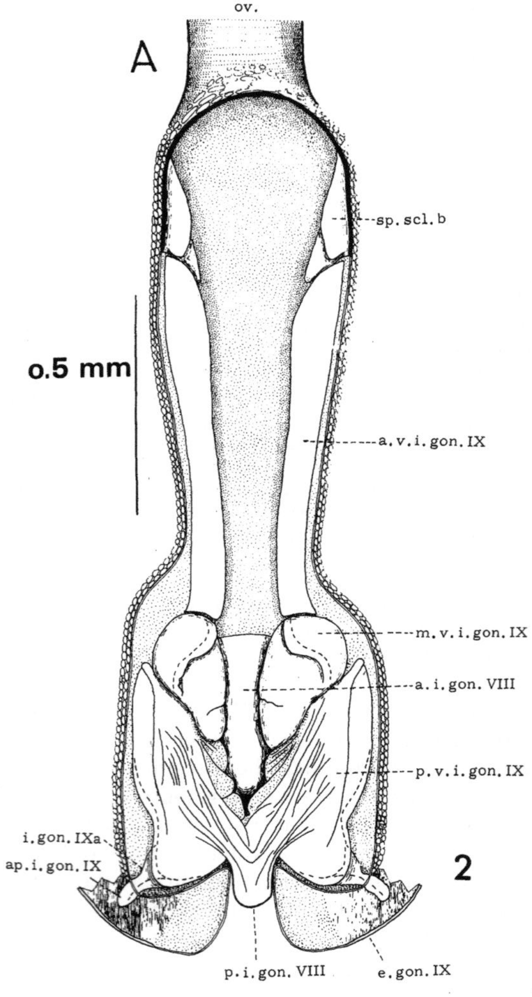

Fig. 1. *Rhyacophila nubila* ZETT.; segments VIII-X (XI) in a ventral view.
Fig. 2. *Sericostoma personatum* KIRBY & SPENCE; ventral wall of the genital chamber.

It is generally stated that the gonopore in Trichoptera is behind segment IX. In *Ecnomus*, however, it is behind segment VIII, bordered ventrally by a tongue, obviously belonging to segment VIII and here interpreted as the united gonopods VIII. In fig. 7 only a rounded keel (e.gon.VIII), lying between the ventral plates, on the ventral side of this tongue is seen. Behind the tongue the ventral side of segment IX is longitudinally cleft. The margins of the cleft are interpreted as gonopods IX, united with segment IX itself. In *Tinodes* (fig. 8) a homologous tongue is found, but it is firmly united with the anterior parts of gonopods IX. In the philopotamids the tongue is longer, almost reaching the posterior end of the abdomen, and in its anterior two thirds united with gonopods IX. In all other forms the gonopods VIII and IX are united for almost their whole length. In hydroptilids, e.g., the gonopore is surrounded by four lips (fig. 5), the dorsal one being segment X (X), the laterals gonopods IX (e.gon.IX), the ventral one gonopods VIII (e.gon.VIII), the latter slightly bilobed as an indication of a paired origin.

The most plesiomorphic condition among the Integripalpia probably is found in *Sericostoma* (fig. 11), in which the external parts of gonopods VIII (e.gon.VIII) and IX (e.gon.IX) are easily recognized. *Molanna* and *Odontocerum* approach this condition. In the limnephilids (fig. 4) the external parts of the gonopods are much reduced, forming the middle lobe (e.gon.VIII) and the side lobes (e.gon.IX) of MCLACHLAN's 'vulvar scale'.

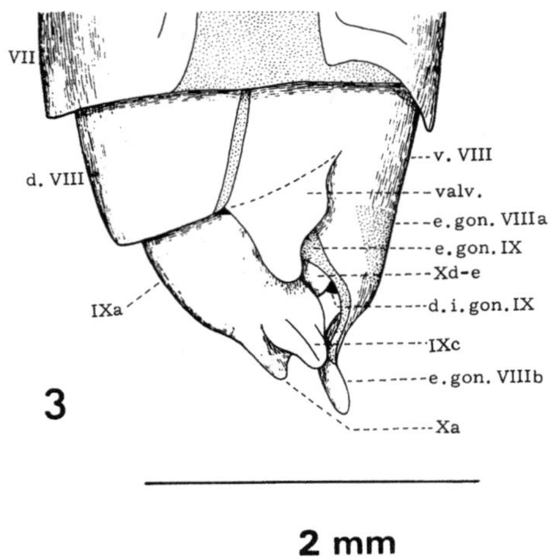

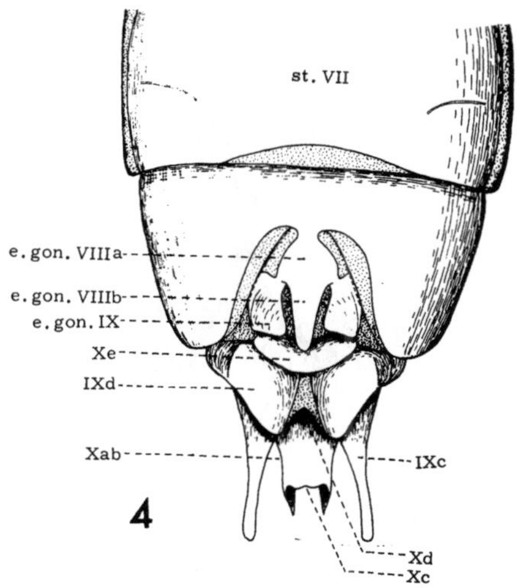

Fig. 3. *Agrypnia pagetana* CURT., posterior end of abdomen as seen from the right side.
Fig. 4. *Limnephilus flavicornis* F.; same, in a ventral view.

In the goerines (fig. 6) the side lobe is a composite structure, formed of part of gonopod IX (e.gon.IX) and part of segment IX itself (IXd).

The apparent ventral side of segment IX actually is the united gonopods VIII and IX, the latter again united with segment IX itself. In many Integripalpia (c.f. above: limnephilids) the external parts of these structures are much reduced in size and so firmly united with the secondary sternum VIII that the gonopore may seem

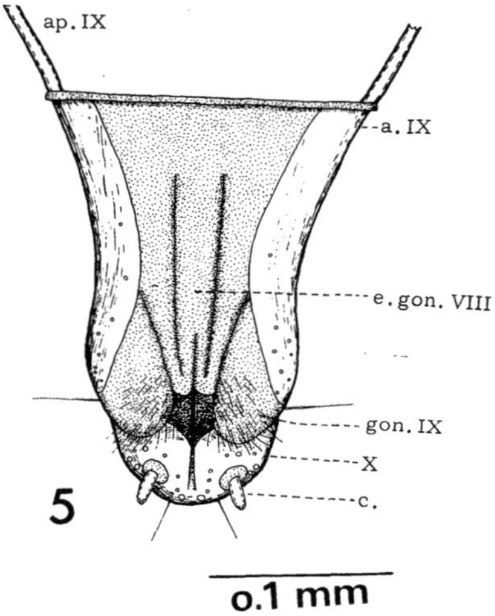

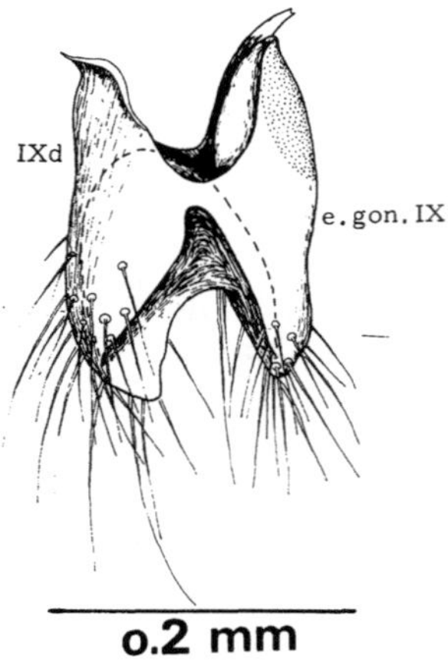

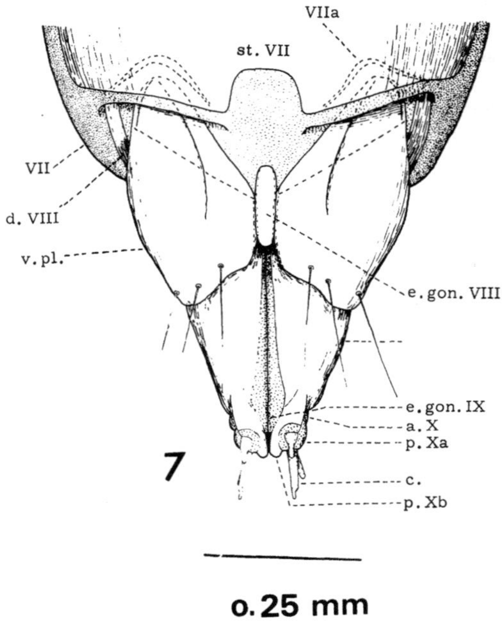

Fig. 5. *Agraylea multipunctata* CURT.; segments IX-X (XI) in a ventral view.
Fig. 6. *Silo nigricornis* PICT.; left side lobe of the 'vulvar scale' in a dorsal view.
Fig. 7. *Ecnomus tenellus* RAMB.; posterior end of abdomen in a ventral view.

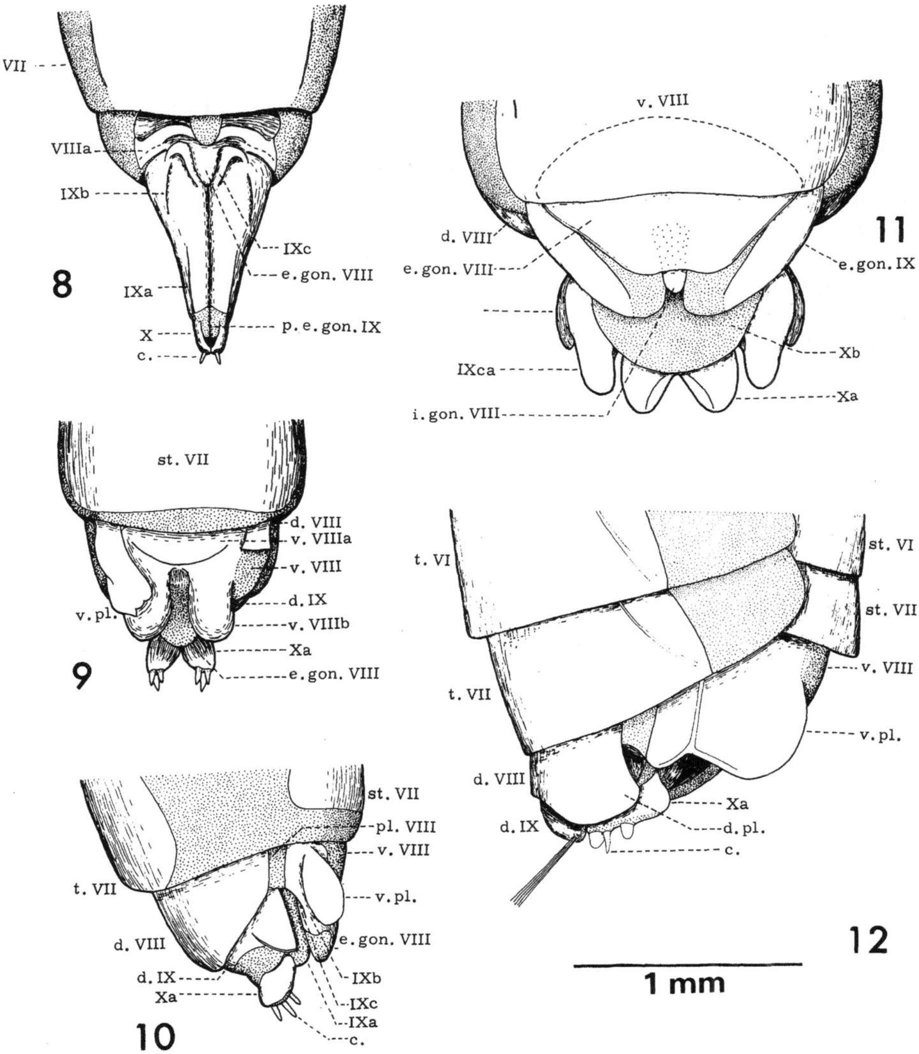

Fig. 8. *Tinodes waeneri* L.; posterior end of abdomen in a ventral view.
Fig. 9. *Holocentropus dubius* STEPH.; same.
Fig. 10. *Polycentropus flavomaculatus* PICT.; same, as seen from the right side.
Fig. 11. *Sericostoma personatum* KIRBY & SPENCE; same, in a ventral view.
Fig. 12. *Hydropsyche* sp.; same, as seen from the right side.

to be situated just behind the latter. In the polycentropines (fig. 10) and the hydropsychids, conditions are very complicated, an atrium with an upper and a lower lip being formed.

In the Integripalpia the dorsal part of segment IX is firmly united with segment X. Together they form a structure of an exceedingly variable and often extremely complex shape.

The true ventral side of segment IX together with part of segment X form the dorsal wall of the genital chamber. The lateral walls of the latter are formed of gonopods IX, the ventral wall of gonopods VIII. The internal parts of the gonopods in most Trichoptera form various sclerotic ridges (fig. 2). The dorsal wall, too, including the processus spermathecae, may have a rather elaborate shape. The complexity of the genital chamber in Trichoptera no doubt is unrivalled among insects.

The theories outlined above are corroborated by a study of the musculature.

In most Trichoptera the genital chamber is provided, besides a spermatheca, with a bursa copulatrix, which probably is homologous with that of Mecoptera and Lepidoptera.

Proc. of the 2nd Int. Symp. on Trichoptera, 1977, Junk, The Hague

The use of ventral sclerites in the taxonomy of larval hydropsychids

P.J. BOON

Abstract
The use of the ventral posterior prosternites by STATZNER (1976) as a character to separate larvae of *Hydropsyche angustipennis* CURTIS and *H. pellucidula* CURTIS is noted and this paper considers its significance in relation to hydropsychid taxonomy. Larvae of the following British species have been examined and the appearance of the posterior prosternites described: *Cheumatopsyche lepida* PICTET, *Diplectrona felix* McLACHLAN, *H. pellucidula, H. angustipennis, H. siltalai* DÖHLER, *H. contubernalis* McLACHLAN, *H. fulvipes* CURTIS, *H. instabilis* CURTIS. It is suggested that key characters such as sclerite shape are preferable to more variable head pattern markings. The separation of *H. angustipennis* and *H. pellucidula* larvae by use of posterior prosternite shape is confirmed for British material. In addition, this character may be used to distinguish *H. fulvipes* from *H. instabilis*. (At present they are separated on head markings — HILDREW & MORGAN, 1974). As a further demonstration of the use of sclerite characters in larval hydropsychid taxonomy, a key is given to the eight species described by HILDREW & MORGAN.

Introduction

This paper was originally to have been entitled 'A new taxonomic character for separating larvae of *Hydropsyche angustipennis* CURTIS and *Hydropsyche pellucidula* CURTIS', and sought to discuss differences between the two species in the shape of certain ventral sclerites. It was only in the late stages of research that the author's attention was drawn to a recent publication by STATZNER (1976) in which this same character was described. The work was therefore extended to examine the same structures in *H. instabilis* CURTIS, *H. siltalai* DÖHLER, *H. fulvipes* CURTIS, *H. contubernalis* McLACHLAN, *Diplectrona felix* McLACHLAN, and *Cheumatopsyche lepida* PICTET, as well as confirming the reliability of the character for separating *H. angustipennis* and *H. pellucidula.* It should be pointed out that this paper does not seek to examine the current taxonomic status of any species within the Hydropsychidae, and the nomenclature conforms to that of HILDREW & MORGAN (1974).

STATZNER (1976) described the appearance of the 'rhomboidal sclerites' on the larval prosternum and showed that their shape in both species was sufficiently different to distinguish easily between the two. There is little mention of these structures in the literature and recent workers such as HILDREW & MORGAN

(1974), SZCZESNY (1974), and VERNEAUX & FAESSEL (1976) have made no reference to them, concentrating rather on head characters. ROSS (1944) described these sclerites as 'prosternal plates' and used their presence or absence to distinguish between the genera *Hydropsyche* and *Cheumatopsyche*. The sclerites were not used in species differentiation within the genus *Hydropsyche*. LEPNEVA (1964) has also described the same structures but refers to them as the 'prothoracic sternite' which appears to include the 'rhomboidal sclerites' posterior to a larger anterior sclerite. However, these are only mentioned in larval descriptions and do not figure in the species key.

The aim of this paper is to describe the rhomboidal sclerites (here termed 'posterior prosternites') for eight species of hydropsychid larvae, and to show how these and other ventral sclerite characters may be used in species separation.

Table 1. A list of the larval material examined.

Species	Number examined	Instar	Collection areas
H. instabilis	42	5	Yorkshire; Cheshire; Carmarthenshire; Breconshire.
H. pellucidula	60	5	Yorkshire; Northumberland; Co. Kerry, Ireland; Toulouse, France; Cheshire.
	9	4	Northumberland; Yorkshire; Cheshire.
H. fulvipes	12	5	Cheshire; Glamorgan.
	1	4	Cheshire.
H. contubernalis	14	5	Northumberland; Yorkshire.
	17	4	Northumberland; Yorkshire.
H. angustipennis	48	5	Yorkshire; Glamorgan; Essex; Cheshire; Northumberland; Flintshire.
	19	4	Yorkshire; Essex.
	8	3	Essex.
H. siltalai	63	5	Northumberland; Pembrokeshire; Carmarthenshire; Yorkshire.
	7	4	Northumberland; Yorkshire.
C. lepida	17	5	Cardiganshire; Yorkshire.
	2	4	Cardiganshire.
D. felix	9	5	Carmarthenshire; Denbighshire.

Methods

The locations from which larvae were obtained are given in Table 1. Identification was carried out in a number of ways. In many cases, distinctive larval characters already described (HILDREW & MORGAN, 1974) were sufficient e.g. *D. felix*, *C. lepida*, *H. siltalai*. Of the other species, attempts were made to rear *H. instabilis*, *H. pellucidula*, *H. contubernalis* and *H. fulvipes*. From this, the identification of *H. pellucidula*, *H. contubernalis* and *H. instabilis* was confirmed (HILDREW & MORGAN, 1974) but rearing of *H. fulvipes* has proved unsuccessful so far. In some instances specimens were provided as named larvae by other workers.

Results and discussion

a. Description of posterior prosternites

The anterior and posterior prosternites are situated on the rear of the prosternum (Fig. 1). The anterior prosternite covers the width of the larva and is characterized by bands of pigmentation along the anterior and posterior margins. The posterior prosternites lie immediately behind this and usually consist of a medial region and a lateral region. The shape and the degree of pigmentation varies with the species, but in general the medial region is darker than the lateral and is frequently separated from it by a pale area (Fig. 2H). Sometimes the posterior prosternites are not immediately visible, and it is necessary to pull back the folded cuticle of the mesosternum to expose them.

The appearance of the prosternites in each species is as follows:

1. *D. felix* (Fig. 2B). The posterior prosternites are absent. The anterior prosternite has a characteristic bi-lobed appearance due to the presence of pale, lateral regions.
2. *C. lepida* (Fig. 2A). The posterior prosternites consist only of two small lateral flecks situated adjacent to the postero-lateral angles of the anterior prosternite.
3. *H. contubernalis* (Fig. 2C). The lateral and medial regions of each posterior prosternite form one continuous structure of uniform, pale colouration.
4. *H. angustipennis* (Fig. 2D). The lateral regions are usually similar in colour to the medial regions and are always in direct conjunction with them, giving the posterior prosternites a uniform, elongated appearance.
5. *H. fulvipes* (Fig. 2E). The medial regions are approximately triangular, frequently tapering to a point in the ventral midline. The lateral regions, which are much paler than the medial regions, are often indistinct and may be separated from the medial regions by pale areas.

6, 7 & 8. *H. pellucidula* (Fig. 2F), *H. instabilis* (Fig. 2G) & *H. siltalai* (Fig. 2H). In all three species the medial regions are irregularly square, oblong or oval. They are much darker than the lateral regions and may or may not be separated from them by pale areas. From samples of 20 animals per species 25% of *H. instabilis*, 50% of *H. pellucidula* and 75% of *H. siltalai* showed separation of lateral and medial regions.

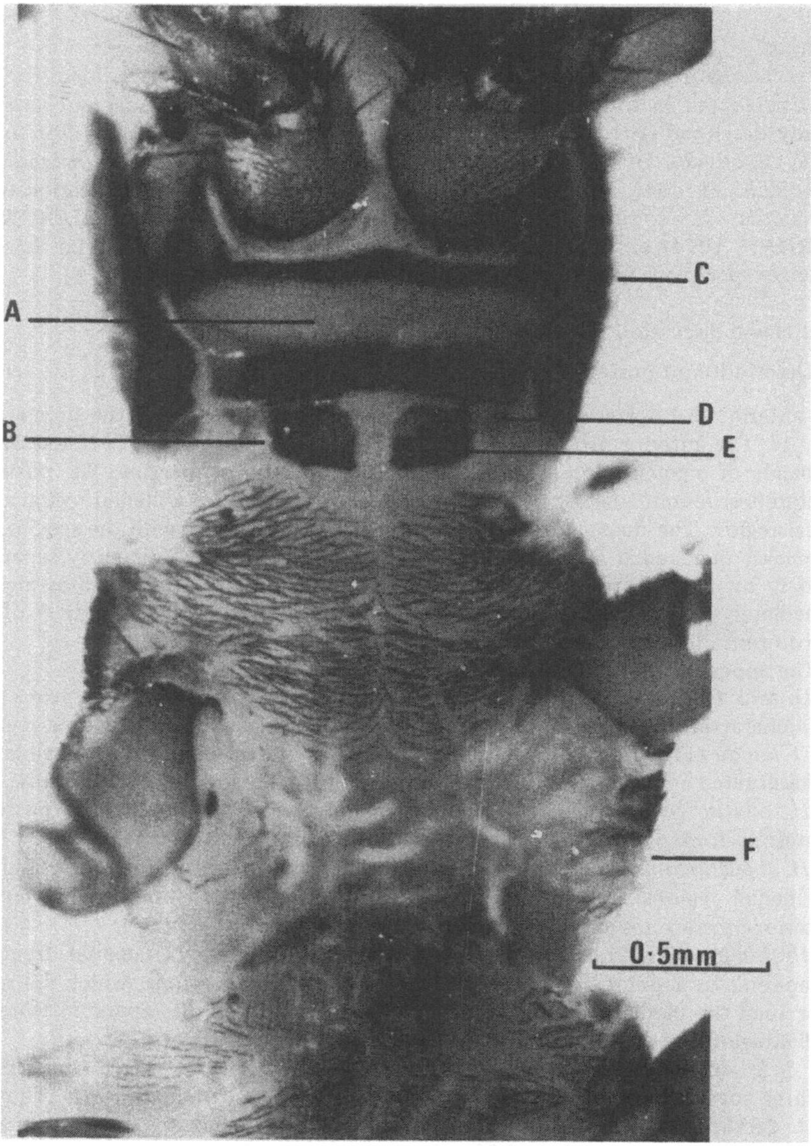

Fig. 1. Ventral view of *H. pellucidula* showing position of prosternites. A — Anterior prosternite, B — Pale area between lateral and medial regions of posterior prosternite, C — Prosternum, D — Lateral region of posterior prosternite, E — Medial region of posterior prosternite, F — Mesosternum.

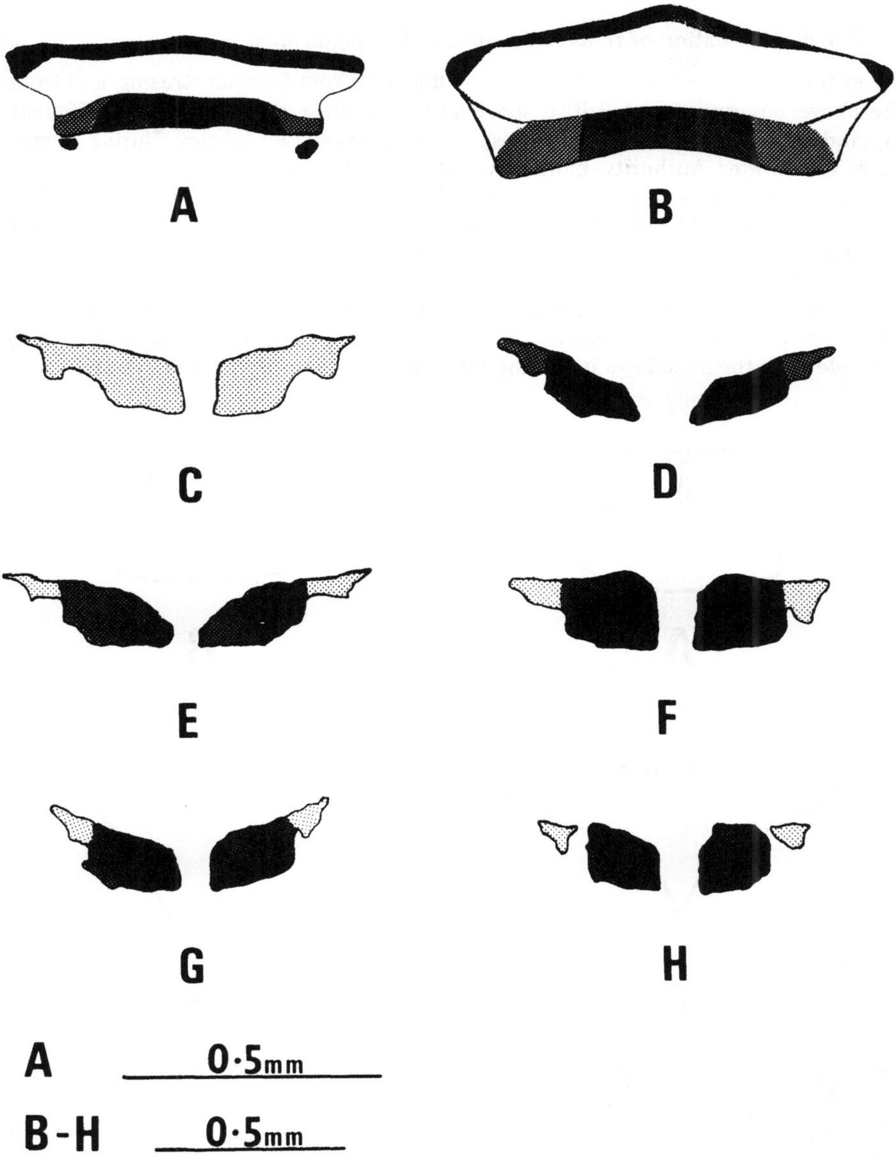

Fig. 2. A — Anterior and posterior prosternites of *C. lepida*. B. — Anterior prosternite of *D. felix*. C-H — Posterior prosternites of *H. contubernalis* (C), *H. angustipennis* (D), *H. fulvipes* (E), *H. pellucidula* (F), *H. instabilis* (G), *H. siltalai* (H).

b. The differentiation of *H. angustipennis* and *H. pellucidula*

From the work of STATZNER (1976) on material from German streams, and from the above descriptions of British specimens, it is clear that a rapid and efficient method is now available to distinguish larvae of these two species. During discussions with Water Authority biologists and other workers, it has become apparent that separating the two species on head differences (HILDREW & MORGAN, 1974) frequently proves difficult. The details given here have been circulated to a number of such workers, all of whom have expressed satisfaction at the clarity of the sclerite character in species separation. It may be used to distinguish larvae down to the third instar (as for HILDREW & MORGAN's key) and it has also been found that with practice live specimens (5th instars) of both species can be separated in the field with the naked eye by noting the shape of the posterior prosternites.

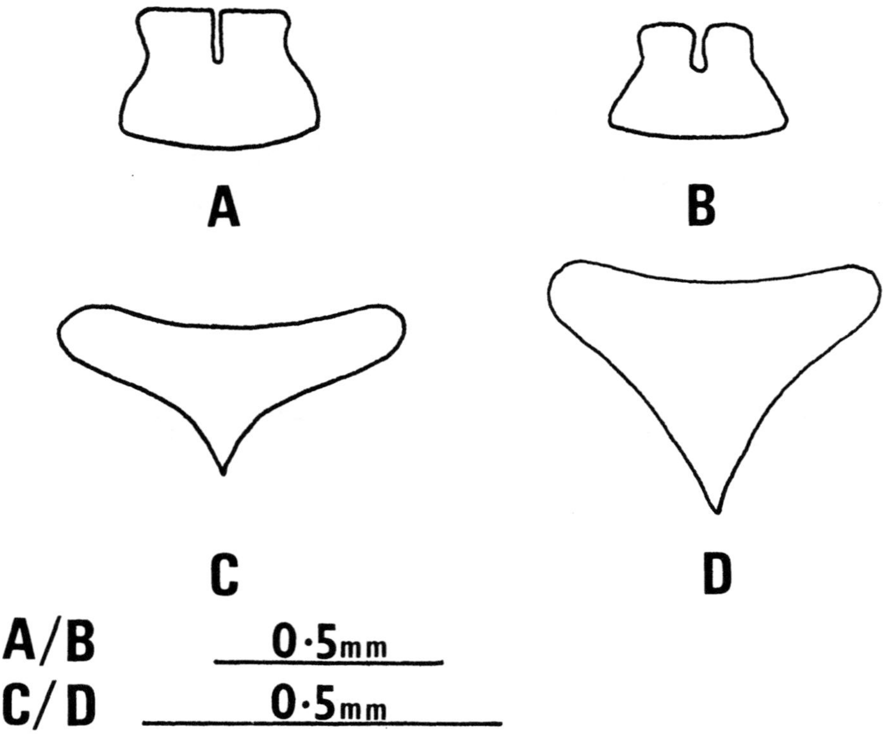

Fig. 3. A — Mentum of *H. pellucidula*, B — Mentum of *H. instabilis*, C — Submentum of *H. instabilis*, D — Submentum of *H. siltalai*.

c. The differentiation of *H. instabilis* and *H. fulvipes*

HILDREW & MORGAN (1974) distinguished between these two species by means of the markings on the fronto-clypeal apotome. It is suggested that the difference in sclerite shape described earlier provides an additional way of checking the identification, taking into consideration the variability frequently noted in hydropsychid head pattern markings.

d. A key to the larvae of the British Hydropsychidae based on ventral sclerite characters

It is obviously recognized that many of the characters in the existing key (HILDREW & MORGAN, 1974) clearly distinguish easily identified species such as *H. siltalai* and *C. lepida.* Apart from its use in separating the four species discussed in the preceding sections, the following key is included to demonstrate the fact that ventral sclerites can form a satisfactory basis for hydropsychid differentiation. Whilst this may be of use in future work on the Hydropsychidae from here and other areas, it also serves to emphasize, in general terms, the value of such characters in caddis taxonomy. (The last key couplet includes a description of the mentum and submentum — characters that have been used by other workers e.g. HICKIN, 1967; HILDREW & MORGAN, 1974; SZCZESNY, 1974; VERNEAUX & FAESSEL, 1976).

1. Posterior prosternites absent (Fig. 2B)	*D. felix*
— Posterior prosternites present	2
2. Posterior prosternites consisting of small flecks adjacent to the postero-lateral angles of the anterior prosternite (Fig. 2A)	*C. lepida*
— Posterior prosternites consisting of medial and lateral regions	3
3. Medial and lateral regions forming continuous, indistinct sclerites (Fig. 2C)	*H. contubernalis*
— Medial regions very distinct	4
4. Medial and lateral regions forming a continuous, distinct elongated structure. Lateral region usually of similar colour to medial but may be slightly paler (Fig. 2D)	*H. angustipennis*
— Lateral regions paler and less distinct than medial regions	5
5. Medial regions irregularly triangular (Fig. 2E)	*H. fulvipes*
— Medial regions irregularly square, oblong or oval	6
6. Lobes of mentum square (Fig. 3A)	*H. pellucidula*
— Lobes of mentum rounded (Fig. 3B)	7
*7. Submentum broad and shallow, having lateral regions extended into noticeable 'arms' (Fig. 3C)	*H. instabilis*
— Submentum narrower and deeper, without markedly extended 'arms' (Fig. 3D)	*H. siltalai*

* Out of the specimens examined approximately 13% proved difficult to separate on this character, as they appeared somewhat intermediate between the two types. [The species can, of course, be separated on the presence or absence of gills on the 7th abdominal segment (HILDREW & MORGAN, 1974)].

Conclusions

It is recognized that further work is required to confirm the reliability of the data presented in this paper. Ideally, all species should be reared through to adults, and additional study is needed to establish the degree to which sclerite characters can be used to distinguish younger instars. However, it is suggested that this work has demonstrated the value of examining features other than those found on the head, which has been the chief area of study in previous investigations on the taxonomy of the Hydropsychidae.

Acknowledgements

I wish to thank the following people for providing specimens: Dr. A.G. HILDREW, Dr. P.D. HILEY, Mr. R.W. JENKINS, Dr. J.P. O'CONNOR, Dr. D. ROY, Dr. W. TOBIAS, Prof. J. VERNEAUX, and Dr. I.D. WALLACE.

I would also like to thank Dr. I.D. WALLACE and biologists from the Anglian Water Authority, Welsh National Water Development Authority and Yorkshire Water Authority for providing comments on the practical use of sclerite characters in identification.

My thanks are also due to Mr. G. HOWSON for photographic assistance and Dr. G.N. PHILIPSON for reading the manuscript.

References

HICKIN, N.E. 1967. Caddis larvae. London: Hutchinson.

HILDREW, A.G. & MORGAN, J.C. 1974. The taxonomy of the British Hydropsychidae (Trichoptera). J. Ent. (B) 43: 217-229.

LEPNEVA, S.G. 1964. Larvae and pupae of Annulipalpia, Trichoptera. Fauna of the U.S.S.R. Zool. Inst. Akad. Nauk. S.S.S.R. 88: 1-560 (trans. Israel Program Sci. Trans. Inc. 1970).

ROSS, H.H. 1944. The caddis flies, or Trichoptera, of Illinois. Bull. Illinois nat. Hist. Surv. 23.

STATZNER, B. 1976. Zur Unterscheidung der Larven und Puppen der Köcherfliegen-Arten *Hydropsyche angustipennis* und *pellucidula* (Trichoptera: Hydropsychidae). Ent. Germ. 3: 265-268.

SZCZESNY, B. 1974. Larvae of the genus *Hydropsyche* (Insecta: Trichoptera) from Poland. Pol. Arch. Hydrobiol. 21: 387-390.

VERNEAUX, J. & FAESSEL, B. 1976. Larves du genre *Hydropsyche* (Trichoptères: Hydropsychidae) taxonomie, données biologiques et écologiques. Annls. Limnol. 12: 7-16.

Discussion

MACKAY: Have you used any setal characters to distinguish between species of *Hydropsyche*? I have found the prominent seta on the antero-lateral surface of the pronotum to be a useful character, according to length and degree of tapering.

WALLACE: No, but one of the purposes of this paper is to show that there are features useful in hydropsychid taxonomy, other than the dorsal head markings so frequently used in the past.

MACKAY: I have noticed that genera lacking the posterior prosternites, or with only tiny prosternites, tend to have the head and thorax in a straight line rather than curled ventrally as in *Hydropsyche.* I presume that this has something to do with the musculature associated with the prosternites. In North America, where we have many species of *Cheumatopsyche,* I can make a preliminary distinction between *Hydropsyche* and *Cheumatopsyche* by looking at the extent of curling.

BADCOCK: There is a colour distinction in the mentum; in *Hydropsyche instabilis* the central part of the mentum is always dark brown, whereas in *H. siltalai* it is yellow.

Proc. of the 2nd Int. Symp. on Trichoptera, 1977, Junk, The Hague

Taxonomic controversies in the Hydropsychidae

RUTH M. BADCOCK

Abstract

The controversy featuring *Hydropsyche instabilis* (CURT.) and *H. siltalai* DÖHL. is outlined. The lectotype of *H. instabilis*, a female in the CURTIS Collection, has been re-examined and is considered to be distinct from *H. siltalai*. The male imago depicted by McLACHLAN as *H. instabilis* has been found in the British Museum (Natural History). Contrary to the view of DÖHLER, based on McLACHLAN's drawing and subsequently accepted in the literature, this specimen is considered to be conspecific with MOSELY's *instabilis* i.e. with *H. siltalai* and not with the female lectotype of *H. instabilis*. However the lectotype in the CURTIS collection has priority for the name *H. instabilis*.

SVENSSON & TJEDER's claim that the designation by BOTOSANEANU & MARINKOVIĆ-GOSPODNETIĆ of a lectotype of *H. siltalai* DÖHL in the CURTIS Collection is invalid is upheld, but it is not agreed that their Swedish *H. siltalai* is conspecific with the female lectotype of *H. instabilis* or that TOBIAS had mixed the sexes of the two species. *H. instabilis* and *H. siltalai* should stand as discrete species. If the suggestion meets with approval, it is proposed to designate as lectotype of *H. siltalai* DÖHLER a male imago in the British Museum (Nat. Hist.) from which it is thought that KIMMINS' drawings for MOSELY's (1939) description were made.

McLACHLAN's (1874-1880) statement that '*Hydropsyche* has proved itself with justice the puzzle of students and of writers on Trichoptera' has been abundantly justified and not least by the recent controversy featuring *Hydropsyche instabilis* (CURTIS) and *Hydropsyche siltalai* DÖHLER.

Although PICTET (1834) erected the genus *Hydropsyche*, the family Hydropsychidae was established by CURTIS (1835) and it was he who first described many of the species occurring in Britain. At that time type specimens were not designated but CURTIS had his own collection of insects which was bought by the National Museum of Victoria and shipped to Australia in 1863 (NEBOISS, 1963). Since then various workers have referrred to the CURTIS Collection and designated lectotypes from it. NEBOISS (1963) designated as lectotype of *Hydropsyche instabilis* (CURTIS) a female labelled 'May, S°gate' in CURTIS' handwriting. A female was selected because the original description (CURTIS, 1834) referred to the female only. The abdomen of this female was mounted as slide T-116 and NEBOISS also mentions preparing a male of *H. instabilis* as T-101.

As shown in Fig. 1, which summarises the controversy diagrammatically,

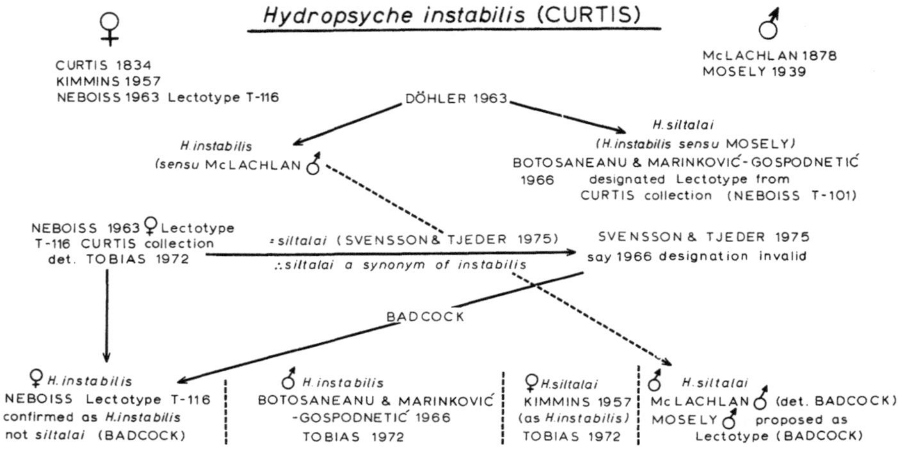

Fig. 1. Diagrammatic representation of the taxonomic controversy featuring *Hydropsyche instabilis* (CURT.) and *H. siltalai* DÖHL

DÖHLER (1963) indicated that two species had been confused in the literature under the name *instabilis*. He listed *H. instabilis* (CURT.) McLACH. and *H. siltalai* n.sp. (*instabilis* auct. pt. syn *instabilis* MOS. 1939). His reasons were published by BOTOSANEANU & MARINKOVIĆ-GOSPODNETIĆ (1966) who quoted a letter in which DÖHLER stated that he possessed two species, one corresponding to the *instabilis* of MOSELY (1939) and because it was the species for which SILTALA had first described the larva as lacking gills on the seventh abdominal segment he proposed to name it *H. siltalai*. The other he maintained was the species depicted by McLACHLAN (1878, Pl. 39) as *H. instabilis* and he proposed to retain the name *instabilis* for that. According to him its larva would have gills on the seventh abdominal segment. BOTOSANEANU & MARINKOVIĆ-GOSPODNETIĆ (1966) recognised DÖHLER's new species, attributing the name *siltalai* to DÖHLER, described diagnostic features for the male genitalia of both *H. siltalai* and *H. instabilis* and designated as lectotype of *H. siltalai* DÖHLER a male (NEBOISS' slide T-101) in the CURTIS Collection the locality being given as England, ? Southgate. NEBOISS had regarded this male as conspecific with his female lectotype of *H. instabilis*. Had that been so, *siltalai* could have become a junior synonym of *instabilis* and a new name might have been found for the other species. However TOBIAS (1972) maintained that the female lectotype did not relate to *H. siltalai* but to the other species which, like DÖHLER, he considered to be conspecific with the *instabilis* of Mc LACHLAN.

As the name *H. instabilis* properly belongs to the female lectotype in the CURTIS Collection, if TOBIAS is correct in his determination, the *instabilis* of MO-

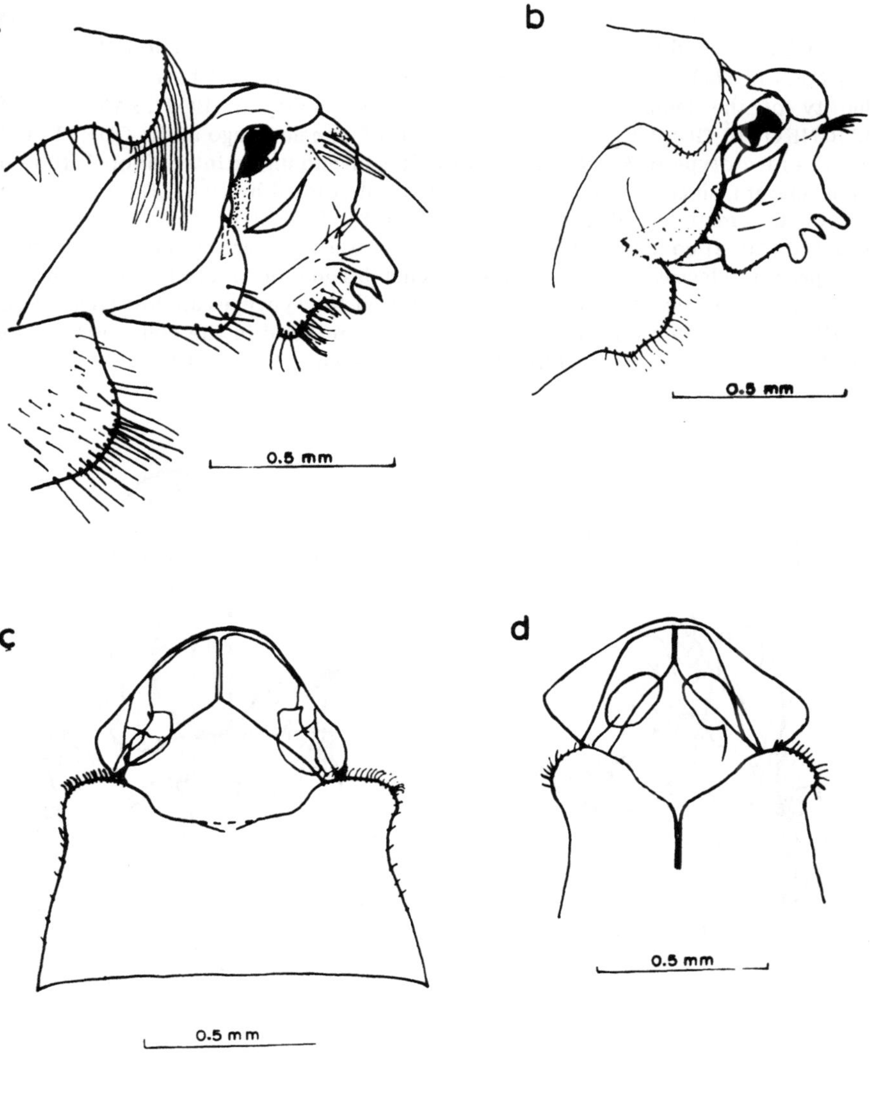

Fig. 2.a, c. *Hydropsyche instabilis*, ♀ lectotype from CURTIS Collection (NEBOISS' T-116), genitalia in lateral (a) and dorsal (c) view.
b, d. *H. siltalai* (from R. TEME, Staffs., England), ♀ genitalia in lateral (b) and dorsal (d) view.

SELY requires another name and could be recognised as *H. siltalai* DÖHLER SEDLAK (1971) adopts this nomenclature in his larval key and so do HILDREW & MORGAN (1974). However SVENSSON and TJEDER (1975) have challenged the identity of the female lectotype and also the validity of BOTOSANEANU & MARINKOVIĆ-GOSPODNETIĆ's designation of a male imago in the CURTIS Collection as lectotype of *H. siltalai*. They say it is invalid under International Rules of Nomenclature, since CURTIS' specimen was not studied by DÖHLER and was not part of the material allegedly misidentified by MOSELY. They also discount TOBIAS' (1972) comment that the designation was incorrect because it concerns a neotype, not a lectotype, maintaining: 'Such a designation would be equally invalid under the Rules because it is not proved that all the syntypes are lost (Art. 75 (C) 3). MOSELY's material is almost certainly to be refound in the collections of the British Museum (Nat. Hist.).' Further, they have seen only one of the two species in

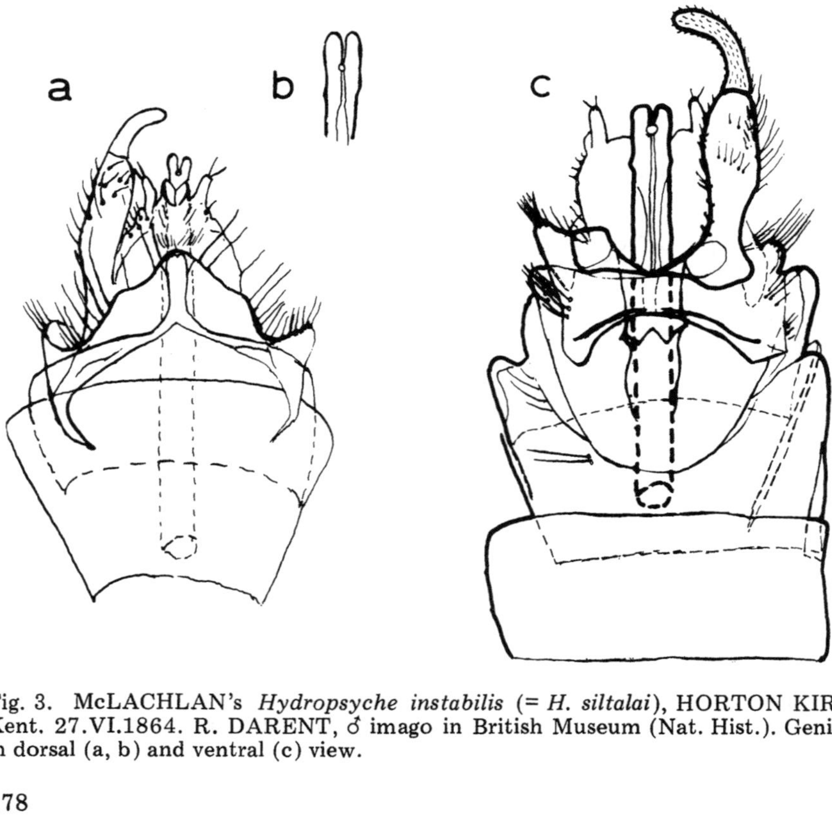

Fig. 3. McLACHLAN's *Hydropsyche instabilis* (= *H. siltalai*), HORTON KIRBY, Kent. 27.VI.1864. R. DARENT, ♂ imago in British Museum (Nat. Hist.). Genitalia in dorsal (a, b) and ventral (c) view.

Sweden. NEBOISS considered females of it to be conspecific with his lectotype of *instabilis*; males from the same area have been determined by SVENSSON & TJEDER as *H. siltalai*. They think it unlikely that there would be males of one species and females of another in the same place at the same time and that this should appertain in both Southgate and the Swedish localities. So they regard their material, the female lectotype and the male imago (T-101) of *H. siltalai* in the CURTIS Collection as all conspecific and attributable to *H. instabilis*, with *H. siltalai* as a synonym. They maintain that a new name is required for the other species of which TOBIAS (1972) has described the female as *H. siltalai* and the male as *H. instabilis*.

I have reared a number of imagines of the two species from the distinctive larvae and am convinced that TOBIAS (1972, II) has not mixed his species. SVENSSON & TJEDER (1975) give a diagram by NEBOISS featuring the opening of the clasper receptacle of the female lectotype. NEBOISS' conclusion of conspecificity was based on the shape of this aperture. Yet the value of it as a main diagnostic feature is doubtful, since TOBIAS (1972, I. Abb. 5) has shown that it may be somewhat variable, although it may be useful in association with other characters. In my experience the form of the dorsal lobe of the ninth abdominal segment, its position relative to the ventral lobe and the size and position of the clasper receptacles, together with the shape of the tip of the abdomen, are more significant features. It was the form of the dorsal lobe in NEBOISS' diagram which first led me to suspect that the lectotype could not be conspecific with *H. siltalai*. Dr. NEBOISS very kindly sent the female lectotype of *H. instabilis* to me and detailed examination of the genitalia (NEBOISS' preparation T-116) convinced me that it was not conspecific with *H. siltalai* and agrees with TOBIAS' description of *H. instabilis*.

In the lectotype (Fig. 2a) while the dorsal lobe of the ninth abdominal segment is narrow and so appears to be elongated, its anterior margin is not inserted beneath the posterior margin of the ventral lobe. However in *H. siltalai* (Fig. 2b) the dorsal lobe is deeper i.e. more lobate and its anterior margin is inserted beneath the ventral lobe. Further, the clasper receptacle has a tooth-like projection on the anterior margin of the aperture resulting in a beak-like slit which is usually much more pronounced in *H. siltalai* than in *instabilis*. The diagrams (Fig. 2b, d) were prepared from an imago reared from a larva lacking gills on the seventh abdominal segment. The aperture of the clasper receptacle agrees with KIMMINS' (1957) diagram of the species which may now be called *H. siltalai*.

In dorsal view the clasper receptacles seem to be smaller in the lectotype (Fig. 2c) than in *H. siltalai* (Fig. 2d). Further the distance between them is greater (at least twice the width of one receptacle) in the type specimen than in *H. siltalai*, where they are closer together. In *H. siltalai*, the two halves of the anterior margin of the ninth abdominal tergum meet in an acute angle whereas in *H. instabilis* the angle is wider, at least a right angle. Associated with this is the narrower dorsal silhouette of the tip of the abdomen (apart from the flanging lobes) in *H. siltalai* than in the lectotype which is more broadly rounded, but in *H. siltalai* the deeper dorsal lobes of the ninth segment constitute a more extensive flange than do the narrower ones of the lectotype. In all these respects, NEBOISS' lectotype of *H. instabilis* differs from *H. siltalai*. There can be little doubt that the female lectotype of *H. instabilis* (CURT.) designated by NEBOISS (1963), is *H. instabilis* in the sense used by DÖHLER (1963), BOTOSANEANU & MARINKOVIĆ-GOSPODNETIĆ (1966) and by TOBIAS (1972) and is not *H. siltalai.*

Swedish material kindly supplied by Dr. SVENSSON has been examined; both male and female imagines are undoubtedly *H. siltalai*, so there are no problems of mixed species. As regards imagines in the CURTIS Collection, there is a question mark associated with Southgate on the label of the male *H. siltalai* (NEBOISS T-101) so the two species may not have been taken in the same area. However in my experience there can sometimes be an overlap in their sequential distribution downstream, so it might well be feasible to take both species in the same locality.

DÖHLER considered that the male imago depicted by McLACHLAN (1878 Pl. 39) differed from MOSELY's *instabilis* and therefore was not *H. siltalai*, presumably because of rather pronounced phallic teeth. However McLACHLAN's *H. instabilis* has been examined in the British Museum (Nat. Hist.). It bears the label 'Horton Kirby, Kent, 27, VI, 1864, River Darent. *H. instabilis* Det. D.E. KIMMINS'. The phallic teeth are small (Figs. 3b, c) like those of *H. siltalai* and it would seem that McLACHLAN may have exaggerated them slightly in his drawing. The ratio of harpago to coxopodite (measured on the specimen) is 1:2.38 which comes within the range recognised for *H. siltalai* (1:2.1, TOBIAS 1972. 1:2.4, BOTOSA-

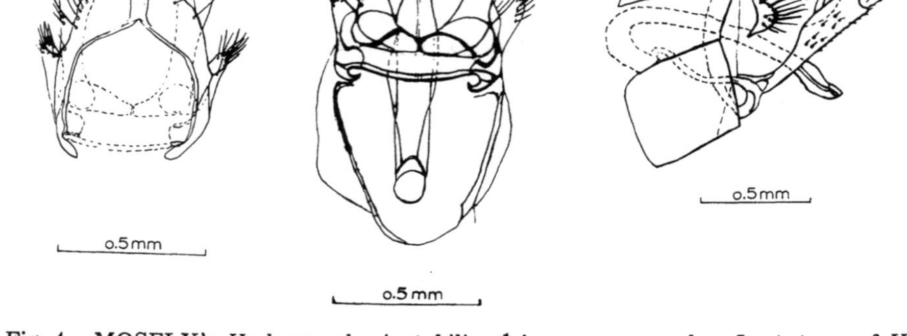

Fig. 4. MOSELY's *Hydropsyche instabilis.* ♂ imago proposed as Lectotype of *H. siltalai* (R. Test, 11.8.1913), genitalia in dorsal (a), ventral (b) and lateral (c) view.

NEANU & MARINKOVIĆ-GOSPODNETIĆ 1966) and is in contrast to that for *H. instabilis* (1:3.2 or 1:3.3). Further, the harpago tapers gently (Figs. 3a, c) and is not hooked as in *H. instabilis*. In my opinion McLACHLAN's *instabilis* is conspecific with that of MOSELY i.e. it is *H. siltalai*. However the female lectotype in the CURTIS Collection still has priority for the name *instabilis*. It is sometimes difficult to be certain of specific identity from drawings made before species were split and it might be desirable to re-examine MARTYNOV's specimen before stating categorically which species his drawings (1934) represent.

MOSELY's specimens have indeed been found in his collection in the British Museum. It is thought that KIMMINS' drawings accompanying MOSELY's 1939 description of *H. instabilis* (= *H. siltalai*) were made from male imago or imagines bearing the label *H. instabilis* R. Test 11.8.1913. One is a pinned specimen with the tip of the abdomen cleared and in an attached tube. It is likely that the lateral view was from this specimen. The other is a slide with the tip of an abdomen mounted dorsal side up and bearing a similar label. The dorsal drawing may have been from either but probably from the slide. These are the drawings to which DÖHLER refers and states to be synonymous with *H. siltalai*. Recent drawings of the specimens (Fig. 4) may help to emphasis that they are. At the moment DÖHLER's name, *H. siltalai*, is strictly speaking incorrect because no type has been correctly designated, but it has been unchallenged. If the members of the Symposium agree, I would like to validate it by designating MOSELY's material with KIMMINS' drawings of it (Figs. 400-402 in MOSELY, 1939). As a single specimen is required, it is suggested that the pinned specimen, with abdomen cleared and in an attached tube had better be the lectotype of *Hydropsyche siltalai* DÖHLER, rather than the prepared slide of parts of a male from the same locality on the same date.

Previously, *H. instabilis* had sometimes been confused with *H. fulvipes* (CURT), since on using MOSELY's key, imagines of *H. instabilis* were determined as *H. fulvipes*. I did this after rearing larvae, hence my note in 1955 alleging the widespread distribution in Britain of *H. fulvipes*; many of the records were really *H. instabilis* of the new nomenclature, a note differentiating between *H. fulvipes* and *H. instabilis* records has been in press since 1974 and should be published shortly.

I thank Mr. P.E.S.WHALLEY and Dr. P.C.BARNARD of the British Museum (Nat. Hist.) for their helpful advice and for access to specimens, also Drs. A. NEBOISS and B.W. SVENSSON for sending material, and Mr. G. Burgess for stencilling Fig. 1.

References

BADCOCK, R.M. 1955. Widespread distribution in Britain of our allegedly rare caddis, *Hydropsyche fulvipes* (CURTIS) (Trich., Hydropsychidae), Ent. mon. Mag. 91: 30-31.

BOTOSANEANU, L. & MARINKOVIĆ-GOSPODNETIĆ, M. 1966. Contribution à la connaissance des *Hydropsyche* du groupe *fulvipes-instabilis*. Étude des genitalia males. Annls Limnol. 2: 503-525.

CURTIS, J. 1834. Descriptions of some hitherto nondescript British species of May-flies of Anglers, Phil. Mag. 4: 212-218.
——. 1835. British Entomology.
DOHLER, W. 1963. Liste der deutschen Trichoptera, NachBl. bayer Ent. 12: 17-22.
HILDREW, A.G. & MORGAN, J.C. 1974. The taxonomy of the British Hydropsychidae (Trichoptera), J. Ent. (B) 43: 217-229.
KIMMINS, D.E. 1957. Notes on some British species of the genus *Hydropsyche*, Entomologist's Gaz. 8: 199-210.
McLACHLAN, R. 1874-1880. A monographic revision and synopsiss of the Trichoptera of the European fauna. (Reprint) Hampton: Classey.
MARTYNOV, A.V. 1934. Trichoptera Annulipalpia, Leningrad.
MOSELY, M.E. 1939. The British Caddis Flies (Trichoptera); a collector's handbook. London: Routledge.
NEBOISS, A. 1963. The Trichoptera types of the species described by J.CURTIS, Beitr. Ent. 13: 582-635.
PICTET, F.J. 1834. Recherches pour servir à l'histoire et à l'anatomie des Phryganides. Geneva.
SEDLAK, E. 1971. Bestimmungstabelle der Larven der haufigen Tschechoslowakischen Arten der Gattung *Hydropsyche* Pictet (Trichoptera), Acta ent. bohemoslov. 68: 185-187.
SVENSSON, B.W. & TJEDER, B. 1975. Taxonomic Notes on Some European Trichoptera, Ent. scand. 6: 67-70.
TOBIAS, W. 1972. Zur Kenntnis europaischer Hydropsychidae (Insecta: Trichoptera) I, II. Seckenberg. biol. 53: 59-89, 245-268.

Proc. of the 2nd Int. Symp. on Trichoptera, 1977, Junk, The Hague

Emergence of caddisflies from the Roseau River, Manitoba

J.F. FLANNAGAN

Abstract

Four, 1m² box emergence traps and 8 Hamilton type stream traps were positioned in sets of 1 box trap plus 2 Hamilton traps over 4 substrates: boulders, cobble, gravel and sand, in a section of riffle and an associated pool in the Roseau River. The traps were in position from 10 May to 26 October 1976 and were emptied every two days. Caddisfly emergence totalling 10,314 specimens and representing 10 families, 22 genera and 42 species were collected over the period 12 May-1 October.

Emergence densities were directly related to substrate size and the boulder substrate was the site of emergence of almost half the caddisflies. Total emergence was found to be related to absolute and/or changes in water temperature. Periodicity of emergence varied among and within species and in most cases was found to be related to illumination but modified by unknown factors. Some species were found to select a specific substrate while others are likely present at a particular station because of water flow or food requirements or a combination of a number of environmental requirements.

Introduction

A proposed irrigation scheme in the North Central U.S.A. would divert water from the south-flowing Mississippi — Missouri river system to the north-flowing Churchill-Nelson system. The possibility of introduction of exotic aquatic animal species via this diversion water necessitated a comparative distributional study of the fauna of these two systems. The caddisflies (and other aquatic insects) of the North Central U.S.A. are quite well known, largely from the work of the Illinois Natural History survey (ROSS, 1944). However, the fauna of the Churchill-Nelson System is rather poorly studied and thus a survey of the aquatic insects of Southern Manitoba was initiated. As part of this survey an area of rapids and an associated pool on the Roseau River, in southern Manitoba, was sampled over the ice free season of 1976.

The Roseau River rises in Northern Minnesota, flows northwest through southern Manitoba (Fig. 1) into the Red River system which eventually drains via the Nelson river into Hudson Bay. The river, which runs through an area of glacial outwash, is of rather an unusual type on the Canadian prairies, since it is permanent, clear, fast and boulder strewn. Most of the other rivers and streams of the area

Fig. 1. Position of the Roseau River in the Churchill Nelson drainage system.

being either temporary (spring and early summer) or slow flowing, silty water bodies. Thus the river contains a larger than normal diversity of habitats — a situation which according to THIENEMANN (1954) should bring about a large diversity of species.

This study included investigation of the organism/substrate relationship, phenology and general biology of the caddisflies of the river.

Materials and methods

During the open water season, 5 April-30 October 1976, emerging insects were collected at a site 2 km downstream from Provincial Trunk Highway 59 (Fig. 1). From 6 April to 10 May collections were made with a hand net. During this period no caddisflies were observed to emerge, although three species of 'winter' stoneflies were collected in relatively large numbers (FLANNAGAN, 1977). On 4 May, four 1 m^2 box traps and 8 stream emergence traps (HAMILTON, 1969) were pinned in sets of 1 box trap and 2 stream traps to each of four substrates — boulders, cobbles, gravel and sand. The box traps (similar to those used by IDE (1940) and SPRULES (1947) were built from wood frames covered on three sides with 400 μ m nitex nylon netting, the fourth side being a nitex covered door. The top was covered with transparent vinyl plastic (light transmission properties discussed by FLANNAGAN & LAWLER, 1972). The stream emergence traps sampled an area of 0.1 m^2. From 10 May to 26 October the traps were emptied every second day except during the periods 13, 14, 15 June; 12, 13, 14 July and 10, 11, 12 August when, for a 48 hour period, the box traps were emptied every two hours. The boulder substrate for the box trap was a 90 cm x 40 cm emergent boulder which was completely enclosed by the trap. The two stream traps in this set were pinned among large emergent boulders. The remaining substrates were completely submerged and were essentially identical for the two kinds of emergence traps — cobble — (5-15 cm diam), gravel (1-5 cm diam) sand (50-1,000 μ m diam). Daily water flow information was provided by Water Surveys Canada. Water temperature was measured using max. and min. thermometers set in the water and read every two days. Dissolved oxygen, pH, and conductivity were measured weekly during the normal sampling schedule and every two hours during the 48-hour sampling period. Chemistry samples were analysed in June, July, August and October by the Freshwater Institute's water analyses laboratory.

Caddisflies were preserved in 75% ethanol and identified using BETTEN & MOSELY (1940), DENNING (1943), ROSS (1944) and GORDON (1974).

Results

Physical and chemical

River flow reached a maximum simultaneously with ice out on 6-7 April and thereafter declined through the summer except for two small peaks associated with heavy rainfall in mid June and early July (Fig. 2). These periods of high water are associated with drops in water temperature and decreases in total numbers of caddisflies emerging.

Although the water chemistry results are not complete it is likely that the dissolved solids, suspended solids and dissolved phosphorus are directly related to water flow. The total dissolved nitrogen and carbon are independent of water flow presumably because they are available both from the atmosphere and from waste products of the fauna and flora of the river.

Biological

The standing crop of algae and macrophytes in the river is visibly high with *Nostoc* sp. forming a carpet over the gravel and larger substrates and long trailing mats of *Spirogyra* sp. filling most of the water space in the riffles. A number of macro-

Table 1. Emergence data for the 42 species of caddisflies collected in the emergence traps.

	Emergence Period and Peak(s) (May, June, July, Aug, Sept, Oct)	Voltinism	No/M² Boulder	No/M² Cobble	No/M² Gravel	No/M² Sand	No/M² Total	Sex Ratio ♂♂:♀♀
RHYACOPHILIDAE								
GLOSSOMOMATINAE								
[2]Protoptila erotica Ross		Univoltine	89	18	57	3	166	1:3
[2]P. maculata (Hagen)		Bivoltine	128	77	64	8	277	1:3
PHILOPOTAMIDAE								
[2]Chimarra socia Hagen		Univoltine	384	114	117	43	658	1:3
PSYCHOMYIIDAE								
Psychomyia flavida Hagen		Univoltine	23	6	4	1	34	-
Plectrocnemia cinerea (Hagen)		Bivoltine	12	3	10	3	28	1:3
Neureclipsis crepuscularis (Walker)		Bivoltine	151	56	88	17	311	1:5
HYDROPSYCHIDAE[1]								
[2]Hydropsyche bifida Banks		Univoltine	20	6	3	3	32	1:20
[2]H. bronta Ross		Univoltine	3	3	2	0	8	1:20
H. recurvata Banks		-	1	0	0	0	1	1:20
H. scalaris Hagen		Univoltine	7	3	1	0	11	1:10
[2]H. slossonae Banks		Univoltine	8	30	13	3	54	1:5
[2]H. walkeri Betten & Mosley		Univoltine	1	4	2	0	7	1:7
H. sp ♀♀			1061	78	133	13	1285	
Cheumatopsyche compyla Ross		Univoltine	318	299	288	28	933	1:3
C. gracilis Banks		Univoltine	28	573	26	0	627	1:4
HYDROPTILIDAE								
[2]Hydroptila ajax Ross		Multivoltine	290	278	223	73	864	1:1
[2]H. albicornis Hagen		Multivoltine	531	313	422	84	1350	1:1
[2]H. angusta Ross		Multivoltine	3	2	3	2	10	1:2
[2]H. armata Ross		Bivoltine	3	1	0	0	4	-
H. consimilis Morton		-	0	2	0	0	2	-
[2]H. grandiosa Ross		Univoltine	2	0	0	1	3	-

..continued

Table - continued

	Emergence Period and Peak(s) (May–Oct)	Voltinism	No/M² Boulder	Cobble	Gravel	Sand	Total	Sex Ratio ♂♂:♀♀
HYDROPTILIDAE (continued)								
[2]H. perdita Morton		Multivoltine	21	15	9	24	69	1:5
[2]H. spatulata Morton		Multivoltine	238	116	131	403	887	1:2
[2]H. virgata Ross		-	0	0	1	0	1	-
H. sp.		-	3	0	0	4	7	-
[2]Tascobia palmata (Ross)		Univoltine	13	0	0	0	13	-
[2]Ochrotrichia tarsalis (Hagen)		Univoltine?	148	6	5	3	169	2:1
[2]Neotrichia okopa Ross		Univoltine	3	0	0	0	3	-
[2]Mayatrichia ayama Mosley		Bivoltine	80	19	30	30	160	1:2
Agraylea multipunctata Curtis		Bivoltine?	0	0	0	2	2	-
PHRYGANEIDAE								
Agrypnia straminea Hagen		-	0	1	0	0	1	-
LIMNEPHILIDAE								
Anabolia bimaculata (Walker)		-	0	0	1	0	1	-
Pychopsyche guttifer (Walker)		-	1	0	0	0	1	-
P. subfasciata (Say)		-	0	1	1	1	3	-
LEPTOCERIDAE								
Leptocella albida (Walker)		Univoltine	1	3	3	7	14	1:1
[2]Leptocella diarina Ross		Univoltine	3	3	33	11	50	1:1
Ceraclea transversa (Hagen)		Bivoltine	2	1	1	0	4	3:1
Oecetis avara (Banks)		Univoltine	69	88	71	40	268	1:2
[2]Oecetis inconspicua (Walker)		Univoltine	0	1	1	8	10	1:1
BRACHYCENTRIDAE								
Micrasema rusticum (Hagen)		Univoltine	7	9	0	0	16	4:1
[2]Micrasema wataga Ross		Univoltine	76	15	31	3	125	1:2
[2]Brachycentrus numerosus (Say)		Univoltine	0	3	11	1	15	-
HELICOPSYCHIDAE								
Helicopsyche borealis (Hagen)		Bivoltine	29	44	27	17	117	1:6
Total			3157	2191	1812	836	8595	

[1] Emergence period of the various species of *Hydropsyche* represents male emergence only. Sex ratio for these species is derived from emergence curve results.
[2] Species not previously recorded from Manitoba.

phytes including *Potamogeton* spp. and *Elodea* sp. are associated with the slower flowing areas.

This large plant biomass is reflected in the animal production since 10,315 caddisflies emerged from a sampling area of 4.8 m^2 over the summer giving a mean density of 2,149 animals/m^2. The emergence period was 10 May to 1 October (Fig. 2) with the peak emergence in mid July. A total of 10 families, 22 genera and 42 species of Trichoptera were represented in this part of the river (Table 1). Densities of animals collected varied directly with substrate size and almost half of the total emergence occurred at the boulder substrate. Sex ratios varied considerably among the various species, the extremes being around 1:20 (males:females) for some *Hydropsyche* species and 4:1 for *Micrasema rusticum*. In general, however, the females usually dominated due perhaps to the presence of females returning to lay

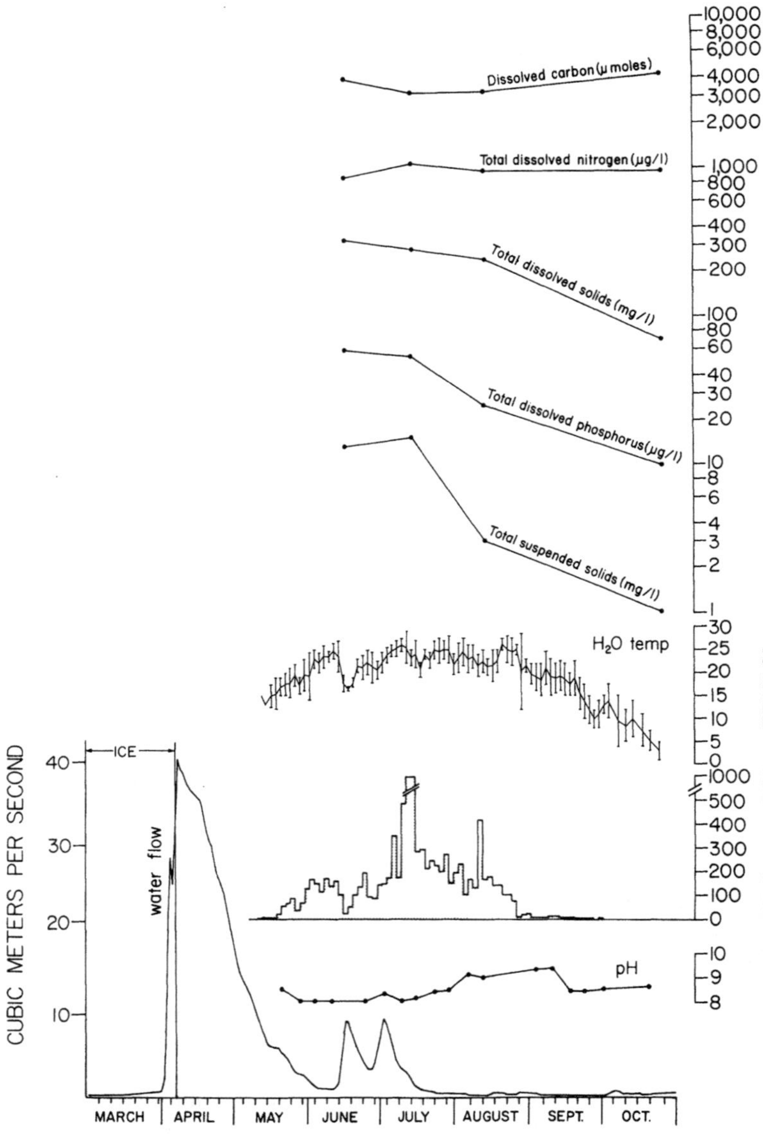

Fig. 2. Physical, chemical and total caddisfly emergence results from the Roseau River, 1976.

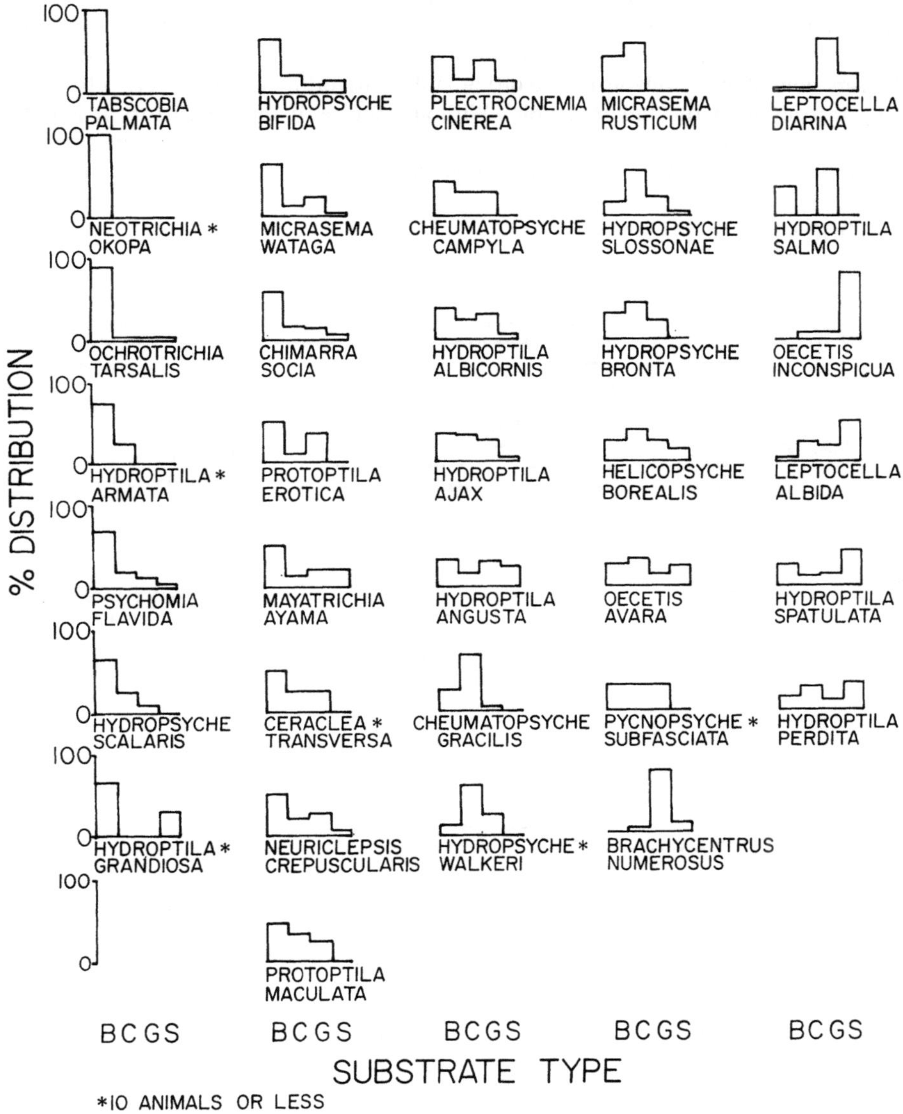

Fig. 3. Percent distribution of the various species collected over the four substrates sampled.

eggs (CORBET, 1966; FLANNAGAN & LAWLER, 1972). Box traps, since they are pinned to the substrate provide less opportunity than floating traps for this source of error in estimating densities. The Trichoptera emerging from this stretch of rivers consisted of univoltine, bivoltine and multivoltine species. Multivoltinism being restricted to the 'micro caddis'. Without studies of the life history it is not possible to say whether the flight periods are as a result of more than one generation occurring during the summer or several cohorts emerging at different times yet

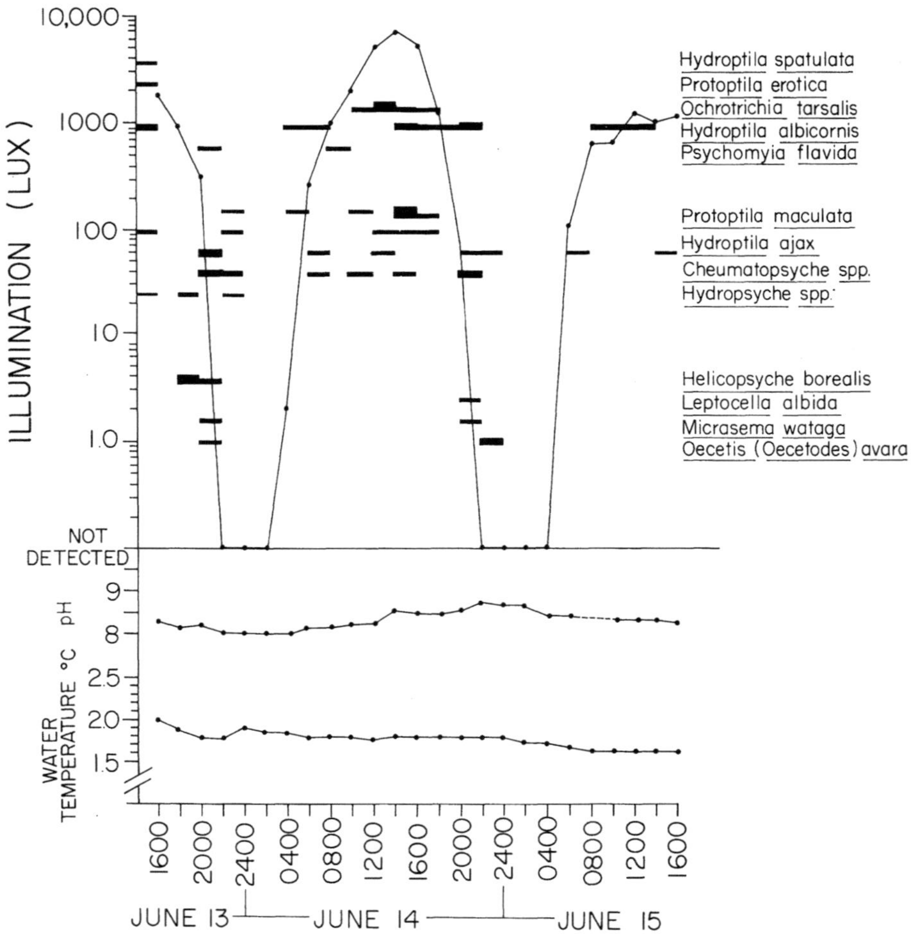

Fig. 4. The relationship between emergence time of caddisflies and illumination, pH and temperature during the June 48-hour samples.

occurring simultaneously (as was found for *Ceraclea transversa* by RESH, (1976). The distribution of the various species over the four substrates sampled (Fig. 3) indicates that, at least in their pupal emergence stages, few species are totally confined to one substrate type. However, these distributions of the various species do show that there is considerable preference — *Tascobia palmata*, *Neotrichia oko-*

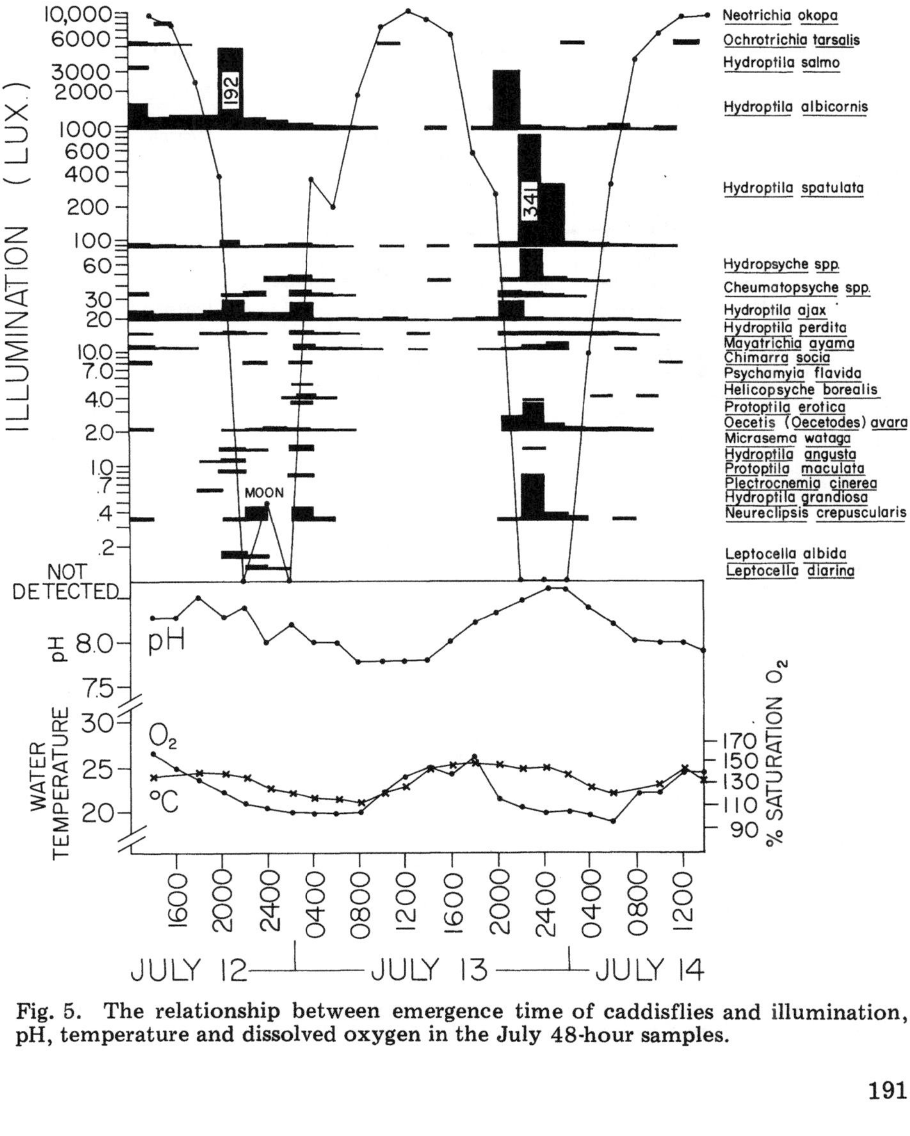

Fig. 5. The relationship between emergence time of caddisflies and illumination, pH, temperature and dissolved oxygen in the July 48-hour samples.

pa, *Ochrotrichia tarsalis*, *Hydroptila armata* and *Psychomia flavida* having a 70% or more occurrence at the boulder substrate station, *Cheumatopsyche gracilis* 70% over the cobble substrate, *Brachycentrus numerosus* over 70% at the gravel station and *Oecetis inconspicua* over 80% at the sand substrate station.

The three series of 48-hour continuous samples (Figs. 4, 5 and 6) although

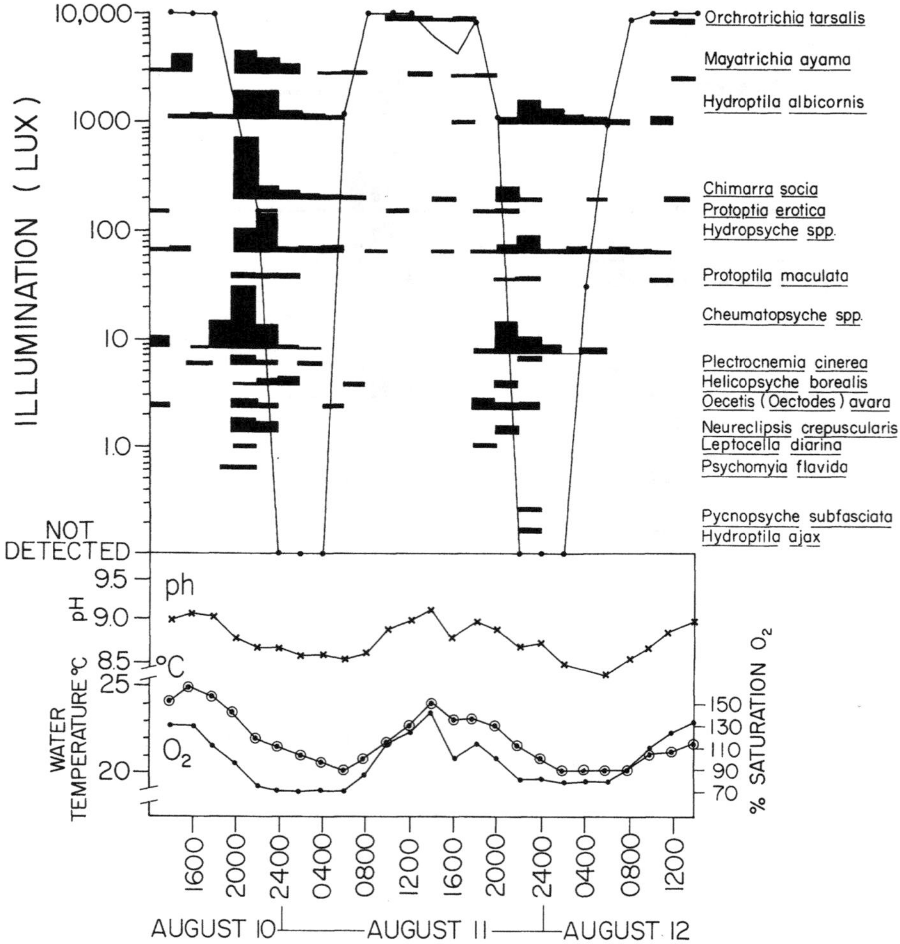

Fig. 6. The relationship between emergence time of caddisflies and illumination, pH, temperature and dissolved oxygen in the August 48-hour samples.

suffering from small sample size for some species, appear to show some interesting phenomena. The species emerging in June seem to emerge largely during the day or evening, while in July emergence is restricted during full daylight and peaks often occur in full darkness and in August peak emergence occurs largely in the evenings. These emergence pattern changes throughout the summer are partly due to the species succession and partly to changes in specific emergence time for individual species.

Emergence of some species (eg. *Ochrotrichia tarsalis*) was almost completely limited to full daylight, some such as *Helicopsyche borealis*, emerged in reduced light while the Hydroptilidae contained most of the species which emerged continuously throughout the full range of light intensities. It should be noted, however, that many of these 'continuously' emerging species had quite definite peaks in reduced light or full darkness.

Discussion

About two thirds of the species caught were univoltine. In view of the lack of evidence to the contrary, at least at these latitudes, we must assume that these species have at least a one year life cycle. Eight species were bivoltine but in most cases the two flight periods were separated by only a short interval (usually less than a month) which does not appear to be sufficient time for the animal to go through a complete generation, in these cases it seems reasonable to assume that we have two simultaneous cohorts similar to that described by RESH (1976) for *Ceraclea transversa*. *Hydroptila armata* appears to be an exception to this situation since it has its first flight period in late May/early June and its second in the middle of September. This is perhaps sufficient time for growth of a second generation as was shown for *Glossosoma penitum* by ANDERSON & BOURNE (1974). The remaining five species (all *Hydroptila*) showed three distinct flight periods, each with a distinct peak and with a more or less normal distribution within each period, and all had distinct gaps of almost a week to a month between flight periods. NIELSEN (1948) reported two successive generations of *Hydroptila femoralis* per year. In this study the three emergence periods are very close together and it is not likely that the three peaks represent three complete life cycles. Possibly the peaks are related to the abundance of *Spirogyra* sp. which appeared (although no biomass measurement were made) to go through three or four cycles of growth with intervening periods of decay throughout the summer.

HYNES (1970) showed that results of attempts to measure densities of aquatic invertebrates in streams varied enormously both with the substrate type and a large series of other factors. Two studies in Canada similar to this one are available for density comparisons; those of IDE (1940) and SPRULES (1947). In general these streams in eastern Canada produced considerably fewer caddisflies than the present river. An exception is SPRULES' (op. cit.) station four, from which 12,423 caddisflies were collected. This station, however, had a substrate of 'rubble' in which stones were piled on top of other stones; this provided a much larger surface area than any of the substrate types in the Roseau river.

MACAN (1961) and HYNES (1970) in reviews of factors influencing the distribution of aquatic animals, suggested that substrate, food availability, water flow, temperature and chemistry could largely explain the species distribution within a given system. In the sampling area under study the chemical parameters and the water temperature showed no detectable differences among the four stations. The substrates are obviously different, but are also partially interrrelated since all stations had some sand, three of the four had gravel, two of the four had cobbles and only one had boulders. Thus if only substrate were responsible for the distribution of caddisflies we might expect all the species collected in the river to be present at the boulder substrate and for their presence at any particular station to be proportional to the amount of their specific substrate at that station, i.e. if a species was limited to boulder substrates it would only be taken in that trap which was set over boulders while species which preferred sand might be expected at all four stations but most commonly at the sand substrate station. An examination of the substrate/emergence results (Fig. 3) indicates, for some species, substrate alone may well be the discriminating environmental factor, eg. *Tascobia palmata* and *Neotrichia okopa* only at the boulder substrate; *Micrasema rusticum* only present at the boulder and cobble stations and most common over cobble, thus a 'cobble' species, etc. However, it is apparent from this figure that substrate alone will not explain all of the distributions.

We must therefore seek other factors to explain their distribution. NIELSEN (1948) found that *Hydroptila* larvae feed by sucking cells of filamentous algae and in the study area this occurred in large amounts at the boulder, cobble and gravel stations. Similarly, RESH (1976) reported that *Ceraclea transversa* was a sponge feeder, sponges occurring in this river only at the three substrates mentioned above and most abundantly on the boulder. Thus food dependancy, either on the algae, the sponges or on the animals which feed on them may well explain the limitation of many of the species to the three larger-sized substrates.

Water current speed obviously plays a large part in the distribution of aquatic invertebrates within a river system. Many workers such as MORETTI & GIANOTTI (1962) with *Agapetus fuscipes* and *Silo nigricornis* and GRENIER (1949) with several species of *Simulium* have shown that species are often very current-specific in their larval instars. However, fine selection of current speeds does not, at least in the Roseau River, appear to be very important in the pupa to adult stages. The boulder substrate, although situated in the fastest part of the river, because it causes a frontal wave before and a sheltered area behind, had a current speed of less than 0.1 m/sec. similar to that at the sand substrate. The two intermediate substrates being in much faster water (gravel 0.3 m/sec, cobble 0.45 m/sec with a river flow of 10 m^3/sec). The peculiar distribution of *Hydroptila grandiosa* (sand and boulders only) is possibly related to the current speed at these two stations.

The distribution of the caddisflies collected, while in some cases apparently related to the individual environmental factor discussed above is likely to be largely controlled by interrelations of these factors with a number of other biotic and abiotic factors. Further research into the biology of North American caddisflies would perhaps aid in the interpretation of their distribution both within and between water systems.

Water temperature and emergence of adult aquatic insects have been shown by IDE (1940), SPRULES (1947), FLANNAGAN & LAWLER (1972) and others to be closely interrelated. Temperature being involved both in the maturation process and as a releaser of emergence behaviour. A similar situation seems to exist in the Roseau river, a correlation coefficient (r) of 0,78 (80 d.f.) being obtained when the $\log_{10} + 1$ of the number of caddisflies emerging is plotted against the water temperature. This correlation is obviously not complete and from Figure 2 it seems that perhaps a more accurate relationship would be that the numbers of caddisflies emerging are directly dependant on changes in temperature. This relationship seems to hold true except for a short period in the middle of July (Fig. 2) when decreasing temperatures do not bring about a decrease in caddis emergence. It is possible that an upper temperature limitation occurs for some species beyond which temperature changes are not important (see FLANNAGAN & LAWLER, 1972).

Air temperature is obviously of importance in the flight activity of adult insects and the change from daytime and evening emergence in June to evening/night/morning emergence in July and evening in August seems likely to be an adaptation within species with a long emergence period and for individual species themselves to ensure that the imago emerges at the time of day most suitable for flight. Presumably night and morning is too cold in June and August and full daylight perhaps causing dehydration problems in July and August. The releasing agent for this emergence behaviour is difficult to determine. Temperature, dissolved oxygen and pH all show distinctive diel cycles in the July and August samples but not in the June ones. The most likely cue is water temperature since it closely follows air temperature.

Conclusions

The Trichoptera fauna of Roseau River consists of at least 42 species, some of which occur in very large numbers. The density of caddisflies varies considerably with substrate, the largest sized substrates having both the most species and largest numbers of caddisflies.

Total emergence of caddisflies was found to be related to absolute and/or changes in water temperature. Periodicity of emergence varied among and within species and in many cases appeared to be dependent on light intensity and modified by time of year. Distribution of caddisflies at the various stations can be correlated, in some species, with substrate type, water flow or food, distribution of other species cannot be easily explained but is probably related to combinations of these factors.

Twenty-two species not previously recorded in Manitoba were collected from the river, all of these species were previously recorded from the Mississippi-Missouri region except one species of *Hydroptila* which needs further investigation.

Acknowledgements

I am much indebted to Misses P.M. LAUFERSWEILER and V. BEAUBIEN and Mr. D.G. COBB for assistance in collection of the samples, sorting the specimens and compilation of the data, Mrs. L.M.B. HIDSON for preparing the figures, Drs. D. ROSENBERG and K. PATALAS for reviewing the manuscript.

References

ANDERSON, N.H. & BOURNE, J.R. 1974. Bionomics of three species of glossosomatid caddis flies (Trichoptera: Glossosomatidae) in Oregon. Can. J. Zool. 52: 405-411.

BETTEN, C. & MOSELY, M.E. 1940. The Francis Walker types of Trichoptera in the British Museum. London: Br. Mus. (nat. Hist.).

CORBET, P.S. 1966. Diel periodicities of emergence and oviposition in riverine Trichoptera. Can. Ent. 98: 1025-1034.

DENNING, D.G. 1943. The Hydropsychidae of Minnesota (Trichoptera). Entomologica. am. 23: 101-170.

FLANNAGAN, J.F. (in press) The winter stoneflies *Allocapnia granulata*, *Taeniopteryx nivalis* and *T. parvula* in Southern Manitoba. Can. Ent.

FLANNAGAN, J.F. & LAWLER, G.H. 1972. Emergence of caddisflies (Trichoptera) and mayflies (Ephemeroptera) from Heming Lake, Manitoba. Can. Ent. 104: 173-183.

GORDON, A.E. 1974. A synopsis and phylogenetic outline of the Nearctic members of *Cheumatopsyche*. Proc. Acad. nat. Sci. Philad. 126: 117-160.

GRENIER, P. 1949. Contribution à l'étude biologique des Simuliides de France. Physiologia Comp. Oecol. 1: 165-330.

HAMILTON, A.L. 1969. A new type of emergence trap for collecting stream insects. J. Fish. Res. Bd. Can. 26: 1685-1689.

HYNES, H.B.N. 1970. The ecology of running waters. Toronto: Univ. of Toronto Press.

IDE, F.P. 1940. Quantitative determination of the insect fauna of rapid water. Univ. Toronto Stud. Biol. Set. 47: 5-20.

MACAN, T.T. 1961. Factors that limit the range of freshwater animals. Biol. Rev. 36: 151-198.

MORETTI, G.P. & GIANOTTI, F.S. 1962. Der Einfluss der Strömung auf die Verteilung der Trichopteren *Agapetus* gr. *fuscipes* Curt. und *Silo* gr. *nigricornis* Pict. Schweiz. Z. Hydrol. 24: 467-84.

NIELSEN, A. 1948. Postembryonic development and biology of the Hydroptilidae. Det Kgl. Danske Vidensk. Selskab. Biol. Skrifter. 5: 1-200.

RESH, V.H. 1976. Life histories of coexisting species of *Ceraclea* caddisflies (Trichoptera: Leptoceridae): the operation of independent functional units in a stream ecosystem. Can. Ent. 108: 1303-1318.

ROSS, H.H. 1944. The caddisflies, or Trichoptera, of Illinois. Bull. Ill. St. nat. Hist. Surv. 23: 1-236.

SPRULES, W.M. 1947. An ecological investigation of stream insects in Algonquin Park, Ontario. Univ. Toronto Stud. Biol. Ser. 56: 1-81.

THIENEMANN, A. 1954. Ein drittes biozönotisches Grundprinzip. Arch. Hydrobiol. 49: 421-422.

Discussion

STATZNER: Did you observe mating activities in the box traps?
FLANNAGAN: I observed mating activity in the box traps, but not in the Hamilton stream traps.
STATZNER: Did you find that mating and oviposition caused a reduction in the number of captured adults and a change in the sex ratio?
FLANNAGAN: The sex ratios in the two kinds of traps were similar and so were the densities (allowing for the different sampling areas), so I do not think that mating activities within the traps affected the population estimates.

Proc. of the 2nd Int. Symp. on Trichoptera, 1977, Junk, The Hague

Evolution of the caddisfly genus *Ceraclea* in Africa; implications for the age of Leptoceridae (Trichoptera)

J.C. MORSE

Abstract

A cladistic phylogeny is hypothesized for the African *Ceraclea* subgenus *Pseudoleptocerus* based on morphological characters of adults. The distributions of modern species are included on the cladogram to determine the probable distributions of the various ancestors. The resulting scenario of dispersals and vicariances is outlined to demonstrate the likely origin, age, and intracontinental evolution of the subgenus. Supporting data for the theory of plate tectonics, coupled with this understanding of the evolution of the subgenus, provides evidence that it and its leptocerid ancestors are at least 65 million years old.

In the recent revision of *Ceraclea* (MORSE, 1975), only 78 species in 2 subgenera were actually considered in detail, the species of the subgenera *Ceraclea* and *Athripsodina* which were known at that time. A full consideration of the third and final subgenus, the African subgenus *Pseudoleptocerus*, is now nearing completion.

The *Ceraclea* subgenus *Pseudoleptocerus* contains at least 15 species, all of which are known only from the Ethiopian Biogeographic Region. The other 3 species of *Ceraclea* known from Africa constitute the *spinosa* group in the subgenus *Athripsodina* (MORSE, 1975).

As demonstrated in the 1975 revision, the sister group of subgenus *Pseudoleptocerus* is the subgenus *Ceraclea* (Fig. 1). The male of their common ancestor (ancestor #1) developed a separate, subanal sclerite in the upper membranes of the genital chamber and the ventral apex of the phallobase became very long, serving as a sort of accessory phallic guide for the cylindrical phalicata (or aedeagus; Fig. 2). When speciation subsequently occurred, the ancestor of *Pseudoleptocerus* (ancestor #2) developed a strongly projecting ninth sternum (Figs. 2-5). The descendant species *grandis* probably changed greatly from this ancestor during the course of time, particularly in the configuration of its tenth tergum (Fig. 3), but species representing intermediate steps are unknown. Another descendant of ancestor #2 lost the male harpago (Fig. 4) and became more colorful, acquiring yellowish-brown scales and clear patches on various portions of its forewings and large, white scales regularly spaced along the forewing longitudinal veins. The species *minima* arose from this ancestor #3 and, like *grandis*, underwent much change in the shape of its tenth tergum since that time. The other recognizable descendant of this ancestor acquired iridescent silver hairs on its forewings and, in the hindwings, the position of the primary branch of the sectoral vein shifted apically beyond the origin of S_4 (Fig. 6). The 2 identifiable descendants of this ancestor #4 both underwent several modifications and gave rise to the majority of the known species in the subgenus. The male of the ancestor of the *squamosa* group (ancestor #5) acquired numerous stout setae on the mesal face of the ventro-basal clasper lobe and the mesal ridge of

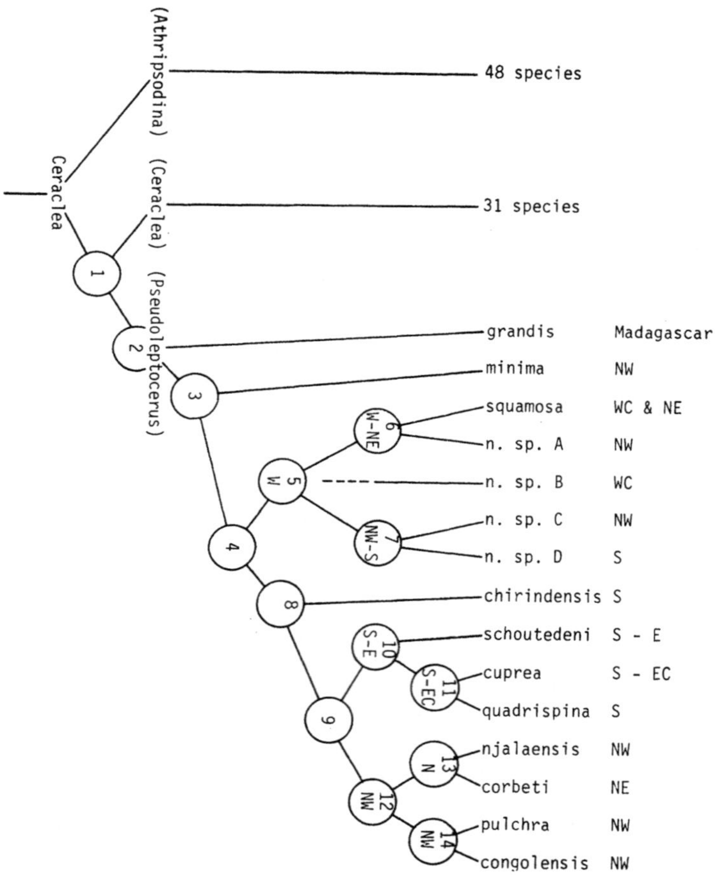

Fig. 1. Phylogeny and distributions of *Ceraclea* (*Pseudoleptocerus*) species and their ancestors in Africa. E = Eastern, EC = Eastcentral, N = Northern, NE = Northeastern, NW = Northwestern, S = Southern, W = Western, WC = Westcentral. Ancestors whose distributions are not indicated were either widespread in tropical and subtropical Africa or their ranges are unknown.

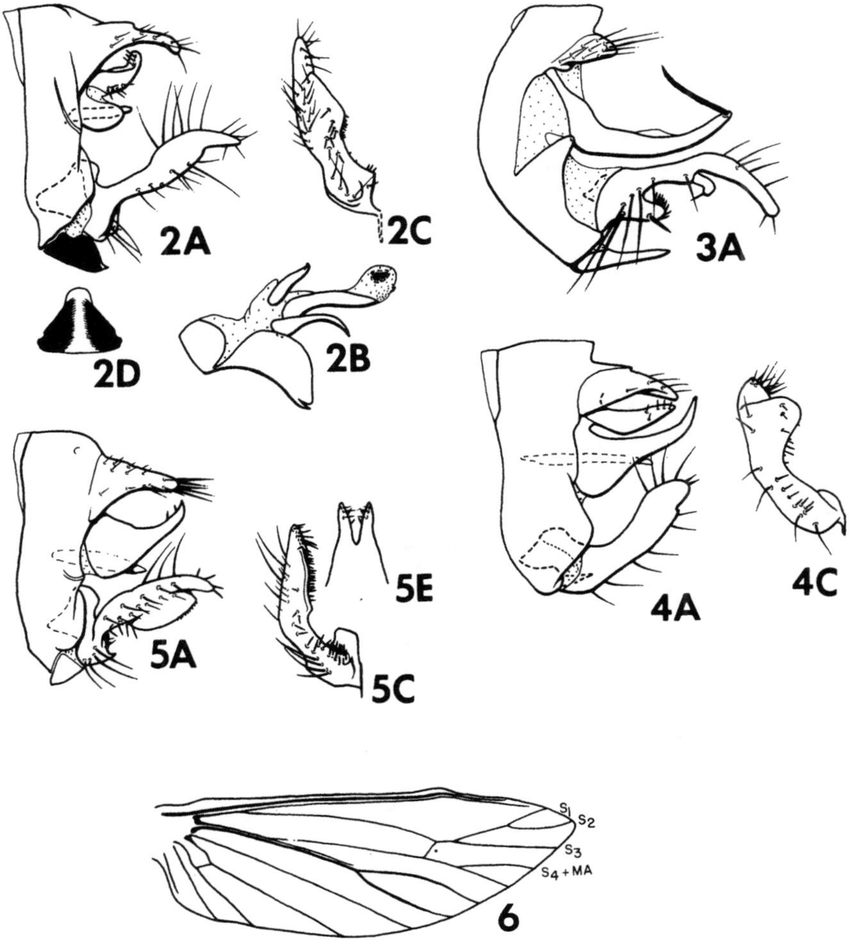

Figs. 2-6. Characters of *Ceraclea* (*Pseudoleptocerus*) species upon which the phylogeny in Figure #1 is based. Figs. 2-5, male genitalia of (2) *C.* (*P.*) *schoutedeni* (NAVAS, 1930), (3) *C.* (*P.*) *grandis* (MOSELY, 1932), (4) *C.* (*P.*) *minima* (KIMMINS, 1956), (5) *C.* (*P*) n. sp. D. A, genitalia, left lateral; B, phallus, left lateral; C, clasper caudal; D, ninth sternum, ventral; E, tenth tergum, caudal. Fig. 6, right hindwig of *C.* (*P.*) sp. female.

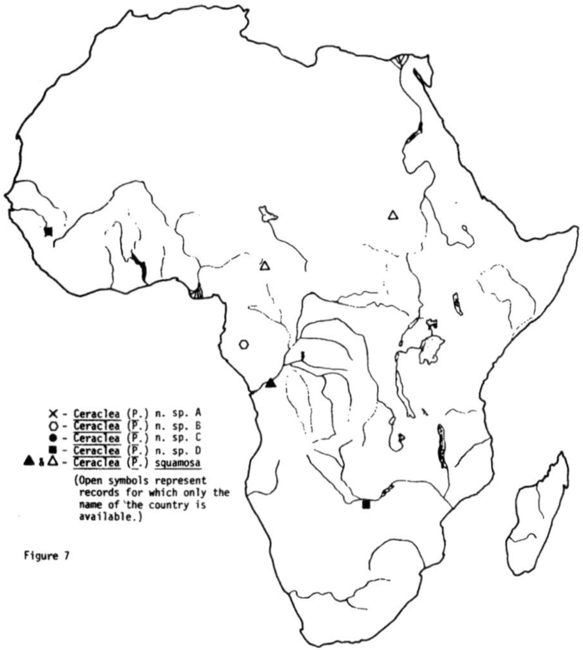

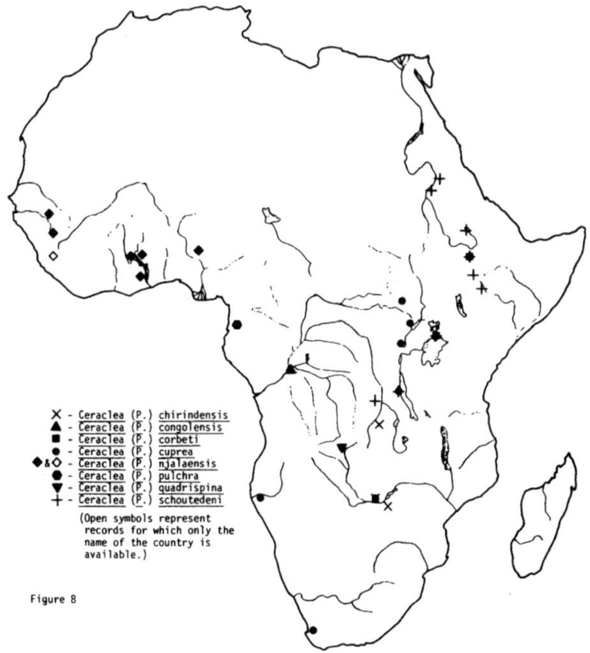

Fig. 7. Distribution of *Ceraclea* (*Pseudoleptocerus*) *squamosa* group species.
Fig. 8. Distribution of *Ceraclea* (*Pseudoleptocerus*) *pulchra* group species.

the clasper became turned and projected posteriorly (Fig. 5). At least 5 species subsequently evolved from it. The male of the ancestor of the *pulchra* group (ancestor #8) gained long superior appendages, an obtuse angle near the base of its mesal clasper ridge, and its ninth sternal sclerite darkened perceptibly (Fig. 2). At least 8 species descended from this ancestor.

Once some idea of the evolutionary picture of the various groups within *Pseudoleptocerus* was obtained, a very interesting story concerning the historical biogeography of the subgenus became possible. From maps of the reported collections of the species (Figs. 7, 8), it is apparent that the subgenus is confined almost exclusively to the tropical regions of the African continent. Furthermore, the composite distribution of the species exhibits a roughly V-shaped pattern, with numerous records in the countries on the west coast and many others along the eastern rift valleys and the Nile drainage, the two lines extending south to a meeting point at the Zambezi River. Only a few collections of caddisflies have been taken in the northern Congo and Ubangi drainages and in the western portions of the White Nile in Sudan, but these produced no records of this subgenus, indicating that its species simply may not occur in the north-central tropics. Their presence or absence in that region does not seriously affect the overall historical picture, however. Looking at the 4 monophyletic subgroups separately, *grandis* is known only from Madagascar, *minima* occurs only in the Northwest, the 5 species of the *squamosa* group are almost entirely confined to the western branch of the V-shaped pattern (Fig. 7), and the 8 species of the *pulchra* group occur in both branches (Fig. 8).

The following reconstruction is based on the principle of logical parsimony in historical biogeography outlined by NELSON (1969, 1973, 1974), ASHLOCK (1974), and ROSS (1974) and put into practice by many authors such as BRUNDIN (1966) in Chironomidae and ROSS (1956) in the primitive mountain caddisflies. Previous historical analysis in the sister subgenus *Ceraclea* has shown that its immediate ancestor probably inhabited North America. We have no evidence to indicate that the immediate ancestor of the subgenus *Pseudoleptocerus* (ancestor #2) occurred anywhere other than Africa. Thus the *Ceraclea-Pseudoleptocerus* common ancestor #1 probably was at one time distributed in the Nearctic and Ethiopian Regions. This rare distribution pattern apparently happened at least one other time in this genus as discussed below. The obvious question presents itself, 'Which direction did dispersal take place; from Africa to North America, or vice versa?' Additional analysis has shown that the ancestor of the subgenus *Athripsodina* occurred somewhere in the Holarctic Region. Logical parsimony thus leads to the hypothesis that the ancestor of the genus *Ceraclea* probably occurred somewhere in the Holarctic Region as well. Therefore, the historical scenario may be related as follows:

1. A specimen or specimens of the *Ceraclea* (*Ceraclea-Pseudoleptocerus*) ancestor #1 dispersed by some, no doubt unusual but unknown, means across the Atlantic Ocean. Notice that this route is perpendicular to the present prevailing trade winds and ocean currents. However, numerous insects have been captured in high altitude wind currents, indicating that means as a possible dispersal mechanism. It is

also possible that some eggs of this insect were inadvertently transported on the feet of some migrating waterfowl or on the exposed portion of floating debris from North America. Given many millions of years, the probability of such unlikely events as these is very high.

2. However it got to Africa, the colonizing insect produced offspring that eventually became sexually and morphologically distinct from the mother population in North America.

3. Over the years, these offspring then spread across the continent of Africa and succeeded in colonizing Madagascar, apparently while it was still attached to the main continent.

4. Subsequent separation of Madagascar from the African continent produced two isolated populations, one on Madagascar and the other on continental Africa.

5. These populations then evolved independently in those two regions. Most of the descendants of the Madagascar ancestor seemingly became extinct. However, the continental population (ancestor #3) eventually gave rise to the species *minima* and to the ancestor with iridescent hairs on its forewings, ancestor #4.

6. Some subsequent situation, possibly unfavorable climatic conditions in the central tropics, reduced the latter ancestor to 2 isolated populations, one western (ancestor #5) and one southern and/or eastern (ancestor #8), which remained separate long enough to become distinct species.

7. The western species (ancestor #5) evolved into the species of the *squamosa* group. Ancestor #7 dispersed southward and gave rise to n. sp. B. The species *squamosa* seems to have invaded the northeastern part of the continent in more recent years.

8. The *pulchra* group ancestor (ancestor #8) and ancestor #9 were probably widespread in their distributions. Ancestor #10 was confined to the southern and eastern section of the continent and ancestor #12 to the northwestern part. One of the descendants of the latter, ancestor #13, dispersed to the Northeast where it gave rise to *corbeti*. The species *cuprea* has advanced southward to the Cape in more recent times, this southern population having been referred to as *cuprea* variety *subfusca* by BARNARD (1934).

This scenario provides an opportunity to place some minimum limits on the ages of the subgenus *Pseudoleptocerus* and its ancestors, despite the fact that no fossils of the group are available with which to calculate the age by more direct means. If we accept the geological estimates of DIETZ & HOLDEN (1970), the *Pseudoleptocerus* ancestor #2 probably colonized Africa and spread eastward over Madagascar before that island separated from the continent about 65 million years ago. Thus the subgenus and its ancestors probably evolved before that time.

Of course, one might argue that the presence of *grandis* on Madagascar may simply represent a dispersal across the Mozambique Channel some time after the end of the Cretaceous Period, but this proposition requires the assumption of still another trans-saltwater dispersal; an unnecessary hypothesis which is contradictory to logical parsimony. Furthermore, the species *grandis* is very different from the other species of the subgenus as is evident upon examining the male genitalia, again suggesting that the species represents a very old lineage.

One other North America-to-Africa dispersal apparently took place in the evolution of the genus *Ceraclea*. The sister group of the African *spinosa* group in the subgenus *Athripsodina* is the Nearctic *tarsipunctata* group (MORSE, 1975). Apparently a specimen or specimens of their common ancestor dispersed by some unusual method from North America to the Lake Tanganyika region and later gave rise to the 3 species *microbatia*, *batia*, and *spinosa*. We presently have no evidence to indicate approximately when that dispersal occurred.

Acknowledgements

This study was supported in part by a Grant for the Improvement of Doctoral Dissertation Research in Systematics from the National Science Foundation (#10-32-RR174-024) and in part by a Grant-in-Aid of Research from Sigma Xi, the Scientific Research Society of North America. Technical Contribution no. 1517, published by permission of the Director, South Carolina Agricultural Experiment Station.

References

ASHLOCK, P.D. 1974. The uses of cladistics. A. Rev. Ecol. Syst. 5: 81-9.

BARNARD, K.H. 1934. South African caddis-flies (Trichoptera). Trans. R. Soc. S. Afr. 21: 291-394.

BRUNDIN, L. 1966. Transantarctic relationships and their significance, as evidenced by the chironomid midges, with a monograph of the subfamilies Podonominae, Aphrotaeniinae and the austral Heptagyiae. K. Svenska Ventenskakad. Handl. 11: 1-472.

DIETZ, R.S. & HOLDEN, J.C. 1970. The breakup of Pangea. Scient. Am. 223: 30-41.

MORSE, J.C. 1975. A phylogeny and revision of the caddisfly genus *Ceraclea* (Trichoptera, Leptoceridae). Contrib. Am. Ent. Inst. 11: 1-97.

NELSON, G.J. 1969. The problem of historical biogeography. Syst. Zool. 18: 243-246.

——. 1973. Comments on Leon Croizat's biogeography. Syst. Zool. 22: 312-320.

——. 1974. Historical biogeography: an alternative formulation. Syst. Zool. 23: 555-558.

ROSS, H.H. 1956. Evolution and classification of the mountain caddisflies. Urbana; Univ. of Illinois Press.

——. 1974. Biological systematics. Reading, Mass: Addison-Wesley Publ. Co.

Discussion

FLINT: The supposition of such unlikely dispersal mechanisms as eggmasses on migrating birds or on floating debris from North America really is not necessary. More recent geological estimates for the time of separation of North America from

Africa are about 180 million years ago (eg., LE PICON & FOX, 1971, J. Geophys. Res. 76: 6294-6308). Thus, your *Ceraclea* (*Pseudoleptocerus*) ancestor #2 may have moved into Africa before that time or it may have arrived by lower altitude wind currents after the continents had separated but while they were still relatively close together. This dispersal mechanism, which seems more likely than those you mentioned, would necessitate placing the age of *Pseudoleptocerus* and its leptocerid ancestors much earlier than you have suggested here.

Proc. of the 2nd Int. Symp. on Trichoptera, 1977, Junk, The Hague

Observations on the larva and pupa of the caddisfly genus *Hagenella* (Trichoptera: Phryganeidae)

I.D. WALLACE & G.B. WIGGINS

Abstract

The literature on Trichoptera indicates that there have been few if any records for the immature stages of species of *Hagenella* since the original descriptions for the European species *H. clathrata* by SILFVENIUS, STRÜCK, and ULMER in the first decade of this century. New data presented here on larval and pupal morphology of this species are therefore of interest, and particularly so because they show some discrepancy from the diagnostic characters traditionally cited for it. Examination of larval exuviae from pupal cases of specimens of *H. clathrata* reared in Finland by SILFVENIUS in 1904 has revealed close concordance with larvae recently reared in England; and larvae with these same general characters recently collected in Canada are probably those of the North American species *H. canadensis*. Discrepancy between available specimens and traditional larval diagnoses may reflect variation in the small local populations in which *H. clathrata* occurs, or colour changes in preserved material may also have some bearing. Additional diagnostic characters are proposed.

Introduction

The Holarctic phryganeid genus *Hagenella* comprises five species: *H. clathrata* (KOL.) in Europe; *H. sibirica* (MART.) in Siberia; *H. apicalis* (MATS.) from Hokkaido, Sakhalin, and north-eastern Asia; *H. dentata* (MART.) from Sakhalin, possibly a synonym of *apicalis*; and *H. canadensis* (BANKS) from north-eastern North America. Larval and pupal stages have been described only for *H. clathrata*, and those by three authors at about the same time: STRÜCK (1904), SILFVENIUS (1904), and ULMER (1909). Although there are in subsequent literature on European Trichoptera repeated references to the information provided by these authors, there is little indication over the past 70 years or so that any worker has had new collections from which the original taxonomic diagnoses could have been expanded.

Information in this paper is derived from several sources, and is of some interest because it includes a re-description of the larva and pupa of *H. clathrata*; this reveals that considerable qualification must be made to the traditional larval diagnosis for this species.

Rearing and description of *Hagenella clathrata*

Two females and five males of *H. clathrata* were captured alive at Whixall Moss, Shropshire, England, on the afternoon of June 25, 1975. Four of the males were taken while in a skipping flight similar to that of certain butterflies, the females while resting on bushes; one of the females was in copula with another male. Placed in a screened cage in a room, the males became active when the sun illuminated the cage for a few hours around mid-day, and initiated copulation; both females were observed in copula, one 2 days and the other 3 days after capture. All males died within the succeeding 16 days; sugar solution with an absorbent wick was provided as food for the adults, but none was seen to use it. Eggs in a gelatinous matrix were deposited just above the water surface on a wet substrate of sedge, grass roots, and sphagnum moss in a shallow bowl of water taken from the site where the adults were collected. The first egg mass was deposited approximately 16 days after the adult insects were collected, and the female died two weeks later; the second egg mass was deposited 37 days after the adults were collected, and that female died shortly afterwards. The egg masses were kept in contact with the water surface, constantly surrounded by a water film. Larvae were seen in the first egg mass 28 days after deposition, and 17 days after deposition in the second; they did not leave the gelatinous matrix until it was completely submerged. Larvae were reared in two polyethylene basins 45 cm in diameter, in 5 cm of water from Whixall Moss; sphagnum and various roots from the site were added. The water was not aerated, and temperature was governed entirely by that of the room, ranging from about 4 to 20°C. Leaves, cracked wheat grains, and tropical fish food were added regularly for food; however, the larvae were retiring in their habits and not often seen feeding. Development of first instars to II and III required approximately 2 weeks for each instar, a period of approximately 12 weeks more to instar IV, with final instars evident by the following March. Pupal cases were first seen in May, the adults successfully emerging in June. On a return visit to the original site on May 9, 1976, two larvae and two pupae were collected after much searching in root and leaf litter from drainage ditches on the peat bog; larvae of *Limnephilus elegans* were common in this habitat.

British adult records of this local species are summarised by BRAY (1966). They are principally from the raised bogs, or 'mosses' of the north-west midlands of England. There are old records from near London and from Wigtownshire, Scotland. It has recently been collected near Aviemore, Scotland (PELHAM-CLINTON pers. comm.).

LARVA (Figs. 1, 3, 4). Length of largest final instar larva 20 mm.

Dorsum of head (Fig. 1) largely uniform dark brown, pale area around base of most primary setae, that of seta no. 5 merging with small irregular light patch along lateral margin of frontoclypeal apotome; frontoclypeal sutures pale and well-delineated; pale area surrounding eye merging with somewhat lighter brown band along side of head; muscle scars numerous on posterior parts of parietals; some

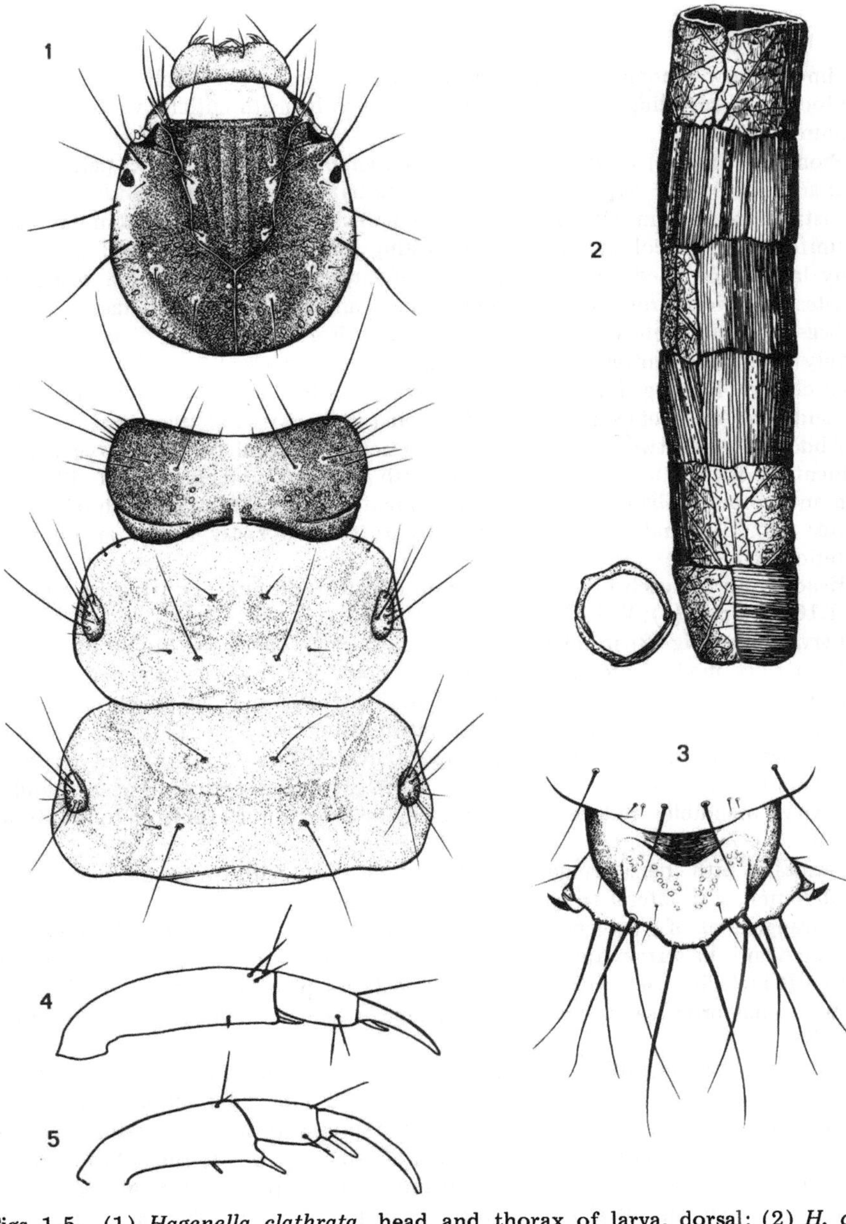

Figs. 1-5. (1) *Hagenella clathrata*, head and thorax of larva, dorsal; (2) *H. clathrata*, larval case, detail of posterior opening; (3) *H. clathrata*, segment IX and anal prolegs, dorsal; (4) *H. clathrata*, fore tibia and tarsus, lateral; (5) *Agrypnia varia*, fore tibia and tarsus, lateral.

specimens show tendency toward lighter brown over central longitudinal band of frontoclypeus revealing longitudinal muscle bands beneath cuticle; venter of head medium brown.

Pronotum (Fig. 1) fairly uniform dark brown, except for indistinct pale median area; anterior border very dark brown to black but transverse dark bands lacking in all instars; prosternum with prosternal horn and sternellum. Meso- and meta-notum of uniform brown colour without contrasting longitudinal bands, and with relatively large sclerotized areas around base of large single seta of *sa1* and *sa2*; *sa3* sclerites of typically rounded phryganeine shape and bearing several setae.

Legs brownish yellow, coxal combs minute. Fore tibiae (Fig. 4) not widened distally and disto-ventral angle not produced into enlarged base for apical seta; tarsal claw of fore leg (Fig. 4) lacking enlarged base for basal seta; both apical tibial seta and tarsal claw seta slightly curved, tending to lie close to leg segment.

Abdomen with two ventral gills on segment I, typical for the Phryganeidae, segments II to VII inclusive each usually with complete complement of gills, VIII with anterodorsal gills only. Sclerite of segment IX (Fig. 3) large, width of sclerite greater than half that of segment IX, two pairs of very stout setae arising from posterior edge.

Head width for each instar: I, 0.44 mm (1); II, 0.68 mm (1); III, 0.86 mm (1); IV, 1.16-1.36 mm (3); V, 1.76-2.28 mm (6).

Larval case (Fig. 2) in all instars of typical ring type, constructed of smooth uniform rings made of rectangular pieces of grass or tree leaves fastened together with silk; case of final instar comprising 5 or 6 rings fastened end to end. Length of final instar case 13-26 mm (5); length of fourth instar case 9 mm (1).

PUPA (Figs. 6 to 9). Length of largest pupal exuvia 21 mm. Front of head produced into rounded prominence (Figs. 6, 7), which in dorsal aspect bears a rounded concavity; mandibles reduced to membranous lobes, much shorter than labrum (Fig. 8).

Mesotarsi with sparse fringe of hairs; metatarsi also with fringe, but hairs fewer.

Abdomen with lateral fringe beginning on segment V, extending posteriorly to VII; arrangement of gills generally similar to that in larva; segment I with upturned dorsomedian lobe terminating in two small knobs; anterior hook-plates on segments III to VII inclusive each with 9-11 hooks, posterior plate on V with 13-17 hooks; quadrate anal processes typical of Phryganeinae each subtended by stout pointed lobe (Fig. 9).

Shortly before pupation, the larva adds up to ten additional ring units to the anterior end of the case giving a length of up to 36 mm. The final preparations for pupation involve sealing the posterior end of the case with a sieve membrane and attaching a loosely constructed tube of silk and plant debris to the anterior end of the case. The largest loose tube was 20 mm. An anterior sieve membrane is not present although the distal end of the loose tube is closed. The pupa can move freely up and down the pupal case proper.

DIAGNOSIS. With all stages now clearly associated with adults of *Hagenella clathrata*,

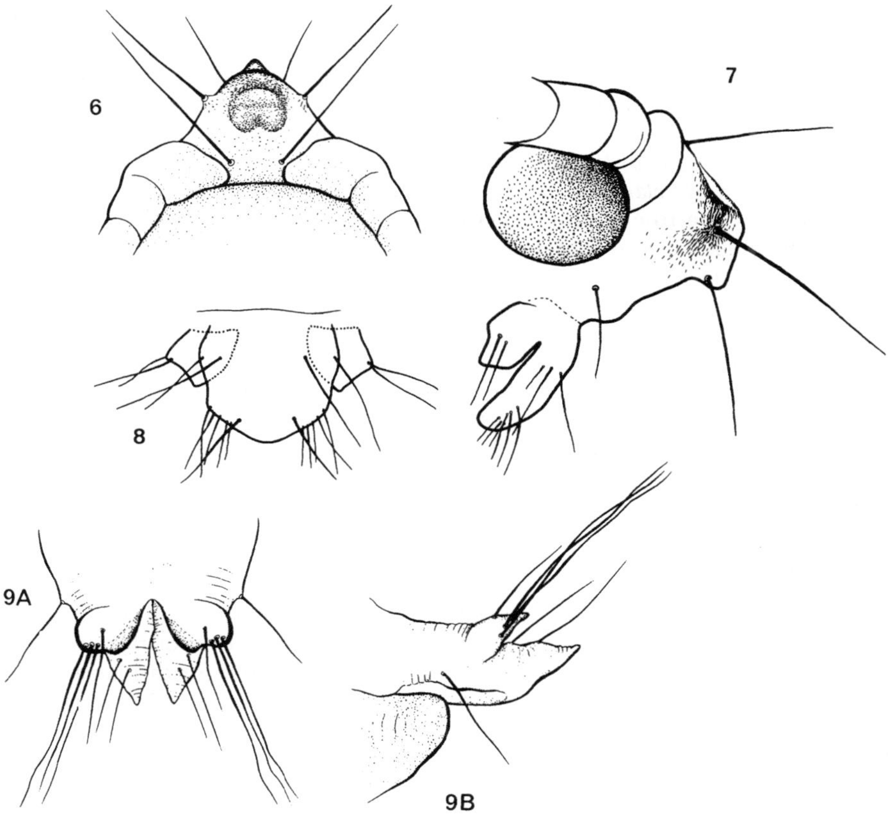

Figs. 6-9. *Hagenella clathrata* pupa. (6) anterior margin of head, dorsal; (7) head, lateral; (8) labrum and mandibles, facial; (9A) anal processes, dorsal; (9B) anal processes, lateral.

it is evident that the larva of the species is most obviously distinctive among other European phryganeids because of a largely unicolorous brown head and pronotum; any confusion in this character with the European *Trichostegia minor* will be resolved by the lack in that species of a dorsomedian hump on abdominal segment I, and also by the random arrangement of the components of the case. A second diagnostic character is the relatively large size of the dorsal sclerite of segment IX, which in dorsal aspect is much greater than half the width of segment IX itself and bears very stout setae posteriorly (Fig. 3); this sclerite is less than half the width of segment IX in most other phryganeids and bears relatively slender setae. The sclerite of segment IX is also relatively large in the genera *Oligostomis* and *Semblis*, but there its anterior border is straight or convex and the ratio of width to length is approximately 2:1; in *H. clathrata*, the anterior border of the sclerite is concave, and the ratio of width to length approximately 3:2 (Fig. 3). A third character is the depressed tibial spur and basal seta of the tarsal claw (Fig. 4), contrasted with the elevated condition of these structures in many other phryganeids (Fig. 5); this character is not unique to *H. clathrata*, and also occurs in *Semblis phalaenoides*.

The pupa of *H. clathrata* can be distinguished from that of other European phryganeids by the frontal protuberance with rounded dorsal concavity (Figs. 6, 7) and by the stout pointed lobe subtending the quadrate anal process (Fig. 9). Although all pupae in our material are males, this same character appears to be present in the illustration from a pupal female provided by SILFVENIUS (1904). *H. clathrata* is one of the few European species, along with those of the genera *Oligostomis* and *Semblis*, to have degenerate pupal mandibles (Fig. 8). Although SILFVENIUS (1904) stated that the anterior end of the pupal case was closed off by a sieve membrane in *H. clathrata*, there is no evidence of it in our material. This aspect of phryganeid biology was discussed at length by WIGGINS (1960), and it seems clear that degenerate pupal mandibles are correlated with omission of the anterior sieve membrane in behaviour leading to pupation.

Larva of *Hagenella canadensis*

Larvae with a unicolorous brown head and pronotum collected in Ontario have been provisionally assigned to the single North American species of this genus, *H. canadensis* (WIGGINS, 1977). Although a positive association for this larva is still lacking, its general similarity to the larva of *H. clathrata* described here and the dissimilarity of these two in colour pattern to all other phryganeid larvae known, leaves little doubt that they are congeneric. In the larva of *H. canadensis*, however, the sclerite of segment IX is not as large as in *H. clathrata*, its width less than half that of segment IX in dorsal aspect.

Discussion

Comparing the characters here assigned to *Hagenella clathrata* with existing keys for identification of European larvae of the Phryganeidae, it is difficult to appreciate the traditional diagnostic character of two dark bands along the lateral margins of the frontoclypeal apotome used by authors for over 70 years for this species. We have noted that there is a slight tendency for the brown colour of the head to be slightly lighter along the mid-dorsal area of the frontoclypeal apotome, and that this area can fade in preserved specimens over the course of a year or so; thus, in the material we have seen the lateral areas of the frontoclypeal apotome can be very slightly darker than the median area. ULMER (1909) mentioned that the frontoclypeal apotome in some specimens is almost entirely dark brown. A similar qualification was made by LEPNEVA (1966), but this is probably derived from ULMER's statement since we are advised by L. ZHILTZOVA (Acad. Sci. U.S.S.R., in litt. to G.B. Wiggins) that LEPNEVA did not have larvae of *H. clathrata*, and based her account on previous descriptions such as that of SILFVENIUS (1904).

In an attempt to ascertain just what material SILFVENIUS had, Wiggins studied larval exuviae from pupal cases from which, according to the label (*Neuronia clathrata* KOL.; Finland, N. TVARMINNE Zoologiska station, 19 June 1904, A.J. SILFVENIUS), adults of *H. clathrata* were reared by SILFVENIUS. From two pupal cases, two sets of larval exuviae were obtained which included the frontoclypeal apotome, both parietals, the pronotum and the sclerite of segment IX; from another case pronotal and parietal sclerites were found; and from a group of three cases sclerites of pronotum, parietals, and segment IX were retrieved. These frontoclypeal and parietal sclerites were entirely uniform brown and identical with the larvae reared and collected in England. Furthermore, the pronotal sclerites were also uniformly dark brown as described for our material, and showed no evidence of the transverse dark band just behind the anterior margin illustrated and described by SILFVENIUS (1904) and again by LEPNEVA (1966).

It seems clear, then, that the material studied by SILFVENIUS which we have seen is consistent in larval characters with our own. His use of lateral banding on the frontoclypeus as a diagnostic character could be attributed to the variation alluded to by ULMER (1909), and that possibility will be tested as more larval material of this species comes to light; or it could have resulted from fading caused by the preservation of the material on which his description was based. The transverse band illustrated on the pronotum by SILFVENIUS (1904) is, however, similar enough to that in *Oligostomis reticulata* (cf. figs. 88 and 97 in LEPNEVA 1966), that the possibility of mixed exuvial sclerites cannot be excluded. But, STRÜCK's (1904) description seems quite definite about lateral banding on the frontoclypeus, and there is little doubt that his material was of this species because his description even alludes to the concave anterior margin on the dorsal sclerite of segment IX. Therefore, variation in expression of dark markings of the frontoclypeal apotome enhanced by the small local populations in which *H. clathrata* seems to occur must be regarded as a distinct possibility. With the broadened taxonomic diagnosis for larvae of *H. clathrata* provided here, additional specimens should become available and provide some answers to these questions.

Acknowledgements

Work on this paper by WIGGINS is part of a general study of the systematics of Trichoptera supported by the National Research Council of Canada (Grant A 5707). Material from the SILFVENIUS collection was made available by the Museum of Zoology, University of Helsinki.

References

BRAY, R.P. 1966. Records of the Phryganeidae (Trichoptera) in northern England 1961-1964, with a summary of the distribution of British species. Trans. nat. Hist. Soc. Northumb. 15: 226-239.

LEPNEVA, S.G. 1966. Larvae and pupae of Integripolpia, Trichoptera. Fauna of U.S.S.R. Zool. Inst. Akad. Nauk. S.S.S.R. 95: 1-700 (Trans. Israel Program Sci. Trans. Inc. 1971).

SILFVENIUS, A.J. 1904. Uber die Metamorphose einiger Phryganeiden und Limnophiliden III. Acta Soc. Fauna Flora fenn. 27: 3-74.

STRÜCK, R. 1904. Beiträge zur Kenntnis der Trichopterenlarven II. Die Metamorphose von *Neuronia clathrata* Kol. Mitt. geogr. Ges. naturh. Mus. Lübeck 19: 3-7.

ULMER, G. 1909. Trichoptera, in BRAUER, A. Die Süsswasserfauna Deutschlands Nos. 5-6. Jena: G. Fischer.

WIGGINS, G.B. 1960. The unusual pupal mandibles in the caddisfly family Phryganeidae (Trichoptera). Can. Ent. 92: 449-457.

——. 1977. Larvae of the North American caddisfly genera (Trichoptera). University of Toronto Press.

Proc. of the 2nd Int. Symp. on Trichoptera, 1977, Junk, The Hague

Probable origins of the West Indian Trichoptera and Odonata faunas

O.S. FLINT, Jr.

Abstract

Upon analysis of the distribution of the West Indian Trichoptera and Odonata the following patterns become clear. The Trichoptera are very highly endemic, over 80% of the species being known from only one island, whereas the Odonata are much less so with only 15% of the species apparently restricted to one island. There does seem to be a correlation between the size of the organism as well as its ecological tolerance and the area it occupies. The small Hydroptilidae and the largest dragonflies, and those species the most ecologically tolerant are the most widely distributed.

The Greater Antilles support a distinct fauna that appears to have been isolated for a long time and which shows three relationships, as follows: 1. a North American element which appears to be rather minor; 2. a Neotropical element, closely related to the fauna of Mexico and Central America, which is the overwhelming percentage; and 3. a very small element that is related to equatorial or southern Africa.

The Lesser Antilles also has a distinct fauna, although seemingly less highly endemic than that of the Greater Antilles. There is a small element that may be called Antillean, but in general the fauna is quite different from that of the Greater Antilles. The primary relationship is with northern South America, from which a part of the fauna seems to have been derived by dispersion.

The entire picture agrees well with the Vicariance Model of Biogeography as proposed by CROIZAT and specifically as developed by ROSEN for the West Indies.

Introduction

The West Indies or Antilles are composed of a vast number of islands and islets ranging in size from a few square meters to the largest, Cuba, covering nearly 115,000 square kilometers. The entire chain of islands lies between the Tropic of Cancer and 12° north latitude. Although the island of Cuba is only about 200 kilometers from either Florida or Yucatan, and Grenada is 150 kilometers from Venezuela, the faunas of the Antilles are those of true oceanic islands.

Many of the small islands are too low to produce sufficient rainfall to support significant lotic communities. However, the islands with elevations approaching 1000 meters have sufficient rainfall to produce permanent streams that may be expected to harbor some caddisflies. Even the Lesser Antillean islands, which might seem to be too small to have adequate habitats, do project high enough into the

Table 1. Antillean islands, showing size, elevation, and numbers of species of Trichoptera and Odonata known from each.

	Jamaica	Cuba	Hispaniola	Puerto Rico	Guadeloupe	Dominica	St. Lucia	St. Vincent	Grenada
Square Kilometers	11,580	114,525	76,071	8,866	1,510	749	616	344	311
Elevation (Meters)	2,252	1,999	3,174	1,338	1,467	1,421	959	1,234	840
Trichoptera	39	36	52	35	3	36	11	12	10
Odonata	62	78	58	43	13	21	16	4	7

prevailing Easterlies to produce tremendous rainfall and many streams. For example, Dominica, a smaller island (Table 1) in one year (1948) had a mean rainfall at 22 stations of 348 cm (137 inches) (HODGE, 1954). The Dominicans claim that the island has one stream for each day of the year (365), which figure I see no real reason to doubt.

For purposes of discussion I have grouped the islands into the Greater Antilles, composed of the four large islands (Cuba, Jamaica, Hispaniola and Puerto Rico) and the Lesser Antilles (Guadeloupe, Dominica, Martinique, St. Lucia, St. Vincent and Grenada). The islands between Puerto Rico and Guadeloupe, the Grenadines, Barbados, and the Bahamas, are small in size and low in elevation and either lack a Trichopterous fauna or at least have none recorded, and the few Odonata listed from them are the most vagile species of little zoogeographic interest. Trinidad has also been excluded from this study as it is on the continental shelf with a continental fauna.

Recent geological studies (NAIRN & STEHLI, 1975; MATTSON, 1977) are generally in agreement over the basic geology of the region. Briefly, and greatly simplified, it appears that volcanic activity began in the Greater Antilles in the Jurassic and reached its peak during the Cretaceous. Although limestone was deposited during this period, this phase did not predominate until the Tertiary. Subsequently these strata have been uplifted and gently deformed. In the Lesser Antilles there have been two phases of volcanic activity. The older phase, occurring from Eocene through Miocene times, involved an eastern arc which has subsequently been eroded and capped with limestone deposits. The westerly arc commenced volcanic activity in the late Miocene and is still active. The two arcs are more or less coincident from Martinique south, but diverge considerably from Guadeloupe north. The development of a coherent and broadly accepted theory of plate tectonics of the region has yet to come in spite of numerous attempts (FREELAND & DIETZ, 1972; MALFAIT & DINKELMAN, 1972; MATTSON, 1977). Beyond the fact that the tectonic history for the region is interrelated with the separation of Europe and Africa from North and South America and the opening of the Atlantic Ocean, the actual relationship of the islands to one another and the continents is wholly uncertain (DONNELLY in NAIRN & STEHLI, 1975).

The nature of the faunas

At this time I know of 165 species of Trichoptera from the West Indies. This includes a number of undescribed species present in the collection at the Smithsonian Institution. I have excluded from this count those species recorded on the basis of larvae or females only, as they may always be the unassociated stages of a species known from another island. The analogous number of species of Odonata for the West Indies is 102.

In Table 1 I have listed the total number of species of caddisflies and dragonflies known from each island. I did include unassociated larvae or females when preparing these figures as this presents a more exact count of the number of species present on any one island.

At present it is clear that the faunas of Cuba or Hispaniola are considerably larger than those of Jamaica or Puerto Rico, and that these in turn are greater than those of the Lesser Antillean islands. Within the Greater Antillean islands, the Trichoptera occurring in Puerto Rico are quite well known and those on Jamaica only slightly less so. Although the caddisfly fauna of Hispaniola may be a bit better known at this time than that of Cuba, I believe that both are relatively poorly known and have much larger faunas than recorded now. I predict that 75-100 species will be found on either Cuba or Hispaniola, with either Jamaica or Puerto Rico supporting only 50-60 species. The Trichopterous faunas of the Lesser Antillean islands, with the exception of Dominica, are almost wholly unknown. For any of the major Lesser Antillean islands I predict a fauna of 40-60 species.

Our knowledge of the insular Odonata is quite different. I believe that the total reported number of species on each of the Greater Antillean islands is correct, give or take one or two. The Lesser Antillean islands are still relatively poorly known, although more fully known than for the Trichoptera. I would expect that the major Lesser Antillean islands will be found to harbor 25-30 species of Odonata.

Although there appears to be a major difference in the size of the fauna between the Greater and Lesser Antilles, this difference appears to be related to the relative sizes of the islands. There also exists a major difference in the composition of the faunas, especially in the Trichoptera, between these two groups of islands. To mention some differences:

The families Rhyacophilidae and Odontoceridae, the subfamily Pseudoneureclipsinae, and the genera (or subgenera) *Curgia*, *Hydropsyche*, *Macronema*, *Orthotrichia* and *Nectopsyche*, are present on the Greater Antilles and absent on the Lesser.

Only at the generic level does the opposite occur: *Wormaldia*, *Polyplectropus*, *Zumatrichia*, *Brachysetodes*, and *Atanatolica* being present on the Lesser Antilles but absent on the Greater.

In other cases a monophyletic group of genera or species on one island group is replaced by another such group on the other island group. This occurs in the family Glossosomatidae, and the genera *Polycentropus*, *Leptonema*, *Smicridea* and *Phylloicus*.

The differences between the faunas of the groups of islands are less marked in the Odonata, tending to be the presence or absence of a species or, rarely, the replacement with a different species on the different island groups.

Endemism

One of the most immediately obvious factors from Table 2 is the degree of endemism of the Trichoptera fauna. For example, 79% of the species of Trichoptera occurring on the Antilles are limited (at our present state of knowledge) to a single island. Only 10% of the Trichoptera have an extensive range that encompasses either both an island and the mainland, or both the Greater and Lesser Antilles. An almost exactly reverse situation occurs with the Odonata, the percentages of island endemic species being only 16% and the widespread species 67%.

If one considers the degree of endemism of the fauna occurring on the two groups of islands, as shown in Table 3, it is clear that both orders show a greater degree of endemism on the Greater Antilles than on the Lesser Antilles. In the Trichoptera 94% of the species occurring on the Greater Antilles are limited to these islands, whereas the comparable figure for the Lesser Antilles is only 70%. In the Odonata the same trend is apparent at 30% for the Greater Antilles and 11% for the Lesser Antilles; only the degree, as discussed above, is much less.

Table 2. Distribution of species of Antillean Trichoptera and Odonata on different island groups.

	TRICHOPTERA		ODONATA	
	No.	%	No.	%
SPECIES LIMITED TO:				
1 GREATER ANTILLEAN ISLAND	106	64	14	14
2 OR MORE G.A. ISLANDS	11	7	15	15
SUBTOTAL	117	71%	29	28%
1 LESSER ANTILLEAN ISLAND	25	15	2	2
2 OR MORE L.A. ISLANDS	7	4	2	2
SUBTOTAL	32	19%	4	4%
G.A. AND CONTINENT	2	1	37	36
L.A. AND CONTINENT	8	5	3	3
G.A. AND L.A. ONLY	5	3	2	2
G.A., L.A., AND CONTINENT	1	1	27	26
SUBTOTAL	16	10%	69	67%
GRAND TOTAL	165	100%	102	100%

In explaining the difference in the degree of endemism between the two orders, two possibilities seem plausible. The first obvious difference is size. The smallest caddisflies, especially some of the Hydroptilidae, are often the most widely distributed. Perhaps these are easily caught up and dispersed by the winds. The average sized caddisflies, 3 to 5 mm long, are almost wholly endemic or limited to a few adjacent islands. These may be too large for easy wind dispersal but too small to have the muscle power and food reserves for extended flight. The dragonflies, however, being much larger, have the necessary powers of flight to readily disperse between islands. Odonata are well known to come aboard ships far out at sea.

The second factor appears to relate to the ecological requirements of the two

Table 3. Distribution of Trichoptera and Odonata species showing differences in degree of endemism between Greater and Lesser Antilles.

	TRICHOPTERA		ODONATA	
	No.	%	No.	%
Of species occurring on Greater Antilles:				
Limited to 1 island	106	85	14	15
on 2 or more islands	11	9	15	16
Subtotal	117	94	29	30
Total on G.A.	125	100%	95	100%
Of species occurring on Lesser Antilles:				
Limited to 1 island	25	54	2	6
on 2 or more islands	7	15	2	6
Subtotal	32	70	4	11
Total on L.A.	46	100%	36	100%
Species occurring on G. A. and L. A. only	5	3	2	2
Species occurring on G. A. and/or L. A. and continent	11	6	67	65
Subtotal	16	9%	69	67%
Total Antillean species	165		102	

groups. Those species that breed primarily in lowland warm-water ponds or slowly flowing streams, seem to be much more widely distributed than those requiring mountain, cool-water streams and springs. In support of this suggestion are the facts that the Odonata with the more restricted distributions tend to be those falling into the second requirement, and that one of the most widely distributed caddisflies, *Oecetis inconspicua* (Walker), although in the most highly endemic size range, is an ecologically tolerant species.

The reason for the lower degree of endemism in the Lesser Antilles may relate to either their small size and therefore their fewer suitable habitats, or the younger age of the island group.

Origins of the fauna

One of the main difficulties in trying to determine the origins of the fauna of the Antilles is the lack of a clear geo-tectonic history of the Region. One thing is clear, however, and that is that since Pangaea began to break up in the Mesozoic, the orientation and placement of the islands and the surrounding continents, including Europe and Africa, have altered greatly and on more than one occasion. Therefore our thinking must not be limited to the present relationship of islands, oceans, and continents.

Because of the high degree of endemism in the Trichoptera, and the fact that many of the endemic elements have speciated to such a degree on the islands that they are not clearly related to any mainland form, I am inclined to believe that to a large degree this is a very old fauna. Perhaps it goes back in its isolation from mainland forms to the early Cenozoic or even the Cretaceous.

HENNIGIAN analysis of all pertinent insular groups and their mainland (including North and South America, Europe and Africa) relatives will be necessary before a true understanding of the various pathways and their relative importance can be reached. Unfortunately, time is not available for such analysis, nor is the Central and South American fauna well enough known now to make such an analysis more than marginally valid. Therefore, I have attempted to look at overall ranges of species in the hope that this will be related to the origin of the fauna as well.

Because the West Indian Trichopterous fauna is so nearly endemic, the overall distribution pattern of the Trichoptera, Table 4, shows nothing meaningful except perhaps from the Lesser Antilles, as will be discussed later. However, the Odonata, with their pattern of much broader distribution (Table 5) do show some interesting trends.

In the Odonata there are, in addition to the endemic element, several other trends clearly shown. The northern element is about 10% and, interestingly, seems to be as well developed on the Lesser Antilles as on the Greater Antilles. A middle American element is rather pronounced in the Greater Antilles and, as would be expected, this is almost three times stronger on the Greater Antilles than on the

Table 4. Distribution of Antillean Trichoptera classified by total range of each species.

TRICHOPTERA	Greater Antilles		Lesser Antilles	
Species on:	125		46	
Endemic Antilles	122	97%	37	80%
Northern	1	1%	0	0
Middle America	1	1%	8	17%
Southern	0	0	1	2%
Widespread	1	1%	0	0

Table 5. Distribution of Antillean Odonata classified by total range of each species.

ODONATA	Greater Antilles		Lesser Antilles	
Species on:	95		36	
Endemic	31	32%	6	16%
Northern	8	8%	3	8%
Middle America	14	14%	2	5%
Southern	37	38%	22	61%
Widespread	5	5%	3	8%

Lesser Antilles. The southern element is very strong, being over a third of the Greater Antillean fauna and almost two-thirds of the Lesser Antillean. Finally, nearly 10% of the species have such a widespread distribution as to be meaningless in terms of origin.

Looking further at the Greater Antillean fauna of the Odonata, it is clear that the related congeneric species of the middle American element are nearly all Neotropical in origin, and this holds for most of the endemic element as well. Thus I am tempted to say that perhaps two-thirds of the Greater Antillean Odonata fauna is derived initially from the south and probably most directly from populations in adjacent Mesoamerica. Up to another 10% is derived from North American progenitors, another 5-10% is so widely distributed as to be uninformative, and perhaps another 15% of the endemic element is presently obscure in its relationship.

The Lesser Antillean Trichoptera with a continental element to their range show an interesting pattern. If one assumes that a species present on a seaward island is present on those islands between it and the continent, then Grenada contains 7 species, St. Vincent 4 and Dominica 3. This pattern, in conjunction with the strong South American element shown by the Odonata, seems to me to indicate a strong dispersal thrust from South America into the Lesser Antilles whose analog seems to be lacking in the Greater Antilles.

It is unfortunate that such a large unanalyzed endemic element remains in the caddisfly and dragonfly faunas, for this group must hold many keys. One pair of examples indicates this point very well. The caddisfly genus *Antillopsyche* is, on the basis of information available from larval, pupal and adult stages, a member of the Polycentropodid subfamily Pseudoneureclipsinae which is known from the Old World tropics from Africa to Indonesia. This group is known in the New World by only the one genus *Antillopsyche* and it is limited to the islands of Cuba, Hispaniola and Puerto Rico. The damselfly genus *Phylolestes* which is known only from Hispaniola is, on the basis of nymphs and adults, clearly a member of the family Synlestidae, which is otherwise known only from Southern Africa and tropical Asia. These two genera, then, point directly to Africa without any New World relationship. It is tempting to suggest that these may be relicts of a time in the Mesozoic when Africa was either in contact with the area or at least closer than now.

Conclusions and summary

There are several points that I would like to emphasize as a result of this preliminary survey.

First, the Greater Antilles has a very distinct fauna. The Trichoptera, at least, appear to have been isolated on the whole for a very long time. The aquatic fauna appears to have three relationships: first, a small North American element; second, a Neotropical element which is predominant and shows a close relationship to Central America; and perhaps a third very small element related to the African fauna.

Second, the Lesser Antilles has a distinct fauna also, but it has a lower degree of endemism. There does seem to be a rather distinct difference between the Trichopterous elements composing the faunas of the two island groups. The primary relationship of the Lesser Antillean fauna is with the adjacent mainland of northern South America. There is also a small percentage of the fauna that is in common with the mainland, and this fauna shows attenuation away from land in a typical dispersal manner.

Third, there seems to be a relationship between size and range. In general the microcaddisflies are more widely distributed than average-sized caddisflies, but the larger Odonata are widely distributed. This may be related to two factors, the first being ease of dispersal. The microcaddis may be easily carried by the wind, but the larger ones not so, whereas the Odonata are large enough to have sufficient powers of flight to fly directly from island to island. The second factor is ecological tolerance. Those species breeding in warmer, lower elevation ponds or slow-flowing streams are more widely dispersed than those requiring cooler, fast-flowing upland brooks.

Fourth, different groups of organisms may show different relationships or degrees of relationship, even among ecologically similar groups such as the freshwater insect orders. For example, almost 80% of the Trichoptera species are known from only a single Antillean island, but this is so for only 15% of the Odonata. Yet, behind this great superficial difference I believe that the basic origins of the fauna of these two orders is the same. This points out the necessity for great care in trying to apply the results of analysis of one group to another group, and this may be even more dangerous when one makes comparison across phyla or kingdoms.

Fifth, and finally, I feel that the distribution of the Trichoptera and Odonata is in basic agreement with the vicariance model of Caribbean biogeography as recently advanced by ROSEN (1975) based on the pioneering work of CROIZAT.

References

FREELAND, G.L., & DIETZ, R.S. 1972. Plate tectonic evolution of the Caribbean-Gulf of Mexico Region. Trans. 6th Carib. Geol. Conf., Venezuela. 259-264.

HODGE, W.H. 1954. The flora of Dominica, B.W.I., Part 1. Lloydia 17: 1-238.

MALFAIT, B.T. & DINKELMAN, M.G. 1972. Circum-Caribbean tectonic and igneous activity and the evolution of the Caribbean Plate. Bull. Geol. Soc. Am. 83: 251-272.

MATTSON, P.H., ed. 1977. Benchmark papers in Geology 33: West Indies island arcs. Stroudsburg, Pa.: Dowden, Hutchinson, & Ross.

NAIRN, A.E.M. & STEHLI, F.G., eds. 1975. The ocean basins and margins, volume 3: The Gulf of Mexico and the Caribbean. New York & London: Plenum Press.

ROSEN, D.E. 1975. A vicariance model of Caribbean biogeography. Syst. Zool. 24: 431-464.

Proc. of the 2nd Int. Symp. on Trichoptera, 1977, Junk, The Hague

Les Trichoptères de Cuba — faunistique, affinités, distribution, écologie

L. BOTOSANEANU

Abstract

A short synthesis of all known aspects of the caddisfly fauna of Cuba and the Isla de Pinos.

Il y a quelques années encore, la faune de trichoptères de Cuba était connue de façon estrêmement imparfaite. La situation s'est notablement améliorée ces dernières années, par suite de l'étude de collections réalisées un peu partout à Cuba, soit par moi-même, soit par des entomologistes cubains.

A présent, un peu moins de 80 espèces et sous-espèces sont connues de Cuba (2 Rhyacophilidae Hydrobiosinae, 8 Glossosomatidae Protoptilinae, 6 Philopotamidae, 1 Psychomyiidae Xyphocentroninae, 6 Polycentropodidae, 6 Hydropsychidae Hydropsychinae et 3 Macronematinae, 31 Hydroptilidae, 3 Leptoceridae, 2 Odontoceridae, 4 Calamoceratidae, 5 Helicopsychidae). Mais on peut s'attendre à une dizaine environ d'espèces supplémentaires. Ce chiffre est passablement élevé si on le compare à ceux obtenus pour les autres îles antillaises (voir les publications de O.S. Flint): pour Jamaica et Puerto Rico, les deux assez bien étudiées et plus petites que Cuba, les chiffres sont 39 et respectivement 32; 27 espèces seulement sont signalées de Hispaniola (mais Haïti, qui possède certainement une riche faune, reste fort mal connue); de l'ensemble des Petites Antilles on a signalé jusqu'à présent 44 espèces. Le pourcentage des taxa endémiques est remarquablement élevé: 61 du total de 76. Il s'agit d'espèces de *Atopsyche* BKS., *Cubanoptila* SYKORA (genre cubain endémique), *Cariboptila* FLINT, *Campsiophora* FLINT (deux genres purement antillais), *Chimarra* STEPH., *Antillopsyche* BKS. (genre purement antillais), *Polycentropus* CURT., *Hydropsyche* PICT., *Smicridea* McL., *Leptonema* GUERIN, *Macronema* PICT., *Alisotrichia* FLINT, *Ochrotrichia* MOS., *Metrichia* ROSS, *Loxotrichia* MOS., *Oxyethira* EAT., *Hydroptila* DALMAN, *Neotrichia* MORT., *Oecetis* McL., *Marilia* F. MÜLLER, *Phylloicus* F. MÜLLER, *Helicopsyche* SIEBOLD.

Cuba représente — tout comme les Antilles en général — une contrée extrêmement intéressante pour l'étude des résultats de l'isolation géographique; l'étude de sa faune de trichoptères permet de distinguer de nombreux cas de vicariance (au niveau générique, spécifique, sous-spécifique): 1. entre les Grandes Antilles considérées comme un tout, et d'autres zones des Amériques; 2. entre les différentes îles

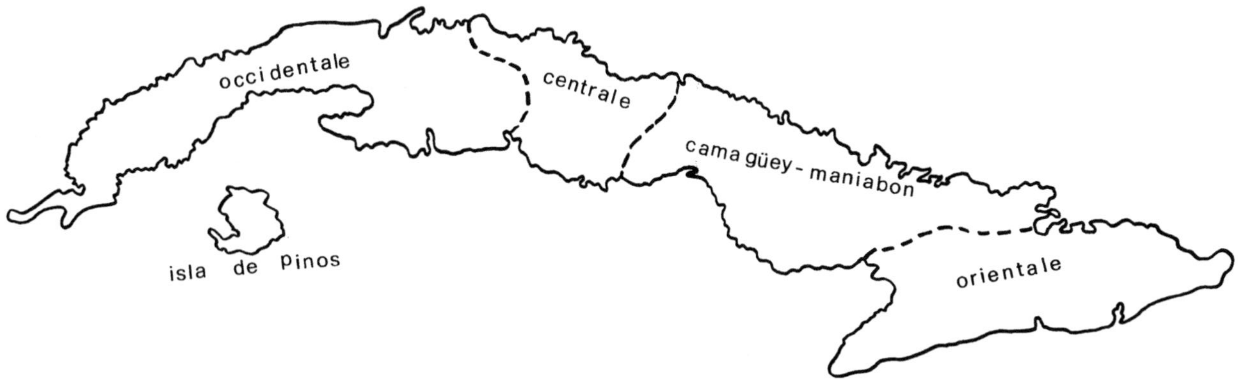

Fig. 1. Cuba et sa division en régions naturelles (selon A. NUÑEZ-JIMÉNEZ)

constituant les Grandes Antilles; 3. entre les différentes provinces naturelles de Cuba; 4. ou même dans les limites d'une seule province naturelle. Nous allons choisir quelques exemples offerts par la faune cubaine, pour illustrer ces 4 cas divers (abréviations: C. = Cuba; Occ. = province Occidentale; Or. = province Orientale; Centr. = province Centrale; I.P. = Isla de Pinos).

1. *Chimarra guapa* BOTS. (C.) — *C. patosa* ROSS (Pérou); *Loxotrichia glasa* ROSS (I.P., Sud des USA) — *L. dalmeria* (Mexique); *Oxyethira alaluz* BOTS. (C.) — *O. maya* DENN. (Georgia, Florida); *Hydroptila medinai* FLINT (C., Puerto Rico) — *H. mexicana* MOS. (Mexique); *Neotrichia iridescens* FLINT (Grandes et Petites Antilles) — *N. olorina* (MOS.) (Mexique).

Comparée à la faune de trichoptères des zones avoisinantes de l'Amérique Continentale, celle de Cuba se distingue nettement par une série de traits positifs, mais surtout négatifs, car c'est (comme celle des autres Grandes Antilles) une faune notablement appauvrie. Les Hydrobiosinae sont représentés seulement par un tout petit nombre d'espèces appartenant au seul genre *Atopsyche*. Les Protoptilinae font un beau bouquet d'espèces, celles-ci appartiennent à 3 genres seulement, tous les 3 endémiques des Grandes Antilles. Les *Chimarra* sont relativement peu nombreuses, mais phylétiquement assez variées. Les Polycentropodidae sont assez bien représentés par les 2 *Antillopsyche* et les 3 *Polycentropus*. Un trait positif des plus remarquables, c'est la présence d'un groupe de 4 *Hydropsyche* étroitement apparentés (avec 2 autres espèces, de Hispaniola, ce sont les seuls représentants du genre aux Antilles, le genre étant d'ailleurs absent d'Amérique du Sud). A l'exception des *Alisotrichia*, présents avec 5 espèces au minimum, les hydroptilides du 'groupe de *Leucotrichia*', si bien représentés ailleurs, sont pratiquement absents. La liste des autres hydroptilides est passablement longue et 'normale', mais les *Hydroptila*, si diversifiés par exemple au Mexique, ne sont présentes que par un petit nombre d'espèces, appartenant au grand 'groupe de *consimilis*'; c'est toujours parmi les hydroptilides que nous trouvons quelques-unes des espèces cubaines les plus isolées du point de vue phylétique, comme *Ochrotrichia insularis* MOS., *O. islenia* BOTS., ou comme les 3 espèces de *Neotrichia* (*alata* FLINT, *pequenita* BOTS., *pinarenia* BOTS.). Les Leptoceridae ne comprennent ni *Brachysetodes*, ni des Triplectidinae. Les Odontoceridae et Calamoceratidae n'appellent pas de commentaire particulier. La faune cubaine acquiert un certain éclat avec les *Helicopsyche*: au moins 5 espèces (mais probablement 1 ou 2 de plus), dont une espèce si isolée que *granpiedrana* BOTS. & SYK.

2. *Atopsyche vinai* SYK. & BOTS (C.) — *A. taina* FLINT (Hispaniola); *Chimarra cubanorum* BOTS. (C.) — *C. spinulifera* FLINT (Hispaniola); *Antillopsyche wrighti* BKS. (C.) et *A. aycara* BOTS. (C.) — *A. tubicola* FLINT (Puerto Rico); *Alisotrichia fundorai* (BOTS. & SYK.) (C.) — *A. hirudopsis* FLINT (Puerto Rico) et peut être *A. argentilinea* FLINT (Jamaica); *Ochrotrichia insularis ayaya* BOTS. (C.) — *O. insularis insularis* MOS. (Jamaica); *Hydroptila selvatica* BOTS. (C.) — *H. ancystrion* FLINT (Jamaica). Plusieurs cas de sous-spéciation à ses débuts ont été découverts (*Hydroptila*, *Oxyethira*).

Les affinités entre la faune de Cuba et celle de chacune des 3 autres îles des Grandes Antilles, sont à peu près du même ordre d'importance.

3. Le géographe cubain A. NUÑEZ JIMÉNEZ propose la division de Cuba en 5 'régions naturelles': Occidentale, Centrale, de Camagüey-Maniabon, Orientale, et Isla de Pinos. Le bien-fondé de cette division est démontré aussi par la faune de trichoptères. A l'exception de la région de Camagüey-Maniabon (presque pas de trichoptères connus et faune certainement très pauvre), chaque région possède une faune comprenant un nombre d'éléments caractéristiques. La région la plus riche et le mieux caractérisée est celle Orientale: 32-33 taxa uniquement connus de cette région. La région Occidentale suit à une bonne distance, avec 8-9 taxa. Il y a 5 taxa connus exclusivement de la région Centrale, et 2 pour Isla de Pinos. Tout ceci reflète fort bien les particularités géomorphologiques, géologiques et hydrologiques des différentes régions. Assez fréquemment, une espèce habite Occ. + Centr. (4 cas), Occ. + I.P. (3 cas), ou ces 3 provinces à la fois (2 cas); les espèces habitant Or. + Centr. sont au nombre de 4; plusieurs espèces ont une large répartition à travers l'île, et on peut considérer les formes suivantes comme étant les plus largement répandues et les plus fréquentes: *Chimarra (Curgia)pulchra* (HAG.), *Smicridea comma* BKS. (sous réserve! voir plus loin), *Neotrichia iridescens* FLINT, *Nectopsyche cubana* (BKS.), *Oecetis inconspicua* (WALK.). Pour revenir aux cas de vicariance, entre diverses régions naturelles cette fois-ci, voici des exemples: *Cariboptila soltera* BOTS. (Occ.) — *C. guajira* BOTS. et *C. poquita* BOTS. (Or.); *Macronema gundlachi* BKS. (Occ. + Centr.) — *M. tremenda* BOTS. (Or.); *Alisotrichia flintiana* BOTS. (Or.) — *A. cimarrona* BOTS. (Occ.); *Ochrotrichia caramba* BOTS. (Occ., Or.) — *O. villarenia* BOTS. (Centr.); *Metrichia espera* BOTS. (Occ.) — *M. cafetalera* BOTS. (Centr.) — *M. munieca* BOTS. (Or.); *Phylloicus chalybeus chalybeus* (HAG.) (Or.) — *P. chalybeus* ssp. (OCC. + I.P.); *Helicopsyche comosa* Kings. (OCC.) — *Helicopsyche n. sp.* prope *comosa* = peut-être *lutea* HAG. (OR.)

4. Un remarquable cas de vicariance non-géographique est fourni par les deux espèces extrêmement voisines *Alisotrichia chiquitica* BOTS. et *A. alayoana* BOTS. Les deux habitent la région orientale mais ne coexistent apparemment jamais dans le même cours d'eau (exclusion compétitive?).

Deux autres aspects systématiques — morphologiques doivent retenir notre attention. Il s'agit d'abord de deux très difficiles problèmes de systématique. Nous avons provisoirement rapporté la plupart des *Smicridea* capturés à Cuba à l'espèce *comma* BKS.; mais il faut dire que pratiquement chaque localité (ou groupe de localités) a fourni un type morphologique légèrement différent, de sorte que l'attribution spécifique — surtout en l'absence d'exemplaires conservés à sec — est sujette à caution; c'est le Dr. O.S. FLINT, jr., qui s'est chargé de trouver la solution de ce puzzle. D'autre part, c'est un problème tout a fait similaire que posent de nombreuses 'populations' de *Helicopsyche*, que je pense pouvoir rattacher au 'groupe *haitiensis*' (d'après le Prof. H.H. ROSS, qui entreprend actuellement l'étude de mes *Helicopsyche* de Cuba: 'They are definitely the most puzzling collections of *Helicopsyche* I have ever studied').

Mentionnons aussi la présence, chez d'assez nombreux trichoptères cubains, de formations androconiales et de soies spécialisées. Ainsi, *Cubanoptila muybonita* BOTS. est, à ma connaissance, le seul Protoptilinae à avoir dans l'aile antérieure du ♂ une importante zone recouverte de petites soies très spécialisées; *Ochrotrichia caramba* BOTS., et *O. villarenia* BOTS. sont probablement les seuls représentants du genre à présenter, sur les deux ailes du ♂, un revêtement d'écailles; *Oecetis maspeluda* BOTS. se distingue par les groupes de fines soies noires de l'aile postérieure (♂, ♀), ainsi que (♂) par un très long penicille de soies androconiales noires sur la base du fémur antérieur; les *Helicopsyche* ♂ du 'groupe *haitiensis*', fournis par au moins une localité de la rég. Orientale, possèdent sur les ailes antérieures une zone androconiale ronde recouverte d'écailles noires; *Helicopsyche comosa* KINGS. et une espèce inédite qui en est très voisine, sont certainement uniques dans le cadre du genre, par le dense revêtement de longues soies des gonopodes.

L'analyse si serrée que possible de la distribution géographique d'une part, des 'sister relationships' d'autre part, permet de formuler les conclusions suivantes.

a. De nombreuses espèces ont un 'cachet antillais' nettement distinct, appartenant à des genres ou à des groupes d'espèces strictement antillais (ou même cubains). Quelques exemples: tous les Protoptilinae, *Chimarra (C.) cubanorum* BOTS. et *C. (C.) garciai* BOTS., les 2 *Antillopsyche*, *Polycentropus nigriceps* BKS. et *P. turquino* BOTS., *Alisotrichia fundorai* (BOTS. & SYK.), *A. chiquitica* BOTS. et *A. alayoana* BOTS., *Ochrotrichia insularis ayaya* BOTS. et *O. islenia* BOTS., *Helicopsyche granpiedrana* BOTS. & SYK.

b. Si les affinités un peu plus éloignées des taxa sont à leur tour prises en considération, on constate que, dans de nombreux cas, les taxa montrent une parenté plus ou moins nette avec des éléments d'Amérique Centrale continentale et/ou du S (surtout du SE) d'Amérique du N (s'il ne s'agit tout simplement d'identité). Quelques exemples: *Polycentropus criollo* BOTS., tous les *Hydropsyche*, *Leucotrichia* cf. *tubifex* FLINT, *Ochrotrichia caramba* BOTS. et *O. villarenia* BOTS., *Orthotrichia americana* BKS. et *O. cristata* MORT., *Loxotrichia glasa* ROSS, les 5 *Oxyethira*, les 4 *Hydroptila*, tous les *Helicopsyche* à l'exception de *granpiedrana*.

c. Au contraire, les taxa appartenant certainement ou éventuellement à des 'trends' d'origine sud-américaine, sont extrêmement rares, mais cependant fort intéressants. Exemples: *Atopsyche cubana* FLINT et *A. vinai* SYK. & BOTS., *Chimarra (Curgia) pulchra* (HAG.) et *C. (Curgia) alayoi* BOTS., *C. (C.) guapa* BOTS. Il est maintenant clairement démontré que la faune de trichoptères de Cuba n'appartient pas au 'faunal circle' représenté par les Petites Antilles + le N. d'Amérique du S. + le S. d'Amérique Centrale continentale, mais bien à celui qui comprend les Grandes Antilles + le S. d'Amérique du N. + le N. d'Amérique Centrale continentale.

On manifeste de nos jours un très grand intérêt pour la zone du Golfe du Mexique — Mer des Caraïbes, et ceci sur le plan des recherches géologiques, géomorphologiques, paléogéographiques, avec toutes les implications de leurs résultats sur le problème de l'origine et de l'évolution des faunes des terres de la zone. Il s'accumule dans ce domaine une bibliographie de plus en plus abondante, et les résultats présentés sont souvent extrêmement différents, contradictoires même, d'un auteur à un autre (en fonction du groupe étudié, du caractère plus ou moins solide de cette étude, des théories géologiques et paléogéographiques prises comme point de départ — et de l'attachement plus ou moins fanatique à une certaine théorie. Ce qui est le plus important à ce stade, c'est de dégager le maximum d'enseignements de l'étude multilatérale de groupes si divers que possible. Les résultats obtenus de l'étude d'un 'excellent' groupe comme les trichoptères, permettent de considérer le schéma suivant comme étant le plus plausible pour l'origine et l'évolution des terres Grand-Antillaises et de leur faune: expulsion en direction de l'Atlantique d'un 'noyau proto-antillais' à partir de l'Amérique Centrale continentale; fragmentation de ce noyau; contacts répétés entre les divers fragments, contacts répétés entre ces fragments — et notamment Cuba — et la Floride et les zones avoisinantes, le Mexique, le N. de l'Amérique Centrale continentale; possibilité de pénétration, à certaines époques, de lignées en provenance du N. de l'Amérique du S., avec utilisation des 'stepping stones' que représentent les Petites Antilles. Migration (dispersion) d'une part, isolement géographique d'autre part, ont joué des rôles également importants dans la constitution de la faune de trichoptères telle qu'elle est aujourd'hui (il serait faux de vouloir opposer ces deux processus, comme certains biogéographes contemporains veulent nous persuader).

En ce qui concerne l'écologie, il faudra considérer seulement comme jalons les quelques idées qui vont suivre. Les massifs montagneux de la région Orientale, avec leurs systèmes hydrographiques abondants et souvent encore protégés par une couverture forestière primaire ou secondaire, sont habités par la faune la plus riche et la plus diversifiée. Malheureusement, on ne sait presque rien sur les trichoptères de la zone comprise entre 1000 et 2000 m. d'altitude, zone souvent pratiquement inaccessible. Les montagnes orientales sont suivies par celles de la région Occidentale, puis par la Sierra Escambray de la région Centrale. Les autres zones de Cuba possèdent une faune considérablement appauvrie, et la région Camagüey-Maniabon est particulièrement pauvre. La pollution industrielle des eaux est pratiquement inexistente, mais c'est la déforestation qui a certainement représenté un facteur défavorable. Une 'bonne rivière' ou un 'bon ruisseau' peuvent être habités par 15 espèces, peut-être même un peu plus. Le Crenon semble pratiquement non habité par les trichoptères, en tout cas il me semble qu'il n'y a pas de formes crénobiontes. Il n'est pas encore possible de distinguer entre la faune du Rhithron et celle du Potamon. Les éléments hygropétriques sont bien représentés dans les différents types d'eaux courantes. La faune des eaux stagnantes est généralement fort pauvre et monotone (mais les grandes étendues marécageuses, comme la Ciénaga de Zapata, sont encore peu connues). La plupart des espèces sont attirées par la lumière artificielle.

Proc. of the 2nd Int. Symp. on Trichoptera, 1977, Junk, The Hague

Species diversity of Trichoptera communities

J.O. SOLEM

Abstract

Species diversity data from 19 light trap samplings are presented. Some data from England and Sweden were extracted from the literature. The extended negative binomial model was used in the calculation of the species diversity. Using α and k as population parameters with standard errors, the data were plotted in a diagram and appeared as ellipses. Species diversity for total flight periods, year to year, and seasonal variations are dealt with. The method seems to give a more detailed separation of communities than Shannon and Simpson indices.

Introduction

The purpose of the present paper is to show species diversity data of Trichoptera communities using the extended negative binomial model elaborated by ENGEN (1974, 1977). The model is a generalization of FISHER's model (FISHER, CORBET & WILLIAMS, 1943) with an additional parameter k. The case $k = 0$ is equivalent to Fisher's model. The present model is two-dimensional and the tabulation used allows a simple comparison of the indices of Simpson, Shannon Weaver and McArthur and Fisher's model (AAGAARD & ENGEN in press).

The data presented here are based on light trap collections which allow optimal identification of the species as the taxonomy of the adults is well known. Therefore, such collections should be well suited for species diversity studies, as also pointed out by RESH, HAAG & NEFF (1975).

Methods

Data on species frequencies from light trap collections in Norway (SOLEM, unpublished), North Sweden (GÖTHBERG, 1970, 1974), and England (CRICHTON, 1960) were computed, and estimates of species diversity were obtained using the extended negative binomial model. The mathematical elaboration of this model to animal populations was made by ENGEN (1974, 1977). The presentation of the data follows AAGAARD & ENGEN (in press), who tabulated the communities in a diagram, where α and k were parameters (Fig. 1). Tables of standard errors of the estimates of α and k are found in ENGEN (1974), and may be applied to obtain standard errors in any direction of an estimate of α and k. The estimated standard errors will appear as ellipses in the α and k diagram.

Calculations were carried cut after the Pseudo Moment Method (ENGEN, 1974), which is rather sensitive to variations in the number of the common species. The variations in the rare species are to a lesser degree taken care of, but that may be done using the Pseudo Maximum Likelihood Method (ENGEN, 1974). The samples were fitted to the equation

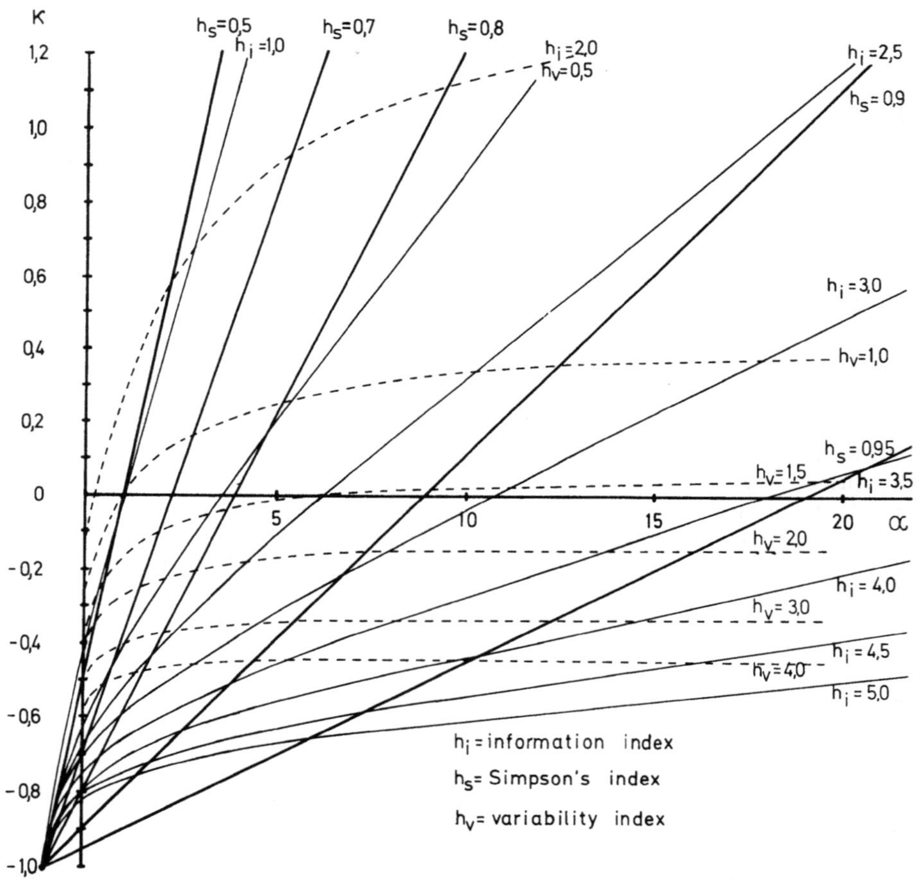

Fig. 1. Values of Simpson index, the information index and the variability (the inverse value of equitability) as functions of α and k. (The species frequencies model is not defined for $\alpha < 0$.) (After AAGAARD & ENGEN, in press).

$$E(N_i) = \alpha\, \omega^k \frac{(k+1)(k+2)\ldots(k+i-1)}{i!} (1-\omega)^i,$$

N_i = Number of species with i representatives in the sample
ω = $\alpha/(\alpha+\nu)$
ν = Expected number of individuals in the sample

The data from light trap collections used seem comparable as nearly all cover the total flight period.

Results

Data from light traps located at standing and slowly flowing water are presented in Fig. 2. Millbarn Pond, England (CRICHTON, 1960), Skalka, North Sweden (GÖTHBERG, 1974), Fiplingvatn, North Norway and Storvatn, Central Norway, all

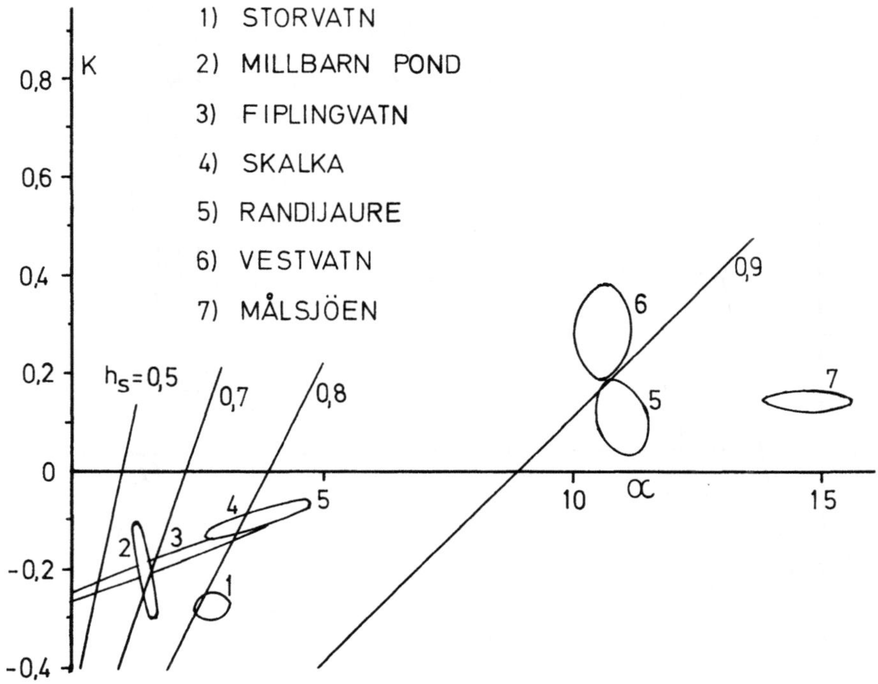

Fig. 2. Estimates of α and k, ± standard errors, from standing and/or slowly flowing waters.

showed low α and k -values. Together they form a group of communities with low diversities in the lower left part of the diagram. Randijaure, North Sweden (GÖTHBERG, 1974), Vestvatn, North Norway and Målsjöen, Central Norway, have fairly high α and k -values. They are as seen with this method definitely separated from the first group of communities, and show higher diversities.

The ellipses drawn in Fig. 3 give results from running water communities. All

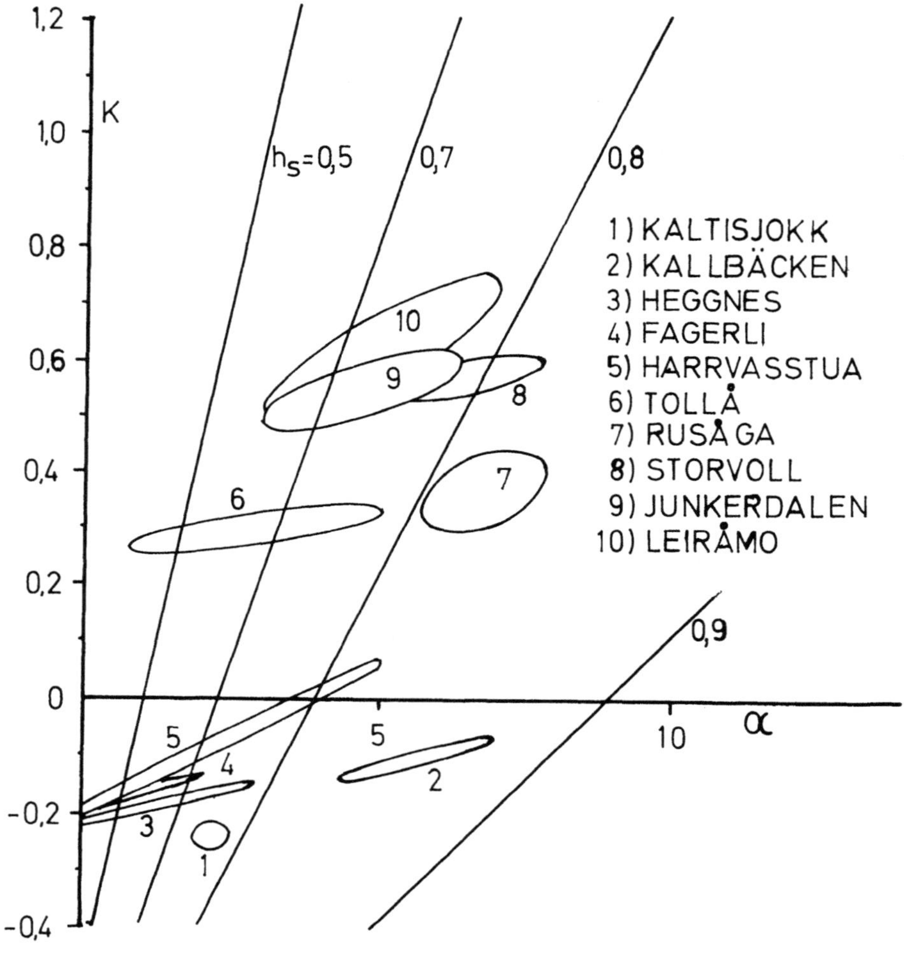

Fig. 3. Estimates of α and k, ± standard errors, from running water communities.

communities show a low α value while the k value perform a greater range. The group of localities with negative k values belong to the same geographical area in North Norway except for two localities that are situated in Northern Sweden. The ellipses with positive k values represent communities belonging to another geographical and geological area in North Norway.

Fig. 4 shows diversity data for three different years at a running water locality in

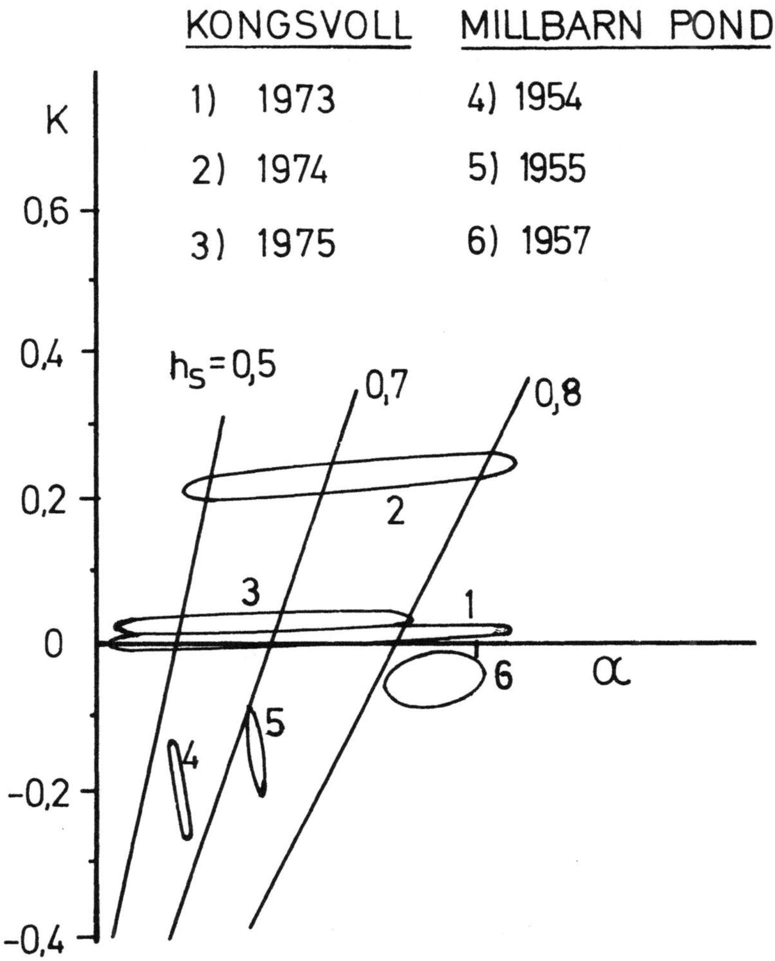

Fig. 4. Year to year variations of estimates of α and k, ± standard errors, at two localities.

Central Norway and at standing water at Millbarn Pond, England (CRICHTON, 1960). At Kongsvoll, Norway, there is good agreement in the 1973 and 1975 estimates. The 1974 estimates is not strictly comparable with the data from 1973 and 1975 as the light trap operated for a shorter time period that year. This discrepancy separates the diversity that year from the two others and the k value was affected more than the α value. A distinctly higher k value was estimated in 1974. I may also mention that the basic data for 1975 had a very low fitness to the model. But, the estimated α and k values and the standard error were almost identical to those of 1973, and the species frequencies of 1973 had an excellent fitness to the model. This strengthens the usefulness of the method.

The Millbarn Pond estimates gave similar trends as the Kongsvoll results in species diversity variations from year to year. In this occasion the 1957 diversity estimate is obtained from a shorter collecting period than in 1954 and 1955. Both

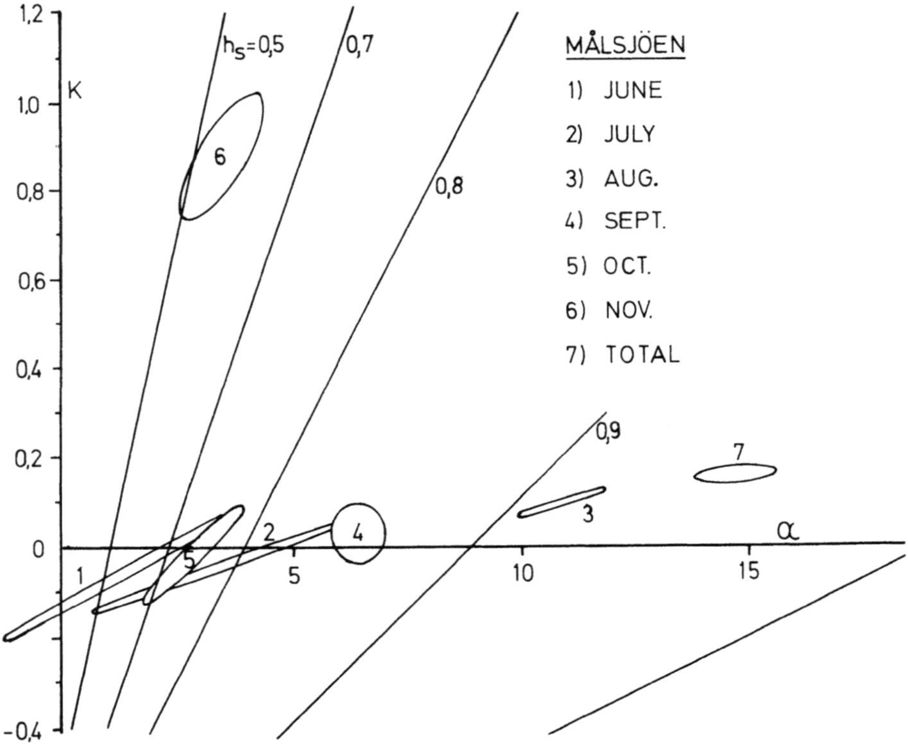

Fig. 5. Seasonal variations of the estimates of α and k, ± standard errors.

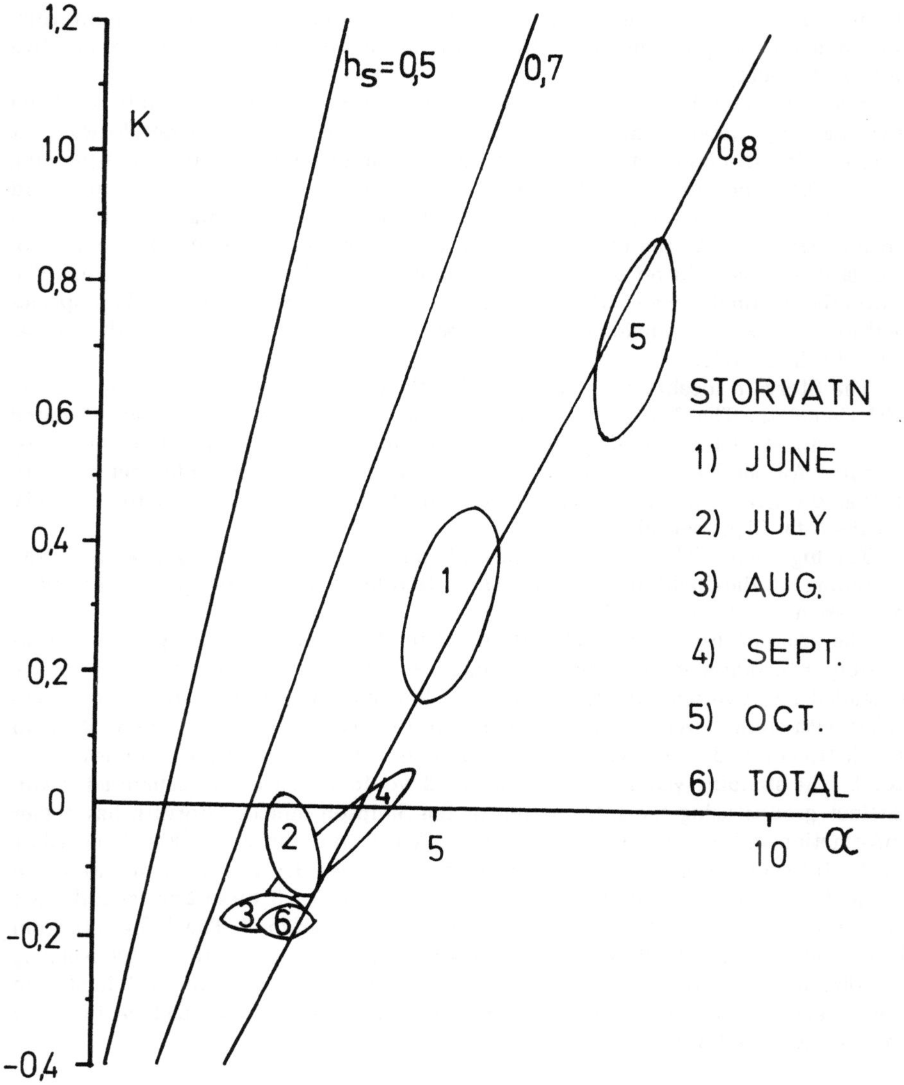

Fig. 6. Seasonal variations of the estimates of α and k, ± standard errors.

higher α and k values were found in 1957 than in the two other years. The Information and the Simpson indices followed the same trend as the extended negative binomial model.

Seasonal trends in the species diversity are reproduced in Figs. 5 and 6, involving two standing water communities Målsjöen and Storvatn, in Central Norway. At Målsjöen the species diversity started with low α and k values in June, the beginning of the flight period. The estimates increased in value throughout summer and reached the highest summer values in August. The August estimate was that with greatest similarity to the estimates of the total flight period. September and October gave successive lower α and k values approaching the July and June data. The November estimates were different from those of the previous months. Few species and low numbers of individuals for each species were obtained, affecting the equitability in the sample.

The Storvatn locality (Fig. 6) showed with some modifications the same as the Målsjöen data. The first and last month, June and October, respectively, deviated from the rest of the sampling months. They showed few species and low numbers, as noted for the November data at Målsjöen. July, August and September had very similar diversities, but as in Målsjöen the August values were nearest the diversity for the total flight period.

The high equitability in June is possibly a result of a slow emergence rate in the beginning of the flight period and that in October because few species and specimens are added to the populations.

The program used in the calculations of the light trap data also gave the Information or Shannon and the Simpson indices. In some cases the α and k values in the extended negative binomial model and the Information and Simpson indices coincided with similar trends, e.g. for the data of the year to year variations at Millbarn Pond. However, discrepancies also occurred, and the most conspicious ones would be: 1. The standing water localities in Fig. 2 would not have been separated in two distinct groups using the Information or the Simpson indices. Storvatn has higher Information index than Skalka. 2. If we rank the communities in Fig. 3 according to the Information and Simpson indices, the localities Heggnes and Leiråmo would be next to each other in the middle of the list. In Fig. 3 these two localities are plotted at the bottom and top, respectively, of the diagram. 3. The 1974 data from Kongsvoll in Fig. 4 would have been between the 1973 and 1975 results. Keeping the dissimilarities in the collecting periods in mind, the separation obtained with the extended negative binomial model seems to give a better picture than the Information and Simpson indices.

Concluding remarks

Oligotrophic communities with few species, e.g. Leiråmo, North Norway, appeared to have low α and fairly high k values, while those with more species had low values of both α and k, e.g. Kaltisjokk, North Sweden.

A high k value in the model indicated few species, and the species richness increases with decreasing k value. Mesotrophic communites, e.g. Målsjöen, Central Norway, showed high α but low k values. These results agree with those from chironomid communities presented (AAGAARD & ENGEN in press).

The area in the diagram with negative k values and low α values seems to be a mixture of communities that are physically controlled, but they have one thing in common: they are strongly dominated by a few species that suppress the equitability greatly. In this respect the organization and structure of the caddisfly community at the eutrophic Millbarn Pond (CRICHTON, 1960) near Reading, England, appeared to be very similar to some oligotrophic communities, e.g. Fiplingvatn and Harrvasstua, in Northern Norway.

References

AAGAARD, K. & ENGEN, S. (in press) Species diversity of chironomid species. *Acta Univ. Carolinae, Praha.*

CRICHTON, M.I. 1960. A study of captures of Trichoptera in a light trap near Reading, Berkshire. *Trans. R. ent. Soc. Lond. 112*: 319-344.

ENGEN, S. 1974. On species frequency models. *Biometrica 61*: 263-270.

——. (1977) Exponential and logarithmic species-area curves. *Am. Nat.* III: 591-594.

FISHER, R.A., CORBET, A.S. & WILLIAMS, C.B. 1943. The relation between the number of species and the number of individuals in a random sample from an animal population. *J. Anim. Ecol. 12*: 42-58.

GÖTHBERG, A. 1970. Die Jahresperiodik der Trichopterenimagines in zwei lappländischen Bächen. *Öst. Fisch. 23*: 118-127.

——. 1974. Trichoptera och Plecoptera från två ljusfällor vid sjöarna Skalka och Randijaure, nordväst om Jokkmokk. *Berichte aus der Ökologischen station Messaure* 14: 1-17.

RESH, V.H., HAAG, K.H. & NEFF, S.E. 1975. Community structure and diversity of caddisfly adults from the Salt River, Kentucky. *Environm. Ent. 4*: 241-253.

Proc. of the 2nd Int. Symp. on Trichoptera, 1977, Junk, The Hague

The undulatory behaviour of larvae of *Hydropsyche pellucidula* CURTIS and *Hydropsyche siltalai* DÖHLER

G.N. PHILIPSON

Abstract
The undulatory movements of the abdomen of *H. pellucidula* and *H. siltalai* larvae are being investigated in relation to water temperature and dissolved oxygen tension.

Recordings of undulatory activity show that periods of quiescence alternate with periods of continuous undulation. The time spent undulating and the rate of undulation both increase with rise in water temperature.

The difference between the oxygen uptake of anaesthetised and active larvae of *H. pellucidula* represents some 70-80% of active oxygen uptake. Estimates of the oxygen uptake per undulation are discussed.

The investigation is continuing.

Larvae of many species of caddisfly make rhythmic undulatory movements of the abdomen. The frequency of these movements is affected by dissolved oxygen concentration and rate of water flow in a way that suggests they have a respiratory ventilatory function (FOX & SIDNEY, 1953; PHILIPSON, 1954; AMBÜHL, 1959).

In previous investigations (PHILIPSON & MOORHOUSE, 1974 and 1976) it has been shown that the undulatory frequency of larvae of various species of Hydropsychidae and Polycentropodidae is also increased by rise in water temperature and that the undulatory frequency of *Plectrocnemia conspersa* (CURTIS) may be increased threefold immediately after feeding. The extent to which such increases are a ventilatory response to increased respiratory requirements is not known. Larvae of *Cyrnus flavidus* McLACHLAN and *Holocentropus picicornis* (STEPHENS) from still water habitats are stimulated to undulate erratically by water flow.

In a consideration of undulation in relation to oxygen uptake in *P. conspersa* and *Polycentropus flavomaculatus* PICTET it has been suggested that undulatory activity may account for some 60-70% of active oxygen uptake. The present paper is a progress report on a similar investigation on larvae of *Hydropsyche pellucidula* CURTIS and *H. siltalai* DÖHLER (formerly identified as *H. instabilis* CURTIS).

Differences in undulatory behaviour have been demonstrated previously by comparing the undulation frequency determined by counting the numbers of undulations during 5-minute periods, under different conditions of water flow rate, temperature and dissolved oxygen tension. To demonstrate further differences in undulatory behaviour, larvae of *H. pellucidula* and *H. siltalai* were acclimated at

10°C for 24 h in air-saturated still water, by which time the larvae were occupying silken retreats which they had spun. The behaviour of the larvae was then observed for 5 minutes during which time the occurrence of individual undulations was marked on a time-scale using an event recorder. The water temperature was then raised to 18°C and after 24 h acclimation under full aeration a further recording was made. The temperature was then raised to 25°C and the procedure repeated. Sections of the recordings obtained are reproduced in Fig. 1. It is seen that periods of quiescence alternate with periods of continuous undulation. The recordings of *H. pellucidula* were all taken from the same larva. At 10°C it was quiescent for 75% of the 5-minute observation time. At 18°C it was quiescent for 55% and at 25°C for 45% of the observation time. The recordings of *H. siltalai* at 10 and 18°C were taken from the same larva, but at 25°C it had left its retreat and was wandering continuously around the experimental vessel, with no pause for undulation. At this temperature the recording was made from another individual which had passed through the whole experimental procedure and still remained in its retreat. The *H. siltalai* larvae recorded were quiescent at 10°C for 50% and at 18 and 25°C for 40% of the observation time.

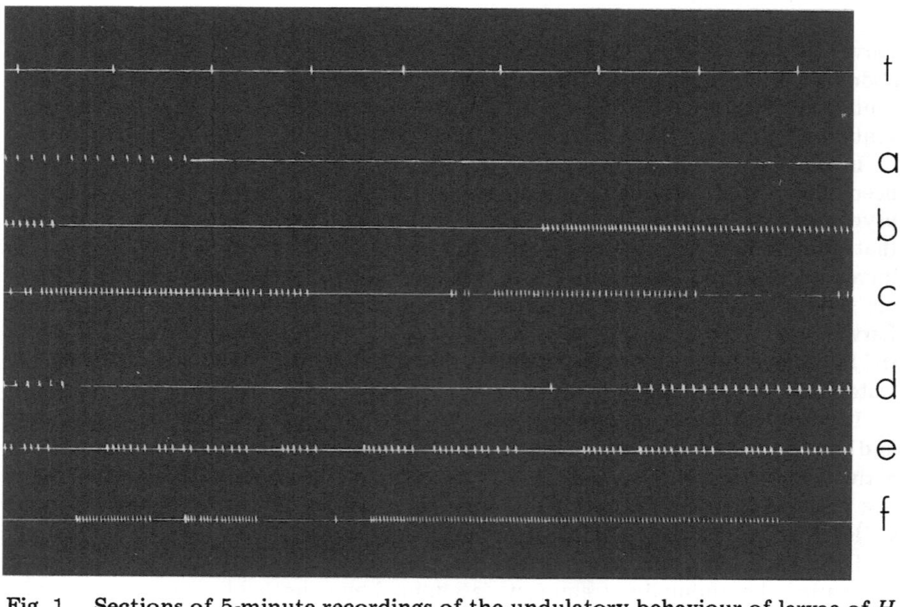

Fig. 1. Sections of 5-minute recordings of the undulatory behaviour of larvae of *H. pellucidula* at (a) 10°C, (b) 18°C and (c) 25°C and *H. siltalai* at (d) 10°C, (e) 18°C and (f) 25°C in air-saturated still water. The time-scale (t) is marked at 15 second intervals.

Determination of undulatory frequency by counting the number of undulations during a 5-minute period does not record periods of quiescence and undulation or rates of undulation within these periods. Inspection of the recordings suggests that the rate of undulation tends to decrease progressively throughout an undulation period. These aspects of undulatory behaviour were examined by measuring the duration of a single period of undulation and its subsequent period of quiescence, together with the associated rates of undulation, at each of the selected temperatures. The data are set out in Table 1. In *H. pellucidula* rise in temperature to 18°C is accompanied by a five-fold decrease in the duration of the quiescent period and a smaller decrease in the period of continuous undulation, while the rate of undulation increases two-fold. These trends in the duration of the periods of quiescence and undulation continue, although less dramatically, with rise in temperature from 18 to 25°C. There is however little change in rate of undulation associated with this further temperature increase. In *H. siltalai* rise in temperature to 18°C is associated with a dramatic decrease in the duration of both quiescence and undulation periods and a smaller increase in undulation rate than in *H. pellucidula*. This trend is not continued in the *H. siltalai* larva recorded at 25°C, where the periods of quiescence and undulation are slightly longer and the undulation rate is much higher than in the larva recorded at 18°C.

Rise in temperature to 18°C is thus associated with a decrease in quiescent time and an increase in undulation rate. Further increase in temperature to 25°C, in *H.*

Table 1. Durations of single periods of continuous undulation and subsequent quiescence in larvae of *H. pellucidula* and *H. siltalai* at 10, 18 and 25°C in air-saturated still water. Undulation rates associated with each period of continuous undulation are given.

	H. pellucidula			H. siltalai		
Temperature	10°C	18°C	25°C	10°C	18°C	25°C
Duration of quiescence (secs) ..	225	46	25	105	15	27
Duration of activity (secs)	73	50	44	102	15	49
Mean undulation rate (unds/min)	35	77	72	38	60	102
Undulation rate, first 15 secs (unds/min) ..	44	92	84	38	-	124
Undulation rate, last 15 secs (unds/min) ..	28	60	64	38	-	88

pellucidula larvae, is associated mainly with further decrease in quiescent time.

In order to investigate the relationship between undulation and oxygen uptake, both standard and active rates of oxygen uptake have been determined. This work, up to the present time, has been restricted to larvae of *H. pellucidula*.

Using the closed bottle method, oxygen uptakes were measured at 10, 18 and 25°C with dissolved oxygen tension (pO_2) at 100% air-saturation level and at 10 and 18°C with pO_2 at 70% air-saturation level. In the determination of active oxygen uptake, larvae were placed singly in glass stoppered bottles in water continuously aerated and maintained at the selected temperature. After 24 h the larvae had constructed silken retreats, in which they remained throughout the subsequent experimental procedure. Following this period of temperature acclimation each bottle was flushed through several times with water at the selected temperature and oxygen tension, sealed and kept at the selected temperature for about 4 h. In determination of standard oxygen uptake larvae were acclimated to the selected temperature in continuously aerated water, after which they were anaesthetised in aerated 0.5% urethane solution at the same temperature. The larvae were then transferred to glass stoppered bottles which were flushed through with 0.5% urethane solution at the selected temperature and oxygen tension. The bottles were then sealed and kept at the selected temperature for about 4 h. In order to achieve a reduction in dissolved oxygen concentration similar to that produced by active larvae, three anaesthetised larvae were placed in each bottle. Dissolved oxygen concentrations were measured using a micro-Winkler method employing the azide modification. The mean reduction in dissolved oxygen concentration was 1.3 ± 0.1 mg/l in experiments with active larvae and 0.9 ± 0.1 mg/l in experiments with anaesthetised larvae. Weights of larvae were determined after drying at 60°C and the rate of oxygen uptake was calculated as mg O_2/g dry wt. h. The results are set out in Table 2. Under each set of experimental conditions the difference between standard and active oxygen uptakes (activity component) is large, amounting to some 70-80% of the active oxygen uptake.

The proportion of the activity component which can be attributed to undulation is not known. However, during the experiments the active larvae were restricted within silken retreats and their visible activity was predominantly undulatory. On the assumption that the entire activity component represents work done on undulation only, the mean oxygen uptake per undulation was calculated, using the mean undulation frequency corresponding to the mean conditions of the experiment, derived from earlier data (PHILIPSON & MOORHOUSE 1974). Similar approximations were made in previous estimates of the oxygen uptake per undulation of *P. conspersa* and *P. flavomaculatus* (PHILIPSON & MOORHOUSE 1976).

The data required for the present calculations and the results obtained are included in Table 2. At 10°C and dissolved oxygen tension at 100% air-saturation level, the mean oxygen uptake per undulation was 2.51 μg O_2/g dry wt. This estimate agrees well with those obtained for the polycentropodid species, which, with the exception of one estimate of 8.6 μg O_2/g obtained for *P. flavomaculatus* larvae, ranged between 3.4 and 4.4 $\mu g\ O_2$ /g dry wt., but it is significantly higher, at the 0.02 level, than each of the other *H. pellucidula* estimates, which range between 0.39 and 0.71 $\mu g\ O_2$ /g dry wt.

The estimates for oxygen uptake per undulation which are available for caddis fly larvae thus appear to form two well-defined groups. The group of higher estimates, derived from *H. pellucidula* and the two polycentropodid species, were all determined at relatively low undulation frequencies ranging between approximately 10 and 40 undulations per 5-minute period. The undulation frequency, of about 20 undulations per 5-minute period, at which the *H. pellucidula* estimate was made lies approximately in the middle of this group. The group of lower estimates, all derived from *H. pellucidula*, were determined at relatively high undulation frequencies ranging between approximately 50 and 330 undulations per 5-minute period. In the polycentropodid species investigated, comparable stress occurs only at low dissolved oxygen tensions and no measurements of oxygen uptake for these species under such conditions are yet available for comparison.

A possible explanation for the difference between these two groups of estimates for oxygen uptake per undulation may be seen in the undulation behaviour of *H. pellucidula*. It has been shown that increase in undulation frequency, measured over a 5-minute period, is accompanied by an approximately two-fold increase in undulation rate within periods of continuous undulation. This increase in undulation rate may be accompanied by a decrease in undulation amplitude which could result in a lower oxygen uptake per undulation.

Table 2. Comparison of oxygen uptake of anaesthetised larvae (standard O_2 uptake) and active larvae (active O_2 uptake) of *H. pellucidula* at 10, 18 and 25° C at 100% air-saturation level and at 10 and 18° C at 70% air-saturation level. Each mean was derived from at least five determinations. O_2 uptake per undulation was calculated from mean undulation frequencies for conditions corresponding to the mean conditions of the experiment (from PHILIPSON & MOORHOUSE 1974) and on the assumption that the difference between standard and active O_2 uptake was wholly due to undulatory activity. Variation is expressed as standard error.

Dissolved oxygen tension ..	100% air saturation			70% air saturation	
Temperature	10°C	18°C	25°C	10°C	18°C
Standard O_2 uptake mg/g/h	0.17±0.01	0.36±0.04	0.46±0.06	0.12±0.02	0.29±0.04
Active O_2 uptake mg/g/h	0.75±0.02	1.55±0.17	1.98±0.15	0.56±0.03	1.48±0.20
Activity component, O_2 uptake mg/g/h	0.58±0.10	1.19±0.08	1.52±0.16	0.44±0.03	1.19±0.21
Activity component, (as % Active O_2 uptake)	77.3	76.8	76.8	78.6	80.4
Mean und. frequency for experiment (unds/5 min)..	19.2±3.5	175.0±27.8	326.7±12.7	52.1±8.2	232.3±3.2
O_2 uptake per undulation µg O_2/g	2.51±0.62	0.56±0.10	0.39±0.05	0.71±0.12	0.43±0.08

A further explananation of the present data is however suggested when further consideration is given to the estimate of 8.6 $\mu g\ O_2/g$ dry wt. per undulation obtained for larvae of *P. flavomaculatus*. It was previously suggested that this high estimate, relating to conditions under which mean undulatory activity amounted to only 4 undulations per 5-minute period, may be due to the larvae being more active in other ways when not under stress (PHILIPSON & MOORHOUSE 1976). It seems probable that the proportion of the activity component of oxygen uptake attributable to undulation will increase progressively with increase in undulatory activity to a theoretical limit when the whole of the activity component is attributable to this activity. Thus the present data, instead of representing two separate groups of estimates with one exceptional estimate, may represent a scale of estimates, albeit somewhat discontinuous, ranging from 8.6 to 0.39 $\mu g\ O_2/g$ dry wt., which tend to decrease with increase in undulation frequency. The association between size of estimate and undulation frequency arises from the assumption in the calculation of the estimates, that the whole of the activity component is attributable to undulation. As the estimates decrease in size with increase in undulation frequency they may approach the true value for oxygen uptake per undualtion more closely.

This problem will not be resolved finally until all activities associated with active oxygen uptake are recognised and quantified, but corroborative evidence may be obtained from estimates over a wider range of conditions.

Future work will involve completion of the programme on *H. siltalai* and will include measurements of undulation amplitude. Further investigations into the relation between undulation and respiration in polycentropodid larvae, including determinations of oxygen uptake of larvae in flowing water, and at low oxygen tensions, is being carried out by my colleague Mr. JOHN GREENWOOD.

References

AMBÜHL, H. 1959. Die Bedeutung der Strömung als ökologischer Factor. Schweiz. Z. Hydrol. 21: 133-264.

FOX, H.M. & SIDNEY, J. 1953. Influence of dissolved oxygen on the respiratory movements of caddis larvae. J. exp. Biol. 30: 235-237.

PHILIPSON G.N. 1954. The effect of water flow and oxygen concentration on six species of caddis fly (Trichoptera) larvae. Proc. zool. Soc. Lond. 124: 547-564.

——. & MOORHOUSE, B.H.S. 1974. Observations on ventilatory and net-spinning activities of larvae of the genus *Hydropsyche* Pictet (Trichoptera, Hydropsychidae) under experimental conditions. Freshwat. Biol. 4: 525-533.

——. 1976. Respiratory behaviour of larvae of four species of the Family Polycentropodidae (Trichoptera). Freshwat. Biol. 6: 347-353.

Discussion

EDINGTON: What do you feel are the ecological implications of your experiments?

PHILIPSON: Differences in undulation behaviour between larvae of certain species will result in differences in energy expenditure on ventilation and thus contribute towards limiting their distribution.

GREENWOOD: My calculations on the cost of ventilation in polycentropodid larvae, when not under extreme stress, are in agreement with those of PHILIPSON. This validation may be of significance since my determinations of respiration rate were made with an oxygen cathode respirometer, which was a different method from that used by PHILIPSON.

Proc. of the 2nd Int. Symp. on Trichoptera, 1977, Junk, The Hague

House building with panels: distribution of building activities by day and night, and intervals between activities, in the larva of *Potamophylax latipennis* (Curt.) Neb. (Trichoptera, Limnephilidae)[1]

H. ZINTL and GISELA STAMMEL

Within a large programme of experiments on the style change (ZINTL, 1970, 1976) we also noted down the building activities separated after the end of the dark and after the end of the light time.

In the laboratory the conditions were: LD 12 : 12, white light (fluorescent lamp Osram Natura 40 W) about 400 lx in front of the insect cases 7 a.m. to 7 p.m.; temperature 289 ± 1 K. February-March the larvae in the open ('76) had approximately LD 12 : 12 but naturally a dawn temperature; 227 ± 1 K.

The larvae mostly build with panels up to the beginning of the 5th instar. They lay a row of large panels (discs) on each other dorsally and ventrally. It is easier to register quantitatively this behaviour of building than that with sand grains after the style change. Therefore in this paper we treat only the building activity with panels.

As units we define: panel-addition activity = addition of 1 disc; panel-removing activity = removal of 1/2 discs (the larva cuts across the disc to tear it off in halves).

Our investigations are concerned with the intervals of these panelbuilding activities and their distribution by day and night.

1. Distribution by day and night

From the histograms (Fig. 1a) we see that panel-addition activity is clearly predominant at night. Removing activity seems to be a little more frequent during the day.

This predominance of addition activity at night (Fig. 2a) is true of the 3rd and 4th instars. It is demonstrable with larvae under the constant conditions of the laboratory and in the open. There are no larvae which build more frequently during the day. If we sum up the addition activities of all larvae of a sample we get 82-86% activities at night. As to the panel-removing activity (Fig. 2b) only during the 4th instar '73, and '76 only in the open, is a predominance during day statistically significant (WILCOXON test: $0.05 > \alpha > 0.0001$). There are only a few larvae having removing activities exclusively during the day: 12-39%.

In order to test whether the shift of the panel-addition activity to the night is triggered internally or externally in '77 we kept larvae under LL regime. The results are seen in Fig. 1a, 2nd row. Even at the beginning of the experiment with constant light maxima of addition activity at night were not noted. Thus an external trig-

[1] To the nuns of Hohenburg.

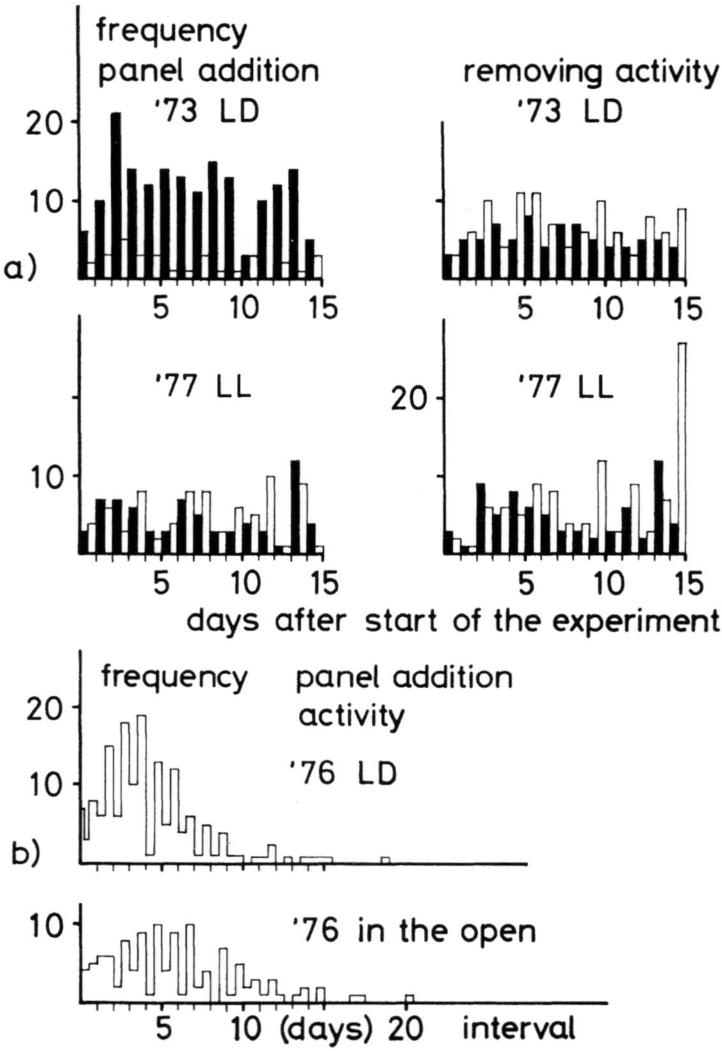

Fig. 1a. Panel-addition and removing activity after 3rd-4th moult under LD and LL. Unit of panel-addition activity: addition of one panel. Unit of panel-removing activity: removing of half a panel. '73: start: 25th January; 42 larvae. '77: start: 21st January; 42 larvae. Black blocks: activity at night respectively during the time which would have been night. b. Histogram of all intervals between two neighbouring panel-addition activities during the 4th instar. Data of 24 ('76 LD) and 25 ('76 in the open) larvae.

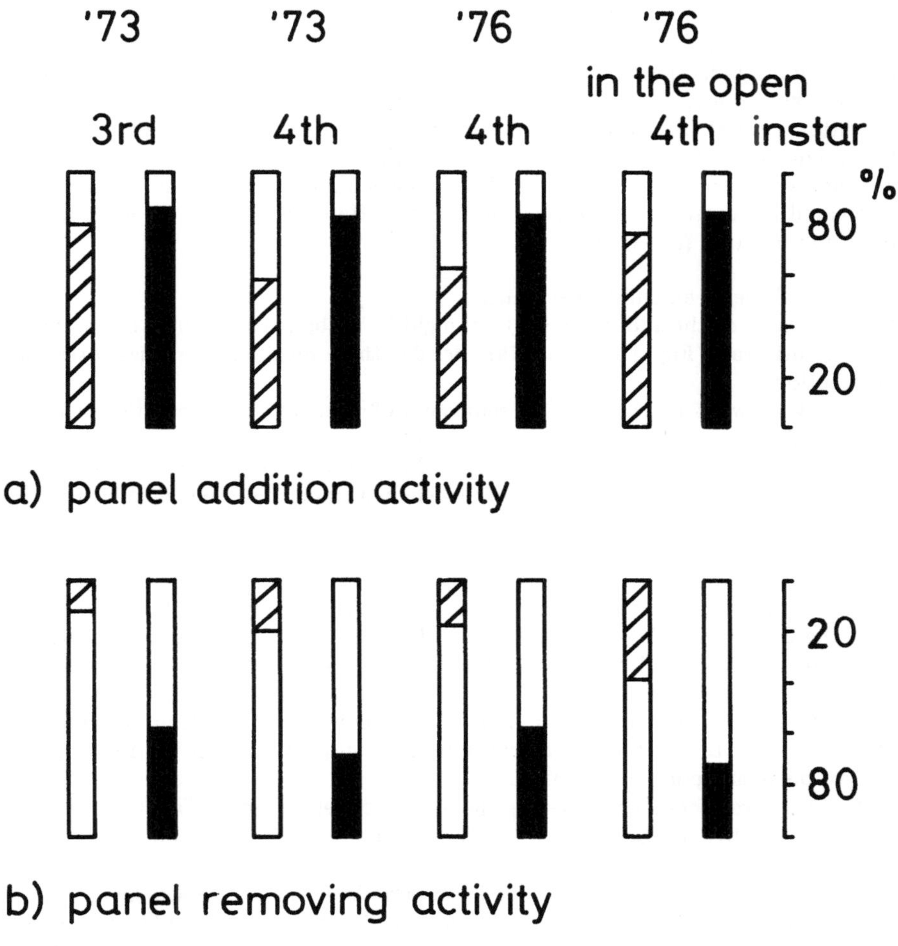

Fig. 2a. Parts (percent) of panel-addition activity at night. Striped: percentage of larvae having built only at night or (in case of four or more addition activities per larva) at the most once during day. Black: percentage of addition activities at night related to the data of the whole sample. Larvae per sample: 23 to 43. b. Parts (percent) of panel-removing activity during day. Striped: percentage of larvae having removed panels only during day or (in accordance with above) at the most once at night. White part of the white/black blocks: percentage of panel removing activity during day related to the data of the whole sample. Larvae per sample: 23 to 43.

gering might be more probable. THORNE (1969) found that 3rd to 5th instar larvae of *Potamophylax stellatus* (= *latipennis*) have activities of feeding and locomotion during darkness. He showed that these periods of activity are triggered mainly exogenously. Noteworthy is the decrease of the addition activity under LL conditions. Apart from the last high maximum (by chance?) the distribution of the panel-removing activity under LL does not differ from that under LD regime. Both are random distributions for day and night (Wilcoxon test: $\alpha >$ and $\gg 0.05$; only calculated for those 15 days). Also there is no difference between the quantities of removing activities ($\alpha > 0.05$).

2. Intervals between building activities

Predominance of the addition activity at night in histograms is turning out as whole number intervals (Fig. 1b, 3 and 4a). Besides these maxima we realize a random distribution.

Fig. 3 shows as an example the sequences of panel-addition activities of larvae '76 LD. Are there more frequently preferred distinct intervals, perhaps different from larva to larva, and its whole multiples? It is difficult to prove such a hypothesis. A single sequence consists only of few elements and so it seemed to be not possible to treat the arrangement of the elements of a single sequence statistically. Therefore, knowing that this is not the best method we summed up the activities of all larvae of a sample.

For the histograms '73 LD and '77 LL (Fig. 4b) we only used the 1st intervals after the 3rd/4th moult of all larvae. In histogram '73 LD there is probably a 2nd maximum at 5 days. With a special procedure for searching for periodical processes (PUSTA: WOLFFGRAMM, 1975; WOLFFGRAMM & THIMM, 1976) besides a period of 1 day a period of 4.9 (5?) days in fact could be found. The superposition histogram '73 LD served for this procedure (Fig. 4c). PUSTA on '77 LL (Fig. 4c) did not prove any continuous period.

Recently we succeeded in testing the longer sequences singly. '76, LD: 10 of 15 tested larvae showed a period of 2 to 6.5 days (period of 5 larvae: 3 days). '73, LD: some insects with periods of 1.5 or *2.5* days (compare: superposition histogram: *4.9* days!). '77, LL: a few single sequencies: periods of 1.5 or 4 or 4.5 days.

Thus we have an indication that the periodic distribution of the addition activity is also triggered externally, provided that a possibly existing internal periodic arrangement (LL: some proved periods) had not been disturbed by the common decrease of the panel-addition activity under LL.

We suggest the following hypothesis to be tested with further experiments: The larvae have periodical tendencies to build. The periods are different from larva to larva. During the time in which there is light these tendencies are repressed and so a shifting into the 'dark time' results.

Under LD the tendencies of addition activity become periodical. These tendencies additionally are shifted into the dark time.

The histograms based on all neighbouring intervals (Fig. 1b and 4a) show the

4	4	3,5	8,5	2	3	7	12	8		
2,5	3,5	6	2	1	11					
5,5	6	1,5								
1	3	5	13							
2,5	3	14	1	1,5	0,5	1	4	6	19	3,5
4	11,5	3,5	2	1,5	10,5	2	4			
1	2	2	1	4	4	9,5	1,5	7	14,5	2,5
6	1	4	4	6	6	4				
3	6	3	3	7	5	5	7			
3	4	5	3,5	3,5	4	6	3,5	3,5		
3	6	3,5	3,5	7	9	10,5				
1,5	2	1	3	4	3	2	4	2		
3	3	3	12	2	9	9				
8	6	5,5	15,5	9	7	2,5				
2	5	8	8	7						
2	4	5	8	4	5	3	3			
4	5	5	5							
3,5	5	10	4	15	4	6	2			
4,5	2									
3	2,5	5	9	1						
0,5	6,5	5,5	6,5	5,5	6,5	7,5	2			

Fig. 3. Sequences of intervals between two neighbouring panel-addition activities. Each row: Data of one larva. '76 LD. 21 larvae. The underlined numbers indicate: After that interval not one but two activities followed (within 12 hours).

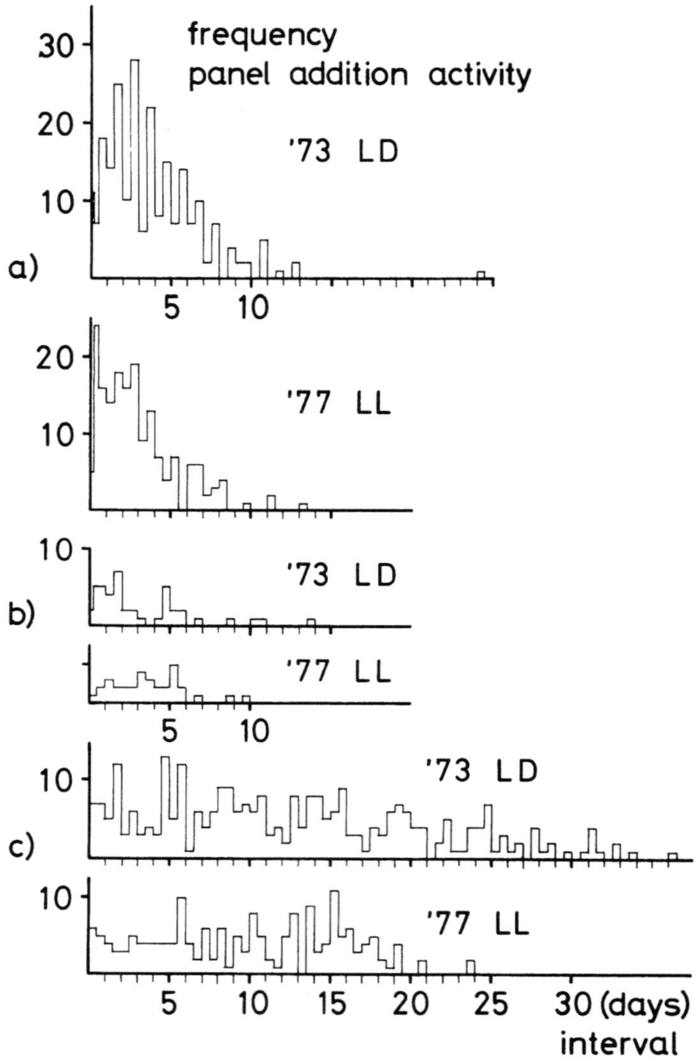

Fig. (4a). Histogram of all intervals between two neighbouring panel-addition activities during the 4th instar. Data of 39 ('73 LD) and 38 ('77 LL) larvae. (b). Histograms of the intervals only between the 1st and the 2nd panel-addition activity. Data of 42 ('73 LD) and 34 ('77 LL) larvae. c. Superposition histograms (all intervals between the 1st and all following panel-addition activities during the 4th instar) '77 experiment interrupted earlier. Data of 39 ('73 LD) and 38 ('77 LL) larvae.

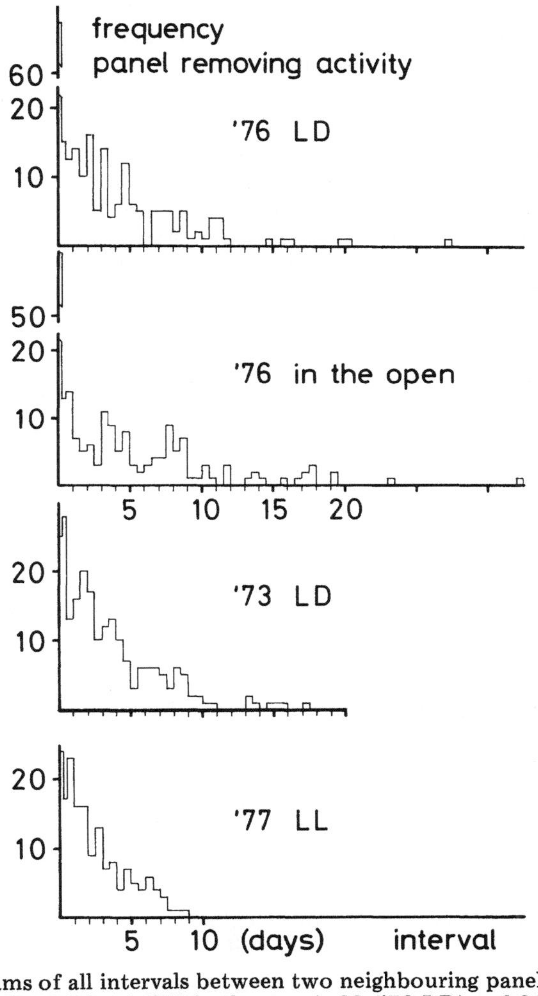

Fig. 5. Histograms of all intervals between two neighbouring panel-removing activities. Data of 23 ('76 LD), 21 ('76 in the open), 39 ('73 LD) and 31 ('77 LL) larvae. Correction of Fig. 7. p. 196 in ZINTL, H. 1976. House building: problems about the spontaneous change of the architectural style in the larva of *Potamophylax latipennis* (CURT.) (Trichoptera, Limnephilidae). — Proc. 1st int. Symp. Trich. 1974. Junk. The Hague.

In Fig. 7. the meanings of the black and white blocks were reversed by mistake and so the results were wrongly interpreted (p. 195 3. and p. 199-200). Therefore I wish to substitute the old Fig. 7. by a new one with the data of '76 additional and connected with a graph about the correlation between the times of the 4th-5th moult and the style change.

expected loss of whole multiples of intervals under LL.

The panel-removing activity under LD exhibits additional maxima (Fig. 5: '76 LD and in the open, '73 LD), especially visible in the histogram '76 in the open. These maxima seem to occur periodically.

Tests of single sequencies showed: '76, LD: 7 of 13 sequencies: periods of 1.5 or 2 or 2.5 days. '76, in the open: 7 of 8 sequencies: periods of 1.5 to 5.5 days, 3 of them of 4.5 days. '73, LD: 8 of 13 sequencies: periods of 1 to 7 days.

Panel-removing activity is characterized by little series with intervals much shorter than 0.5 days following much longer intervals. Histogram '77 LL (Fig. 5, below) shows a random distribution; 2nd maxima were seen no more.

Our team will continue the experiments in order to elucidate the mechanism triggering the distribution of building activities.

Acknowledgements

This work is dedicated to the nuns of Hohenburg, D-8172 Lenggries. Thanks are due to Mr H.J. VERMEHREN for confirmation of the identification of the larvae of *Potamophylax latipennis*, and to Dr F. THIMM and Dr J. WOLFFGRAMM for the PUSTA tests.

References

THORN, M.J. 1969. Behaviour of the caddis fly larva *Potamophylax stellatus* Curtis (Trichoptera). Proc. R. ent. Soc. Lond. (A) 44: 91-110.

WOLFFGRAMM J. 1975. Aufbau, Organisation und zentrale Steuerung variabler Lautmusterfolgen im Gesang des Rollerkanaris (*Serinus canaria* L.). Dissertation, Universität, D-7800 Freiburg.

—. & THIMM, F. 1976. Bearbeitergesteuerte Analyse von Verhaltenszeitreihen am Digitalrechner. Biol. Cybernetics 21: 61-78.

ZINTL, H. 1970. Zum Problem der Wahl des Baumaterials bei der Larve von *Potamophylax latipennis* (CURT.) Neb. (Trichoptera, Limnephilidae). Z. Tpsych. 27: 129-135.

—. 1974. House building: problems about the spontaneous change of the architectural style in the larva of *Potamophylax latipennis* (CURT.) (Trichoptera, Limnephilidae). Proc. 1st int. Symp. Trich: 187-201.

Discussion

NIELSEN: *Potamophylax latipennis* certainly does build cases of this type, but in my research area it is very unusual. In *P. nigricornis* it seems to be obligatory in a certain phase of larval development. In *P. latipennis* it may depend on the availability of dead leaves.

ZINTL: In order to build with panels, dead leaves or adequate material (in experiments discs of sheet) must be available. If dead leaves are lacking, larvae from our population make houses only with small particles. It is yet to be found out if *P. latipennis* larvae from sites where panel houses have never been seen are able to work with panels.

DENIS: Building activity properly consists of movement sequences, and in the end a new particle is not always added. Do you have ideas on the importance and arrangement of these incomplete sequences of building activity?

ZINTL: No, we have not observed the larvae more often than twice in 24 hours.

A correction and short addition to ZINTL, H. (1974): House building: problems about the spontaneous change of the architectural style in the larva of *Potamophylax latipennis* (CURT.) (Trichoptera, Limnephilidae). Proc. 1st int. Symp. Trich.: 187-201. Junk. The Hague.

In Fig. 7 (p. 196) of that paper the meanings of the black and white blocks were reversed by mistake and so the results were wrongly interpreted.
Fig. 7 corrected: Black blocks: start with sand grains. White blocks: start with leaf particles.
Interpretation corrected: P. 195, 3.: From the graphs of Fig. 7 we can gather that connected with short to middle lengths of the 4th instar during the time of ascending frequency and during the maximum of SCs change processes with leaf particles predominate. We might imagine that the shifting of the motivations in the central nervous system in insects with a shorter instar length is more rapid and leads to the building activity with leaf particles as a suboptimal material.
P. 200, above: . . . and the more frequent are changes not at once with sand grains. We suppose the shifting of the motivation then to be more advanced. For this case the hypothetical oscillation of the tendency to build with panels is damped more strongly, compared with conditions of a very long instar when the amplitudes of the tendency to build with panels decrease only slowly and SCs frequently only succeeded in using sand grains at once.

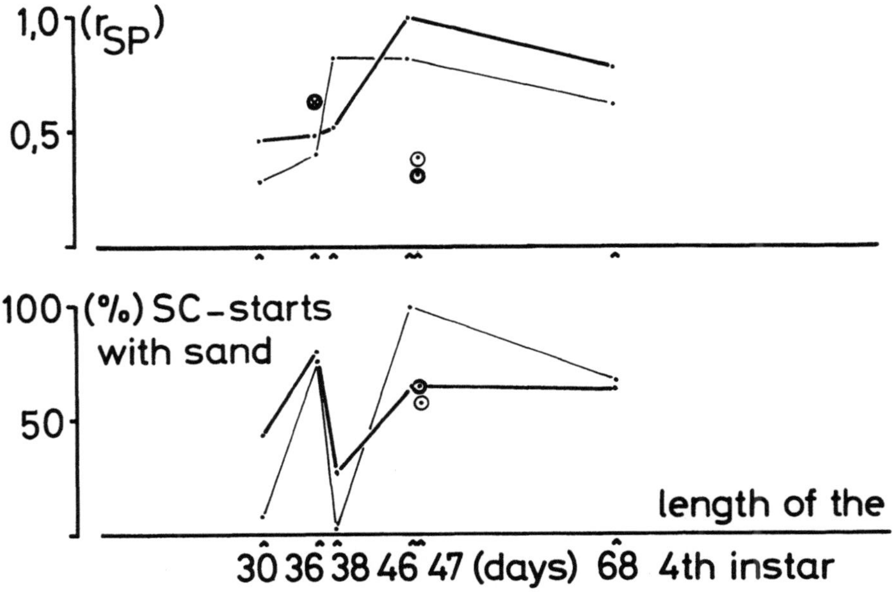

Correlation between the intervals of time '3rd-4th moult — 4th-5th moult' and '3rd-4th moult — style change'. Percentage of the style change starts with sand dependent on the length of the 4th instar. Ord.: above: SPEARMAN's correlation coefficient; below: percentage of the style change starts with sand. Absc.: medians of the 4th instar length. Thinly inked curve and circle: style changes ante 4th-5th moult; thickly inked curve and circle: style changes post 4th-5th moult. Circles with point: '76, experiment in the open. Circle with cross: '73, environment without sand.

Addition: In the meantime it turned out that there is not a simple relation between the frequencies of the style change starts with sand and the length of the 4th instar (see the Fig., graph below).

Strangey enough with a length of 36 days of the 4th instar (median, '73) a maximum of SC-starts with sand is already achieved and two days later (median 38 days, '67) lost again. Then with longer periods of the 4th instar only high magnitudes of SC-starts with sand follow.

That first maximum of SC-starts with sand is not connected with a strong correlation between 'length of the 4th instar' and 'interval of time 3rd-4th moult — style change (see the Fig., graph above).

We realize that there are many problems involved in these relations. We shall try to throw the light upon them in another paper.

Proc. of the 2nd Int. Symp. on Trichoptera, 1977, Junk, The Hague

Stream flow and the behaviour of caddis larvae

N.V. JONES, M.R. LITTERICK and R.G. PEARSON

Introduction

The conditions under which caddis larvae are living are different in different parts of streams and this would be expected to show in the composition of the caddis communities. Conditions at any one point may be variable, which may produce fluctuations in the species compositon of the fauna as well as in the density of the populations. If this is the case, do the larvae respond in any way that is dependent on the environmental changes? If they do, how and when do they do so? What are the implications of stream management practices to populations of caddis larvae?

The aim of this paper is to illustrate these questions and hopefully, to stimulate an interest in the movements of trichopteran larvae at the population level.

The data used come from two studies carried out on the River Hull in N. Humberside (Fig. 1). One of these (PEARSON, 1974) was concerned with a general study of the distribution of invertebrates in the river system and the other, LITTERICK (1973), studied the drifting and upstream movements (USM) of various invertebrates. Both were incomplete in their coverage of the Trichoptera but they do add to the questions raised by workers on drift in general (see WATERS, 1972), upstream migration and recolonization mechanisms (WILLIAMS & HYNES, 1976).

The R. Hull

The R. Hull originates as a series of springs emerging from the chalk of the Yorkshire Wolds (Fig. 1). It flows south for about 36 km and drains into the Humber at Kingston upon Hull. The river is used for a variety of purposes including water abstraction, navigation, sewage disposal, land drainage and recreation. The banks are raised for about half its length, with the surrounding low-lying land drained by a system of dykes. Stations 24, 22, 16, 14, 15 on Figure 1 are typical chalk streams whereas Stations 7, 5, 3, 1 are canalised and affected by the tide. Stations 12 a and b are above and below a weir and 19 and 20 are in a canal.

The channel is maintained by both dredging and weed-cutting activities. A small mill dam collapsed above Station 24 on August 7th 1971 and produced an unusual spate down the part of the stream under study at the time.

Methods

The distribution and density of invertebrates in the river were studied by taking

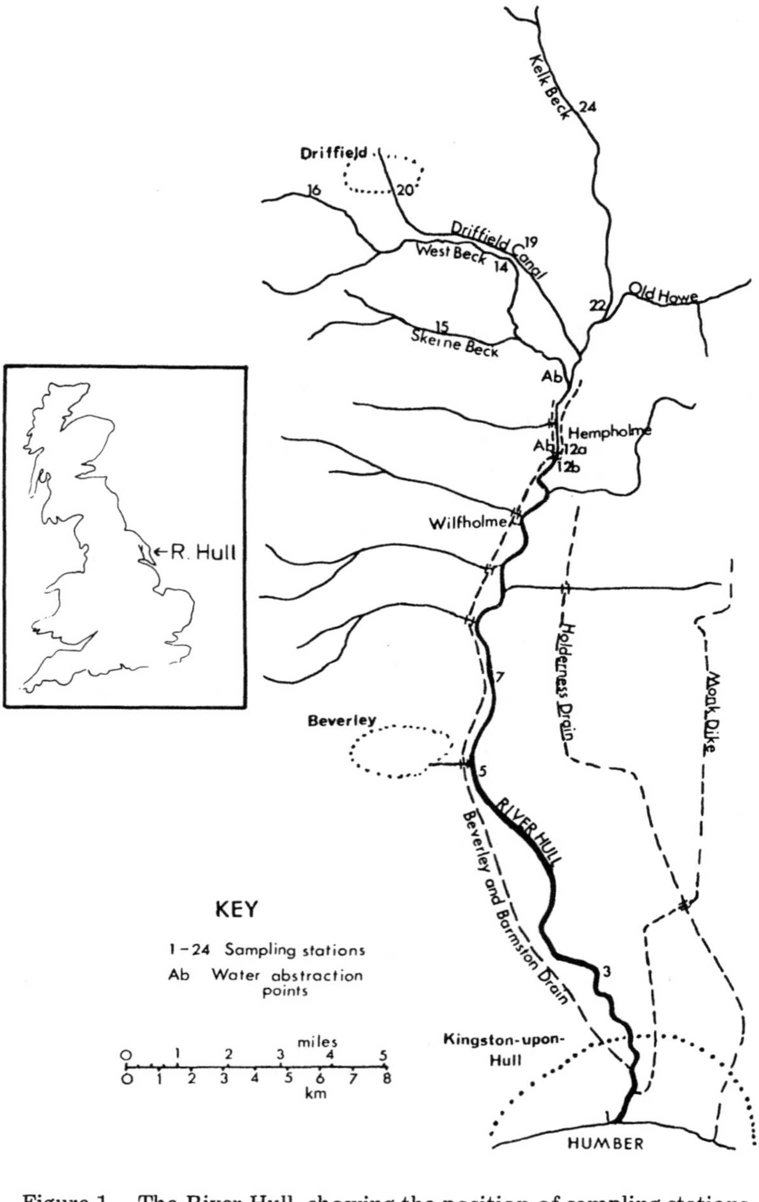

Figure 1. The River Hull, showing the position of sampling stations.

monthly samples with an air-lift (PEARSON, LITTERICK & JONE, 1973). The quantitative benthic samples associated with the drift and USM studies were taken with a Surber sampler.

Drifting animals were collected in drift nets similar to those described by ELLIOTT (1970). The animals moving upstream were collected in gravel-filled traps designed for the purpose (LITTERICK, 1973).

Table 1. The occurrence of Trichoptera at twelve sampling stations on the River Hull. x = < three records, xx = > three records during monthly collections, July 1970-September 1973.

Trichoptera	3	5	7	12a	12b	22	24	14	15	16	19	20
					-Stations-							
Leptoceridae				xx	xx	xx						
Molanna sp.					xx	xx						
Phryganeidae			x	x		xx						
Agapetus fuscipes (Curtis)	x			xx			xx	xx	xx	xx		
Silo nigricornis (Pictet)							xx	xx	xx	xx		
Melampophylax mucoreus (Hagen)							xx	xx	xx	xx		
Halesus radiatus (Curtis)				xx	xx	xx						
Other Limnephilidae	x		xx	xx	xx	xx		xx			xx	
Sericostomatidae				xx	xx	xx		xx			xx	
Rhyacophila dorsalis (Curtis)							xx	xx	xx	xx		
Hydroptilidae			xx	xx	xx	xx				xx	xx	
Polycentropus kingi McL.						xx		xx		xx	xx	
Holocentropus dubius (Ramb)			xx	xx	xx							

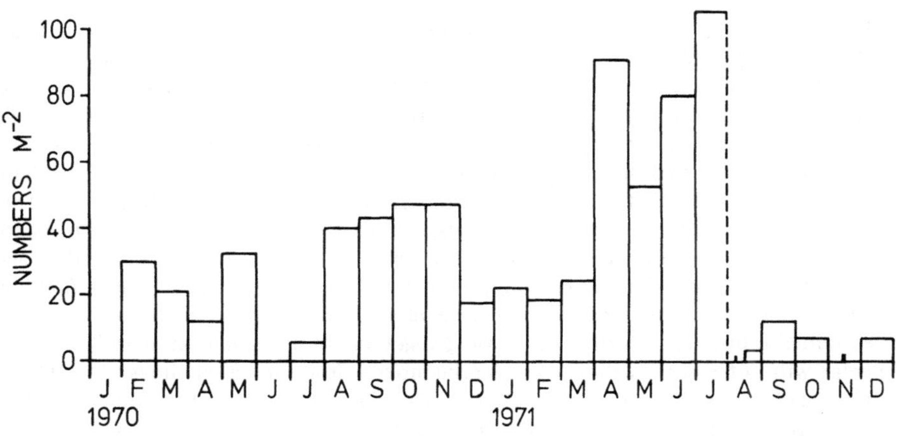

Figure 2. *Rhyacophila dorsalis*. Density of larvae in the benthos near Station 24, based on monthly samples over 2 years. No samples taken in January and June 1970. Broken line indicates time of spate due to collapse of dam (August 7th).

Results

A general picture of the distribution of the commoner Trichoptera can be seen from Table 1. An idea of the month to month and year to year variation in the population density of these insects at one station (12b) is shown in Table 2, which covers two years.

Rhyacophila dorsalis larvae also showed considerable variation in population density as is seen in Table 3 and Figure 2 which summarises the results of sampling the benthos at four sites (I-IV) near Station 24. This Table also shows the small contribution that this species makes to the drift and USM as well as the effect of the spate on the population (A11 & A17 = sampling dates).

Quantitative figures for the cased caddis larvae were not collected but Table 4 shows the collections made in the USM traps. The seasonal pattern of activity is apparent for all three species, being particularly obvious in the case of *M. mucoreus*. The effect of the spate in August 1971 is again apparent although it is possible that these species were pupating and emerging by this time and would, therefore, not be sampled by the USM traps.

The diurnal activity of these cased species was investigated by taking collections from the traps every two hours over a 24 hour period. The results, shown in Table 5, indicate that no cased caddis larvae appeared in the drift nets which were raised off the stream bottom, but that they were represented in the USM traps. Most of the activity of all three species was during the period of darkness but this was not so clearly regulated by changes in light intensity as the behaviour of most of the other invertebrates which were collected in the drift nets.

Discussion

Life in running water is clearly dominated by the strength of the current. This factor, directly and indirectly, affects the distribution of the various species as is reflected in Table 1. We often assume that the heavy cases of stone-cased species help the larvae to maintain station in the stronger currents that they inhabit. It is possible, however, that reliance on such a passive measure to counter the effect of the current would expose the species to the catastrophic effects of spates, particularly if they occur at the time when the larvae are in their early instars. The variability of numbers seen in Table 2 suggests that the populations are subject to frequent and large fluctuations, some of which may be due to environmental factors. Some behavioural measures may, therefore, be essential to give a more flexible strategy than that provided by the case alone. This may also be necessary to maintain the populations of species using vegetable building matter as well as those species without cases, as they may also be moved downstream at times (ELLIOTT, 1968; TOWNSEND & HILDREW, 1976). This flexibility could be achieved by the females flying upstream before laying eggs, as is suggested by MULLER (1954), but it could also be contributed to by the behaviour of the larvae.

It is clear from Tables 3, 4 and Figure 2 that *Rhyacophila dorsalis*, *Melampophylax mucoreus*, *Agapetus fuscipes* and *Silo nigricornis* populations are very susceptible to spates. Even allowing for emergence, the spate produced a dramatic drop in numbers of *R. dorsalis* that persisted until December at least, and caused this species to be caught in the drift nets immediately after the spate. These animals were still drifting out despite the fact that the water level had already dropped before the samples were taken. The lack of cased species in the USM collections following the spate may be, at least partially, attributed to the denuding effect of the unusual flow on the populations. The recovery of such populations after natural or man-made disturbances might be possible due to the larvae moving upstream as they are capable of such movements as is seen in Table 4.

These movements may, however, be relatively unimportant as they may depend on the species being present in the first place and may only cover limited distances. The seasonality of the catches is noticeable, with a general increase in upstream

Table 2. Numbers of Trichoptera m^{-2} estimated from five 0.04 m^{-2} samples in each case. Station 12b.a = 1971-1972 b = 1972-1973

	Jun		July		Aug		Sept		Oct		Nov		Dec		Jan		Feb		Mar		Apr		May	
	a	b	a	b	a	b	a	b	a	b	a	b	a	b	a	b	a	b	a	b	a	b	a	b
Limnephilidae	0	0	0	40	0	40	0	20	0	20			0	6	0	40	0	5					33	0
Sericostomatidae	463	15	100	10	50	0															17	0	33	0
Polycentropodidae	113	0	125	5	0	30	22	90	0	100	0	40	300	44	0	90	0	10	0	90	0	120	0	95
Hydroptilidae											0	5												
Leptoceridae											25	0			100	0	25	0	0	35	0	15		
Agapetus sp.							11	0	17	0							25	15						

Table 3. *Rhyacophila dorsalis.* Numbers of larvae collected by benthos sampling, drift nets and upstream migration traps. (1970-1971 Station 24). Benthos = Nos m^{-2}. Drift. USM = Nos 24 h^{-1}. ¦ = time of spate.

Month	Feb	Mar	Apr	May	Jun	Jul	Aug	Sep	Oct	Nov	Dec	Jan	Feb	Mar	Apr	May	Jun	Jul	A11	A17	Sep	Oct	Nov	Dec
BENTHOS																								
I	33	33	25	108	/	0	25	0	50	37	0	38	50	50	188	150	238	250	/	0	0	25	/	13
II	50	50	13	25	/	25	63	150	88	100	63	38	13	25	175	50	125	150	0	13	0	0	0	0
III	38	0	13	0	/	0	75	25	38	38	0	13	13	13	0	13	38	25	0	0	50	0	0	13
IV	0	0	0	0	/	0	0	0	13	13	0	0	0	13	0	0	0	0	0	0	0	0	0	0
Σ	121	83	51	133	/	25	163	175	189	188	63	89	76	101	363	213	401	425	/	13	50	25	/	26
DRIFT																								
I					/						2	4	2		2		2	8	32					
II	1				/						2				2									
III					/	8													2					
Σ	1	0	0	0	/	8	0	0	0	0	4	4	2	0	4	0	2	8	34	0	0	0	0	0
USM																								
I						0					1					1							1	
II						/	1																	
III																								
Σ						/	1	0	0	0	1	0	0	0	0	1	0	0	0	0	0	0	1	0

movement during March-June. This might be a reflection of an increase in general activity, including searching for food, during the main growth period of spring and early summer or be a specific movement prior to pupation. The larvae of *Silo* sp are more active during, or just after, the period of darkness as shown in Table 5. This diurnal pattern is not so clear with *Agapetus* and *Melampophylax* which were caught in smaller numbers. Most other invertebrates living in the stream show a night-active period, particularly with respect to drift. *Silo* adults are day-active (CRICHTON, 1976) which suggests that this is another example of a change in diel

Table 4. Numbers of three species of cased caddis larvae collected per 24 hrs in USM traps at Station 24 (1970-71) ¦ = time of spate.

Table 4A. MELAMPOPHYLAX MUCOREUS.

Month		Jul	Aug	Sep	Oct	Nov	Dec	Jan	Feb	Mar	Apr	May	Jun	Jul	A11	A17	Sep	Oct	Nov	Dec
TRAP	I	2	1	0	0	0	0	0	0	6	22	37	4	2	4	1	0	0	0	0
	II	/	0	0	0	0	0	0	0	52	112	69	7	0	0	3	0	0	0	0
	III	/	1	0	0	0	0	0	0	10	5	18	2	16	1	1	0	0	0	0
	Σ	/	2	0	0	0	0	0	0	68	139	124	13	18	5	5	0	0	0	0

Table 4B. AGAPETUS sp (fuscipes)

Month		Jul	Aug	Sep	Oct	Nov	Dec	Jan	Feb	Mar	Apr	May	Jun	Jul	A11	A17	Sep	Oct	Nov	Dec
TRAP	I	2	4	2	12	11	3	5	5	15	21	31	15	1	0	0	0	0	0	1
	II	/	5	3	8	17	6	6	15	66	45	17	8	4	0	0	0	0	1	0
	III	/	0	2	1	0	3	4	4	9	3	3	4	1	0	0	0	0	0	0
	Σ	/	9	7	21	28	12	15	24	90	69	51	27	6	0	0	0	0	1	1

Table 4C. SILO sp (nigricornis)

Month		Jul	Aug	Sep	Oct	Nov	Dec	Jan	Feb	Mar	Apr	May	Jun	Jul	A11	A17	Sep	Oct	Nov	Dec
TRAP	I	7	3	13	6	19	15	13	18	48	194	48	25	7	1	3	0	0	2	0
	II	/	2	6	5	17	15	9	43	55	120	29	15	10	0	0	0	0	1	1
	III	/	0	4	10	12	10	4	16	44	74	31	41	11	16	0	0	6	2	3
	Σ	/	5	23	27	48	40	26	77	147	388	108	81	28	17	3	0	6	5	4

Table 5. Drift and USM collections for all species. 16-17th May 1972. Numbers per two hour collection period. —— = Period of darkness

	DRIFT													USM												
Collection number	I	II	III	IV	V	VI	VII	VIII	IX	X	XI	XII	Σ	I	II	III	IV	V	VI	VII	VIII	IX	X	XI	XII	Σ
Silo sp.													0	5	1	6	12	16	12	36	38	35	15	6	10	192
Agapetus sp.													0	2	4	3	6	6	2	3	2	2	1	3	2	36
Melampophylax mucoreus													0		1	0	0	1	2	1	3	1				9
Total Trichoptera													0	7	6	9	18	23	16	40	43	38	16	9	12	237
Total of 13 invertebrate spp excluding Gammarus & Trichoptera	7	10	13	40	39	269	436	140	25	10	9	4	1002	0	0	0	1	0	2	4	6	1	4	4	5	27

activity between the larval and adult stages as observed by SOLEM (1976) in lake-dwelling caddis.

Man-made disturbances such as dredging, weed-cutting and river control resulting in unseasonal flow patterns may produce far reaching and relatively long term effects on caddis populations as well as on other species. Different species of caddis, as well as of other taxa, will employ different strategies to overcome these problems. Recolonisation can be achieved by drifting into affected areas, by upstream movement to such areas, by migration from within the substratum or by aerial stages as discussed by WILLIAMS & HYNES (1976). These authors quote their caddis populations as being mainly recruited from the drift and with smaller and about equal proportions, coming from upstream movement and aerial sources. This study was carried out on a silty stream in the summer, and the only Trichoptera named are *Hydropsyche* sp. which occurred mainly as small larvae.

It seems, then, that some stages of some trichopteran larvae are drifted downstream, even in relatively slow flowing areas. Other species move upstream, at least at some time of the year. Much of this behaviour is exhibited at night, although some evidence exists to suggest that some larvae are day-active (WATERS 1968). Adults have been observed to show a definite upstream flight prior to egg-laying in some studies (ROOS, 1957) but not in others (ELLIOTT, 1968). What are the relative importance of these behavioural mechanisms in maintaining populations of different types of Trichoptera under both natural and man-made stream conditions? Which stages, of which species, exhibit the various behaviour patterns and which stimuli control them?

References

CRICHTON, M.I. 1976. The interpretation of light trap catches of Trichoptera from the Rothamsted Insect Survey. Proc. 1st int. Symp. on Trich. 147-158.

ELLIOTT, J.M. 1968. The life histories and drifting of Trichoptera in a Dartmoor stream. J. Anim. Ecol. 37: 615-625.

—. 1970. Methods of sampling invertebrate drift in running water. Ann. Limnol. 6: 133-159.

LITTERICK M.R. 1973. The drifting and upstream movement of the macro-invertebrate fauna of a chalk stream. Ph.D. Thesis. University of Hull.

MÜLLER, K. 1954. Investigations on the organic drift in North Swedish streams. Rep. Inst. Freshwat. Res. Drottningholm. 35: 133-148.

PEARSON, R.G. 1974. A study of the macro-invertebrate fauna of the River Hull. Ph.D. Thesis. University of Hull.

—. , LITTERICK, M.R. & JONES, N.V. 1973. An air-lift for quantitative sampling of the benthos. Freshwat. Biol. 3: 309-315.

ROOS, T. 1957. Studies on upstream migration in adult stream-dwelling insects. l. Rep. Inst. Freshwat. Res. Drottningholm 38: 167-193.

SOLEM, J.O. 1976. A progress report on diel rhythmicity in Trichoptera. Proc. of 1st int. Symp. on Trich. 205-206.

TOWNSEND, C.R. & HILDREW, A.G. 1976. Field experiments on the drifting, colonization and continuous redistribution of stream benthos. J. Anim. Ecol. 45: 759-772.

WATERS, T.F. 1968. Diurnal periodicity in the drift of a day-active stream invertebrate. Ecology 49: 152-153.
——. 1972. The drift of stream insects. A. Rev. ent. 17: 253-272.
WILLIAMS, D.D. & HYNES, H.B.N. 1976. The recolonisation mechanisms of stream benthos. Oikos 27: 265-272.

Discussion

MACKAY: In Canada I have found a caddisfly larva whose sand-grain case clearly offers no protection from catastrophic drift. During the spring thaw when high discharge in small woodland streams causes catastrophic drift, about 60% of a population of *Pycnopsyche gentilis* (Limnephilidae) may be washed downstream. At this time the 5th-instar larvae are in cases made of beech leaves. After the thaw, larvae begin to transform the leafy case into one of sand grains. I have interpreted this behaviour as protection against leaf-chewing detritivores at a time when few leaves remain in the stream (having been flushed out) and beech leaves are beginning to decompose and are thus becoming palatable (see Can. Ent. 1972 and Ecology 1973).

MORETTI: Is there a relationship between caddisfly movements and those of the other groups?

JONES: To my knowledge, such interactions between populations have not been investigated.

Proc. of the 2nd Int. Symp. on Trichoptera, 1977, Junk, The Hague

Concepts of evolutionary ecology in Nearctic Trichoptera

ROSEMARY J. MACKAY & G.B. WIGGINS

Abstract
There are 144 genera of Nearctic Trichoptera compared with 60 Ephemeroptera, 82 Odonata, and 87 Plecoptera. Trichoptera have the broadest range in habitat and trophic category; the distribution of genera in the Eastern Deciduous Forest Biome shows a good fit to expected numbers according to a model showing functional aspects of an eastern stream ecosystem.

Filter-feeding Trichoptera partition fine particulate organic resources. In the Hydropsychidae, Arctopsychinae filter large particles in mountain streams; Macronematinae filter small particles in large rivers; Hydropsychinae are generalists but are under-represented in the western montane region. The Polycentropodidae feed mainly on fine particles in the East; the only two genera in the western montane region are predacious. The Philopotamidae feed on tiny particles and are uniformly distributed in the East and West.

One-third of Nearctic Trichoptera genera are Limnephilidae. Apataniinae, Neophylacinae, Goerinae and most Dicosmoecinae are confined to lotic habitats where they graze on rock surfaces. Limnephilinae and Pseudostenophylacinae are shredders, and are more diversified ecologically, especially in lentic habitats including transient waters. In the upstream areas of western montage regions, shredder/collector and grazer genera of limnephilids are equally common; grazers are fewer in the East. The evolution of western Trichoptera appears to have been influenced by food resources available in more open streams in largely coniferous forest.

(This paper will be published in *Ecology*, late 1978).

Proc. of the 2nd Int. Symp. on Trichoptera, 1977, Junk, The Hague

Ecological aspects of life history in some net-spinning Trichoptera

A.G. HILDREW

Introduction

Theoretical ecologists have shown that species coexisting stably in the same community may not share the same limiting factors (LEVIN, 1970). The minimum distinction which must occur between two resource-limited species involves differences along niche dimensions (HUTCHINSON, 1957) which determine their use of those resources. The observation of ecological differences in a guild (ROOT 1967) of species is often taken to infer that by such differences competition is avoided and that they serve to divide up a limiting resource. WILLIAMSON (1972) has clearly indicated the weakness of this approach; differences between coexisting species are not necessarily associated with the avoidance of competition now or in the past, nor with limiting resources, though they can indicate situations worthy of further investigation.

A number of workers have made observations on coexisting species of net-spinning caddis larvae. WALLACE (1975) and WALLACE et al. (1977) have stressed that species partition food resources by means of different net-mesh sizes, microdistribution patterns and temporal variations in net construction. WILLIAMS & HYNES (1973) showed distributional differences between a philopotamid and two hydropsychid species coexisting in a single river. EDINGTON (1968) stressed the role of separate flow preferences between *Hydropsyche siltalai* DÖHLER and *Plectrocnemia conspersa* (CURTIS). OSWOOD (1976) found that coexisting Hydropsychidae larvae showed interspecific differences in life history.

From these and other studies two resources emerge as possibly limiting for populations of hydropsychid larvae. WALLACE et al. (1977) postulate that interspecific competition for food has led to partitioning of particles on the basis of their size. ELLIOTT (1968) however, suggests that competition for net-spinning sites in *H. siltalai* may lead to density dependent drifting. The emphasis of the present paper is on the temporal separation of *Hydropsyche siltalai* DÖHLER, *H. instabilis* (CURTIS) and *H. pellucidula* (CURTIS). The pair *H. siltalai* and *H. pellucidula* are probably the most commonly coexisting hydropsychid species in Britain. In the system dealt with here, *H. siltalai* coexists with *H. pellucidula* in most of the main river, while *H. instabilis* occupies the major tributaries with *H. siltalai* (EDINGTON & HILDREW, 1973).

Methods

Most of the observations described here were made on the River Usk, South Wales and on tributary streams. The Usk is a typical salmonid river, unpolluted over most of its length. Two major stations were chosen for study, one on the Caerfanel tributary (Nat. Grid ref. SO 060174) and the other on the Usk itself at Llangynidr Bridge (Nat. Grid ref. SO 152203). Other stations and rivers are described as appropriate in the results section.

The information on life histories was obtained by taking samples from rapids with a long net (mesh size 0.44 mm). Samples were normally in excess of 50 larvae. All samples were preserved in Pampel's fluid in the field.

Quantitative samples were taken with a Surber Sampler of 25 cm side (1/16 m^2) and a net mesh size of 0.44 mm. Any stream mosses falling within the sample area were stripped from the stones and placed in the net. After removal of larvae in the laboratory, the moss was dried to constant weight at 60°C.

Larvae were identified to species and assigned to instar (ELLIOTT, 1968; HILDREW & MORGAN, 1974). First instar larvae, however, are unidentifiable to species. Pupae were identified by means of their associated larval fragments.

The colonisation of artificial surfaces was followed in the field. Pieces of roofing slate (22 × 36 cm) with plastic artificial grass stuck on the upper surfaces were placed in rapids sections. This material has been used as suitable for net-spinning in an experimental study by PHILIPSON (1969). The slates were fixed to the stream bed by long nails and jammed between large rocks. After two weeks exposure the larvae were carefully removed in the field and preserved.

Results

The instar distribution of *H. instabilis* at Caerfanel and *H. siltalai* and *H. pellucidula* at Llangynidr Bridge is given in Fig. 1. The samples were taken in 1970/71 though others taken in 1976/77 revealed no major differences. For each sample the mean larval instar was calculated (excluding instar 1) after BENKE (1970). The life history of each species shows a simple one year pattern, probably with a long flight period. However, *H. instabilis* and *H. siltalai* overwinter mainly as 3rd instar larvae and grow rapidly to full size the following spring, whereas *H. pellucidula* overwinters mainly as 5th instar larvae. In all species there appears to be a winter pause in growth although some increase in weight of *H. pellucidula* larvae may occur (P.J. BOON personal communication).

The life histories of *H. siltalai* and *H. pellucidula* are both modified down the length of the river. Samples were taken on four occasions in 1969/70 from three successive downstream stations on the Usk system: Caerfanel, Llangynidr Bridge and Usk Weir. The results show that the mean instar of *H. siltalai* in the tributary (Station 1 Fig. 2) is always lower than at the river stations (Stations 2 and 3 Fig. 2). The difference between these latter two is, however, slight. These sampling intervals failed to reveal a similar pattern for *H. pellucidula* at the same stations. However, an undergraduate project student at Queen Mary College, London (supervised by P.J.

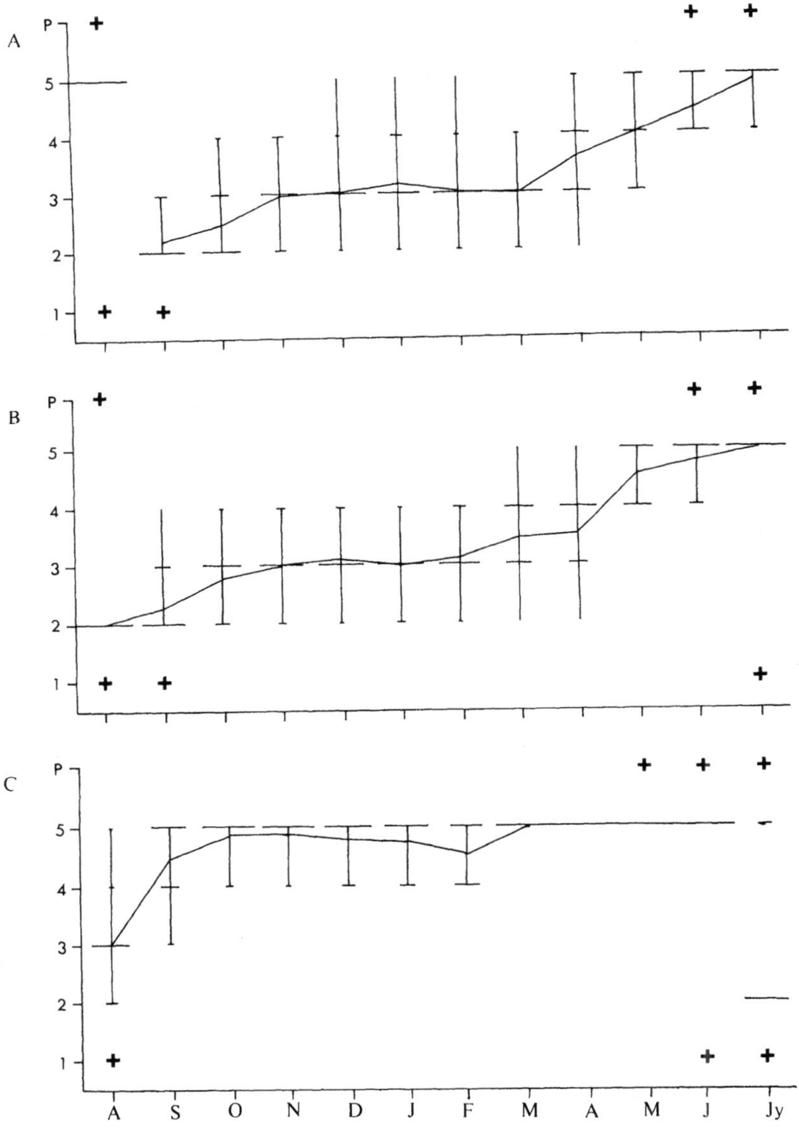

Fig. 1. The life histories of A. *H. instabilis* B. *H. siltalai* and C. *H. pellucidula*. The vertical lines give the range of instars and the horizontal lines the proportion of each instar present in the sample. The mean instar points are joined up for consecutive samples of a cohort. Crosses denote presence of pupae and first instar larvae.

BOON and myself) worked on a similar system, the River North Tyne, in north east England. His samples were taken approximately fortnightly from 26 July to 24 September 1975. His results (MACDONALD, 1976) are given in Fig. 3 and show an acceleration of growth from upstream (Kielder) to the most downstream station (Dunkirk).

Samples from Caerfanel (*H. instabilis* and *H. siltalai*) and from Llangynidr Bridge (*H. siltalai*) taken at fortnightly intervals from April to June 1977 show that the mean instar of *H. siltalai* is greater from the river station (Llangynidr Bridge) than from the stream station (Caerfanel). In this case however, samples were taken separately from stream mosses (mainly *Fontinalis antipyretica*) on the tops of stones and from underneath stones. Clearly the mean instar at any one time is greater below stones than in the moss, both for *H. siltalai* and *H. instabilis* in the stream, and for *H. siltalai* in the river (Fig. 4).

Quantitative samples of hydropsychid larvae were taken from the Usk at Llagynidr Bridge in April 1970. The relationship between the density of larvae of *H. siltalai* and *H. pellucidula* and the dry weight of moss is given in Fig. 5. There is a highly significant correlation ($r = 0.84$ $p < 0.001$, $n = 28$) for *H. siltalai* but not for *H. pellucidula*. In Fig. 5 are also plotted points from similar samples of *H. siltalai* taken from the neighbouring River Wye at Newbridge.

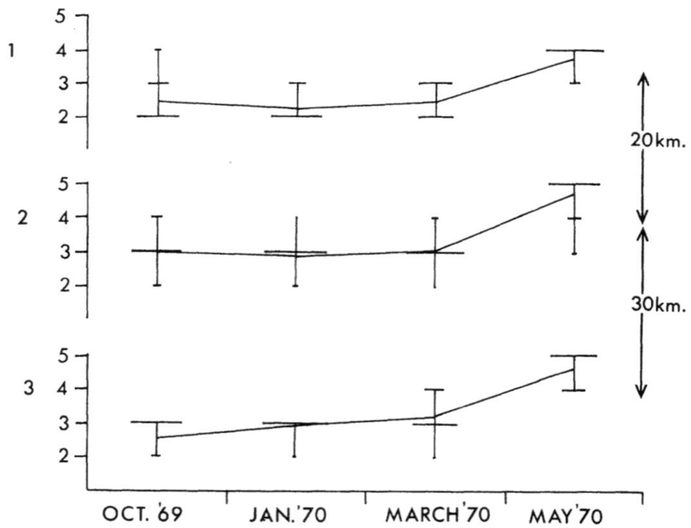

Fig. 2. The seasonal instar distribution of *H. siltalai* at three stations on the Usk system. The range, proportional representation and mean instars in each sample are given as in Fig. 1. The figures on the right indicate the intervening distance between successive downstream stations 1-3.

This procedure was repeated in the tributary stream at Caerfanel in spring 1977. Total numbers of larvae in ten samples and numbers of *H. instabilis* and *H. siltalai* separately, are plotted against moss weight in Fig. 6. There are significant correlations for total numbers ($r = 0.88$, $p < 0.001$), for numbers of *H. siltalai* ($r = 0.89$, $p < 0.001$) and for numbers of *H. instabilis* ($r = 0.67$, $p < 0.05$ with moss weight.

Colonisation experiments on artificial substrates were carried out at fortnightly intervals between 21 April and 11 June 1977 at Caerfanel and Llangynidr Bridge. At least 4 slates were exposed for colonisation on each occasion. At Caerfanel both *H. instabilis* and *H. siltalai* appeared on the slates whereas at Llangynidr Bridge *H. siltalai* colonised them but *H. pellucidula* did not. The total numbers colonising in two week intervals, and separate figures for *H. siltalai* and *H. instabilis*, are given in

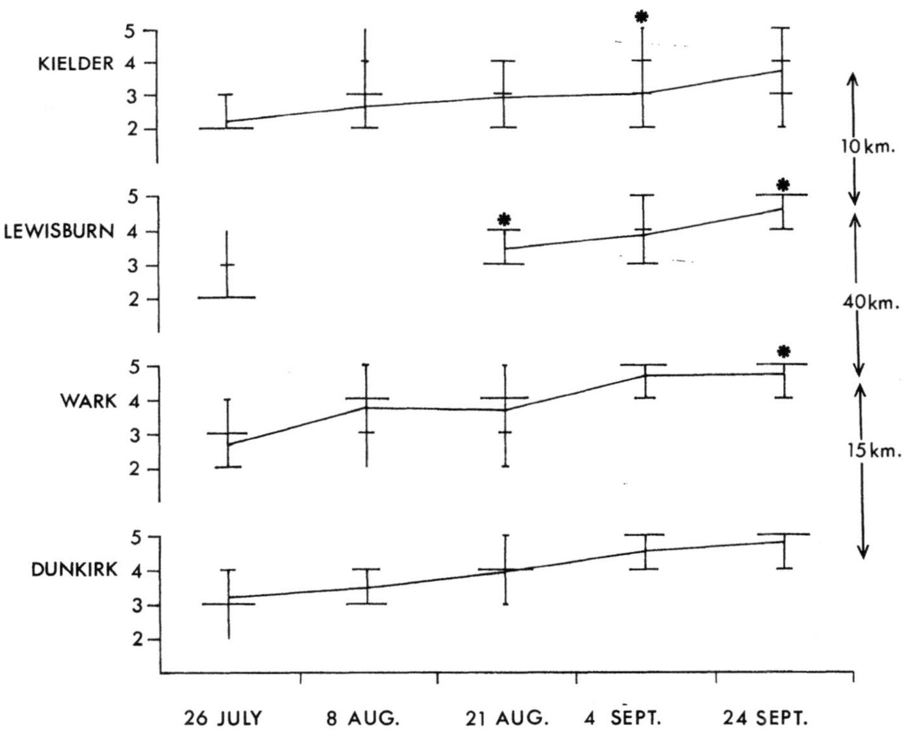

Fig. 3. The instar distribution of *H. pellucidula* in the River North Tyne. The range, proportional representation and mean of instars in each sample are given as in Fig. 1. The figures on the right indicate the intervening distance between successive downstream stations (Kielder — Dunkirk). Asterisks show where samples were smaller than twenty larvae.

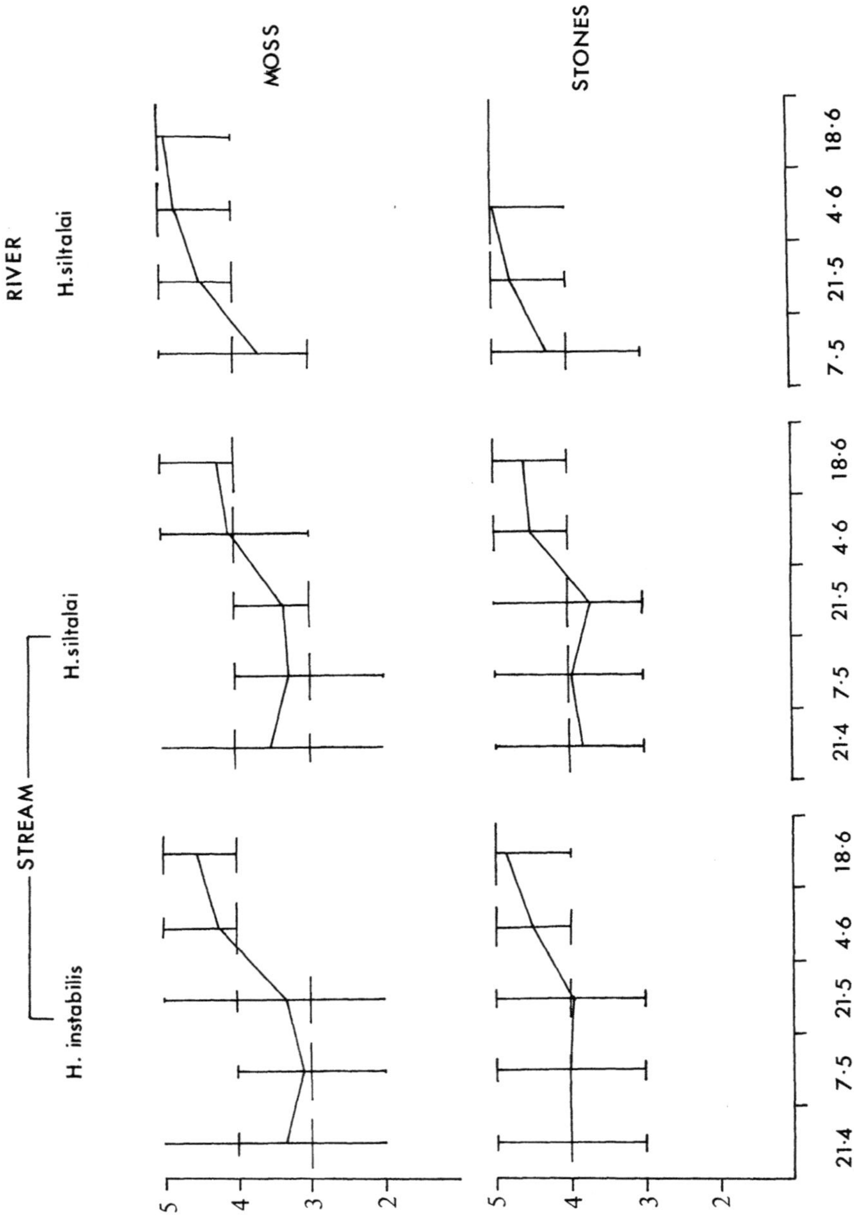

Fig. 4. The instar distribution of *H. siltalai* from the Usk system in a series of samples from moss and from under stones. The range, proportional representation and mean of instars in each sample are given as in Fig. 1.

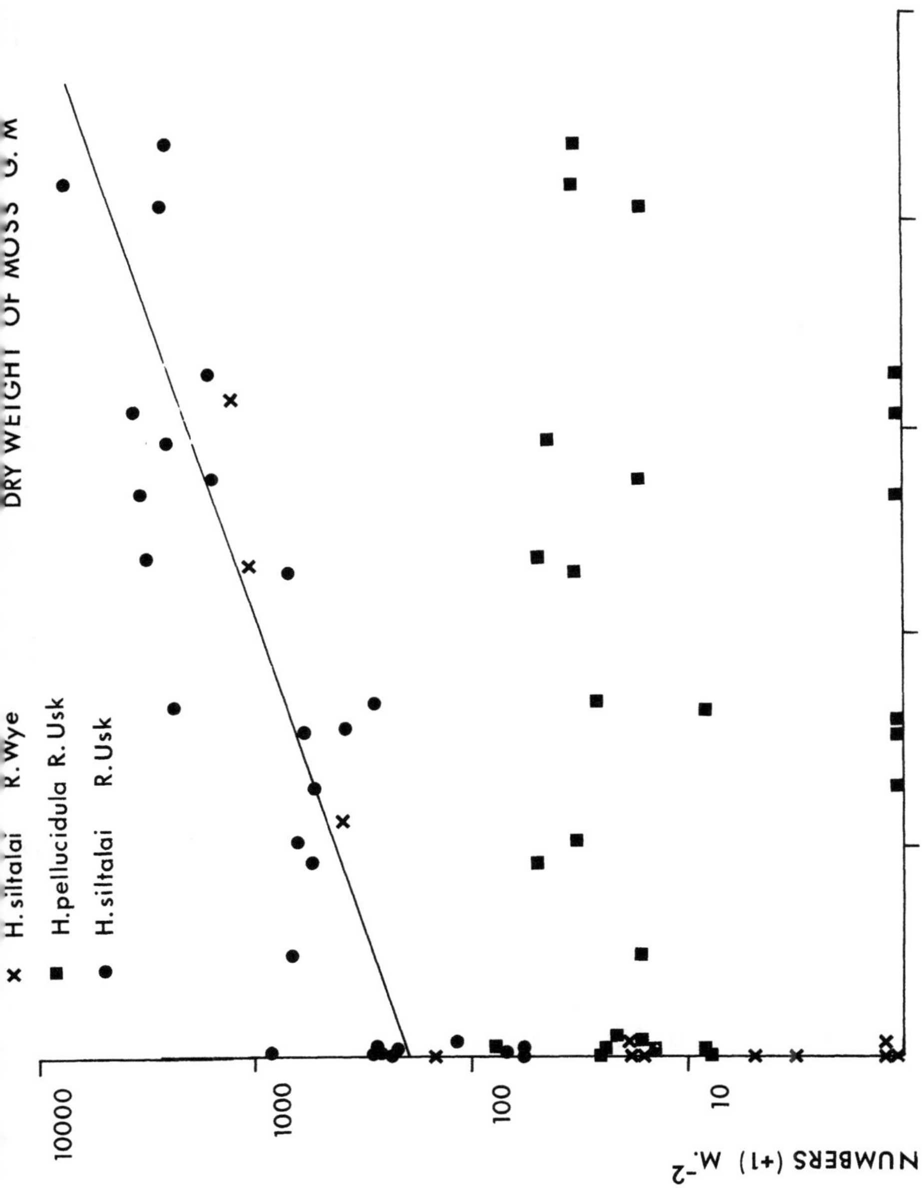

Fig. 5. The relationship between abundance of *Hydropsyche* larvae (numbers + 1 m^{-2}) and dry weight of moss (g, m^{-2}). The regression line is fitted by least squares for *H. siltalai* from the River Usk. Equation of the line is $y = 2.3 + 0.007x$ (where $y = \log_{10}$ numbers of larvae m^{-2} and x = dry weight of moss m^{-2}).

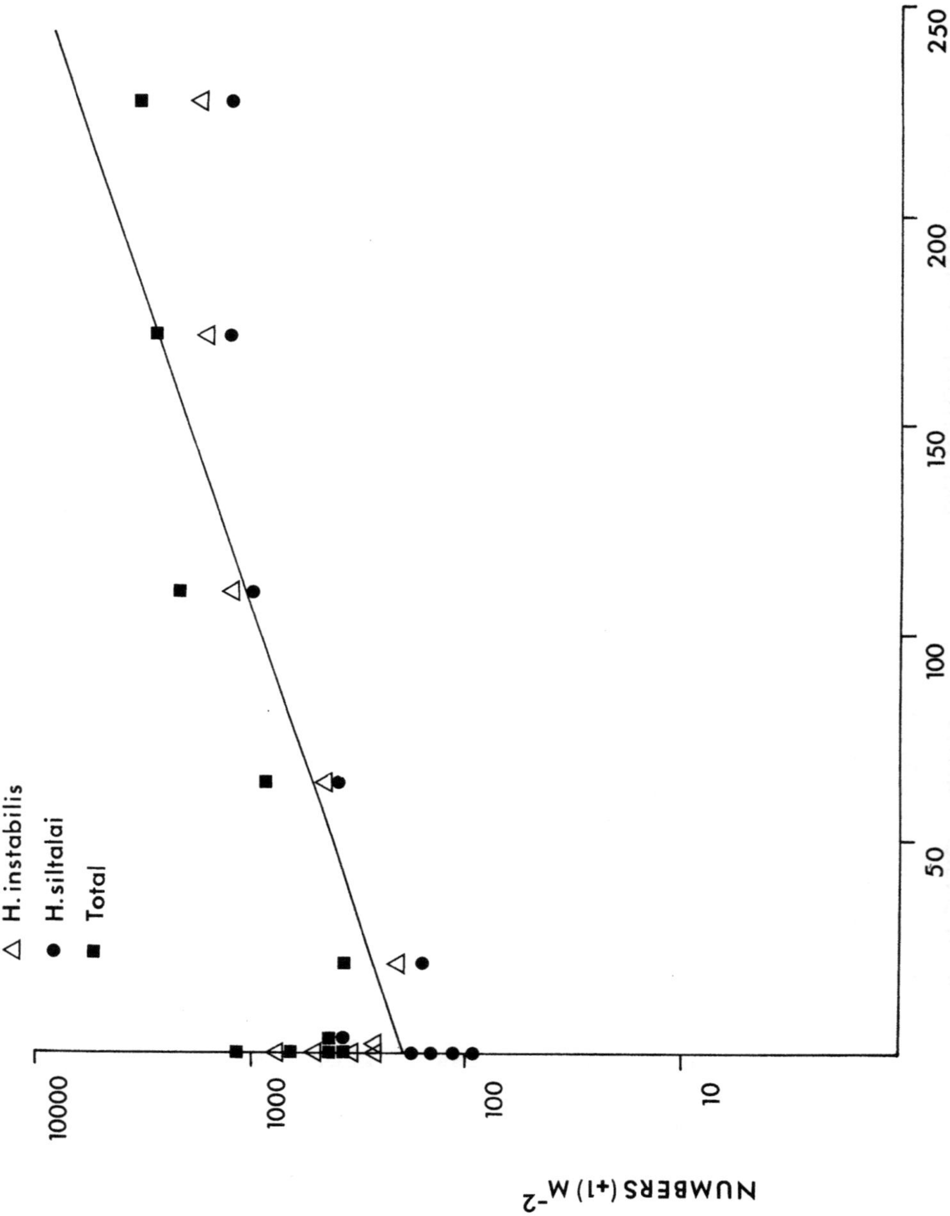

Fig. 6. The relationship between abundance of *Hydropsyche* larvae (numbers + 1 m^{-2}) and dry weight of moss (g. m^{-2}) in the Caerfanel tributary stream. The line plotted is the regression line for *H. siltalai* from the main river (see Fig. 5).

Fig. 7c, b and a. These may be compared with the numbers of *H. siltalai* colonising at Llangynidr Bridge (Fig. 7d). The total numbers of larvae colonising at Caerfanel (two species) is very similar to the numbers of *H. siltalai* alone at Llangynidr Bridge. There is a progressive reduction in the numbers colonising over the experimental period at both stations.

Discussion

The life histories of the three species (Fig. 1a, b and c) are in broad agreement with other descriptions of hydropsychid life cycles (e.g. ELLIOTT, 1968; HYNES, 1961). There are marked differences between *H. pellucidula*, which grows fastest in late summer, and the other two, which grow fastest in spring. An acceleration of life histories downstream and at lower altitudes (Fig.s 2, 3 and 4) is well known for

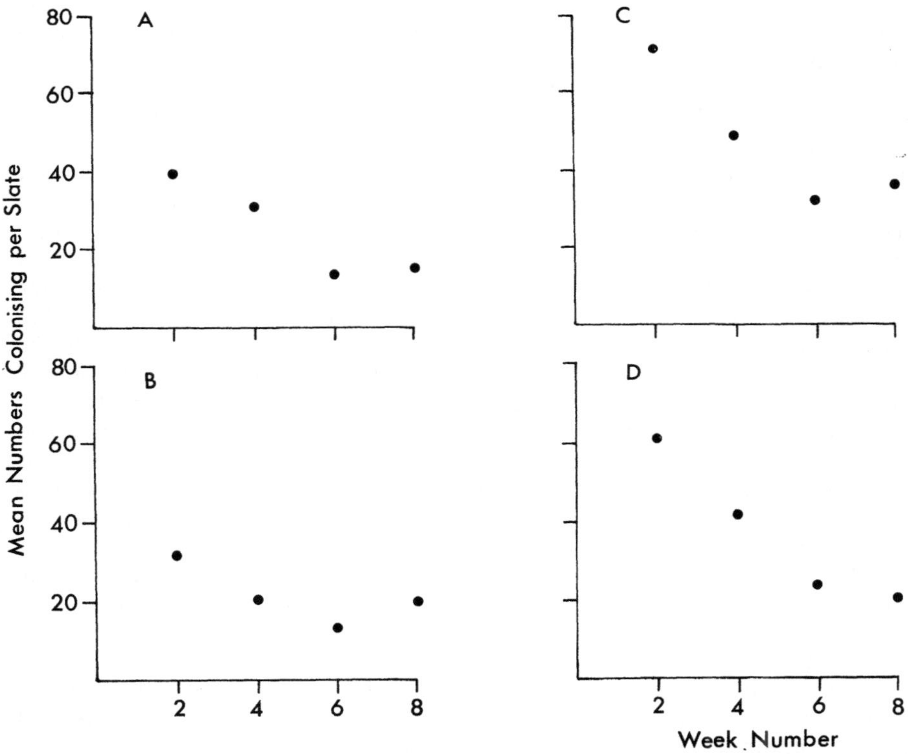

Fig. 7. The numbers of *Hydropsyche* larvae colonising artificial moss in two week intervals, A. *H. instabilis* at Caerfanel, B. *H. siltalai* alone at Caerfanel, C. total numbers at Caerfanel, D. *H. siltalai* alone at Llangynidr Bridge.

many insects (HYNES, 1970), including Hydropsychidae (e.g. MARINCOVIĆ-GOSPODNETIĆ, 1961). This is almost certainly an effect of temperature. The man River Usk has daily maxima in summer as high as those in the Caerfanel tributary but much higher daily minima (unpublished data). Temperature, (see BOON & SHIRES, 1976), similarly could account for the pattern of growth of *H. pellucidula* in the North Tyne. The difference in larval size between moss and stones is less easily explained. KOWNACKA (1971) found young *Rhyacophila* larvae in moss, and there are many other examples of changes in habitat with larval growth (e.g. CUMMINS, 1964). In the case of hydropsychid larvae I suggest it is a result of a limited supply of net-spinning sites.

Life history differences might act as an ecological isolation mechanism for co-existing hydropsychids limited by food supply, since larvae of lower instars spin nets of smaller mesh sizes (WILLIAMS & HYNES, 1973; WALLACE, 1975). However, the evidence from the Usk system is that *H. siltalai* and *H. instabilis* are limited by net-spinning sites. The evidence for a spatial limitation has hitherto been based upon the observation of stridulation by larvae in agonistic encounters (JOHNSON, 1964), spacing behaviour (GLASS & BOVBJERG, 1969) and possible density dependent drift at times of maximum growth (ELLIOTT, 1968).

Several authors have noticed the use of stream mosses for net-spinning by hydropsychid larvae (SPRULES, 1947; TANAKA, 1968; EDINGTON, 1968; WILLIAMS & HYNES, 1973). The density of *H. siltalai* larvae at Llangynidr Bridge is clearly related to the amount of moss present (Fig. 5). Stream mosses are a seasonally available resource; low summer flows frequently leave much of it above the water and it is further eroded by spates in winter. Its growth in spring coincides with the growth of *H. siltalai*. The larvae mostly overwinter under stones, and it may be that interference competition for net-spinning sites under stones in spring leads to the exclusion of the smallest of them into the drift. This would coincide with the peak in drift of *H. siltalai* in a Dartmoor stream observed by ELLIOTT (1968). It would also have the consequence that larvae smaller than the mean size under the stones should be most likely to colonise the moss. This is clearly the case (Fig. 4). The samples of *H. siltalai* from the River Wye showed that in quadrats with no moss they were much less common than in the Usk. However, in the three samples containing moss the points fall almost exactly on the regression line for the Usk samples (Fig. 5). I suggest that differences in numbers between the two rivers in moss-free samples are due to differences in rock shape and substrate stability as this influences net-spinning sites. In samples where substrates are comparable (i.e. where there is moss) abundances are comparable.

H. pellucidula does not grow rapidly in the spring, is not limited by moss supply (Fig. 5) and therefore does not participate in the colonisation of natural moss or artificial substrates. Life history differences between *H. siltalai* and *H. pellucidula* may therefore be one important aspect of niche differentiation. No such differences are shown by *H. siltalai* and *H. instabilis* coexisting in the Caerfanel (Fig. 1a and b). Both appear limited by moss supply (Fig. 6) and share in the colonisation of artificial moss (Fig. 7). In fact, they appear to act as one species in their use of this resource. There is some evidence that their coexistence, unlike that of *H. siltalai* and *H. pellucidula*, is unstable. In 1969, the proportion of *H. siltalai* in the Caerfanel was only about 8%, in 1977 it was nearly 50% (Fig. 6). Coexistence probably depends upon such factors as random environmental fluctuations which may temporarily reverse any competitive advantage of one over another (HUTCHINSON, 1961) or on recolonisation of headwaters from downstream by *H. siltalai.*

Summary

1. All three species described have simple one year life cycles. The two which most commonly coexist, overwinter at different stages (*H. pellucidula* as 5th instars, *H. siltalai* as 3rd instars).
2. The life cycles of *H. siltalai* and *H. pellucidula* are accelerated at downstream stations (probably due to environmental temperature).
3. *H. siltalai* and *H. instabilis* both use stream mosses for net-spinning in spring. The mean larval instar of both species from moss is smaller than that of larvae under stones.
4. The abundance of *H. siltalai* and *H. instabilis* is significantly related to the weight of moss present.
5. Only *H. instabilis* and *H. siltalai* colonised artificial moss, *H. pellucidula* did not. The total numbers colonising where *H. instabilis* and *H. siltalai* lived together were similar to the numbers of *H. siltalai* colonising where it lived alone.
6. It is postulated that the density of hydropsychid larvae in these situations is limited by competition for net-spinning sites. The contribution of life history differences to the stable coexistence of *H. siltalai* and *H. pellucidula* is discussed.

Acknowledgements

I should like to thank my wife PAMELA for her help with the field work. Dr. MIKE WINTERBOURN and Dr. JOHN EDINGTON kindly read the manuscript. Thanks are due to Professor N.B. MARSHALL for provision of facilities at Queen Mary College. This work was partly financed by a N.E.R.C. research studentship and by a N.E.R.C. research grant to Dr. J.M. EDINGTON.

References

BENKE, A.C. 1970. A method for comparing individual growth rates of aquatic insects with special reference to the Odonata. Ecology 51: 328-331.

BOON, P.J. & SHIRES, S.W. 1976. Temperature studies on a river system in north-east England. Freshwat. Biol. 6: 23-32.

CUMMINS, K.W. 1964. Factors limiting the distribution of the caddis flies *Pycnopsyche lepida* (HAGEN) and *Pycnopsyche guttifer* (WALKER) in a Michigan stream (Trichoptera: Limnephilidae). Ecol. Monogr. 34: 271-295.

EDINGTON, J.M. 1968. Habitat preferences in net-spinning caddis larvae with special reference to the influence of water velocity. J. Anim. Ecol. 37: 675-692.

—. & HILDREW, A.G. 1973. Experimental observations relating to the distribu-

tion of net-spinning Trichoptera in streams. Verh. Internat. Verein. Limnol. 18: 1549-1558.
ELLIOTT, J.M. 1968. The life histories and drifting of Trichoptera in a Dartmoor stream. J. Anim. Ecol. 37: 615-625.
GLASS, L.W. & BOVBJERG, R.V. 1969. Density and dispersion in laboratory populations of caddisfly larvae (*Cheumatopsyche*, Hydropsychidae). Ecology 50: 1082-1084.
HILDREW, A.G. & MORGAN, J.C. 1974. The taxonomy of the British Hydropsychidae (Trichoptera). J. Ent. (B) 43: 217-229.
HUTCHINSON, G.E. 1957. Concluding remarks. Cold Spring Harbor Symp. Quant. Biol. 22: 415-427.
—. 1961. The paradox of the plankton. Am. Nat. 95: 137-145.
HYNES, H.B.N. 1961. The invertebrate fauna of a Welsh mountain stream. Arch. Hydrobiol. 57: 344-388.
—. 1970. The ecology of running waters. Liverpool University Press.
JOHNSTONE, G.W. 1964. Stridulation by larval Hydropsychidae. Proc. R. ent. Soc. Lond. (A) 39: 146-150.
KOWNACKA, M. 1971. The bottom fauna of the stream Sucha Woda (High Tatra Mts.) in the annual cycle. Acta hydrobiol. Krakow 13: 415-438.
LEVIN, S.A. 1970. Community equilibria and stability, and an extension of the competitive exclusion principle. Am. Nat. 104: 413-423.
MACDONALD, R.A. 1976. Distribution and life history of *Hydropsyche* species, in the River North Tyne, Northumberland. Unpublished honours project, Queen Mary College, London.
MARINKOVIĆ-GOSPODNETIĆ, M. 1961. The dynamics of populations of *Hydropsyche fulvipes* CURTIS and *Hydropsyche saxonica* McLACHLAN (Trichoptera), Godisnjak biol. Inst. Saraj. 14: 15-84.
OSWOOD, M.W. 1976. Comparative life histories of the Hydropsychidae (Trichoptera) in a mountain lake outlet. Am. Midl. Nat. 96: 493-497.
PHILIPSON, G.N. 1969. Some factors affecting the net-spinning of the caddis fly *Hydropsyche instabilis* CURTIS (Trichoptera, Hydropsychidae) Hydrobiologia 34: 369-377.
ROOT, R.B. 1967. The niche exploitation pattern of the blue-gray gnatcatcher. Ecol. Monogr. 37: 317-349.
SPRULES, W.M. 1967. An ecological investigation of stream insects in Algonquin Park, Ontario. Univ. Toronto Stud. biol. 56: 1-81.
TANAKA, H. 1968. The aggregation of net-spinning caddisworms in the drainage ditch of fish-farm ponds. Bull. Freshwat. Fish. Res. Lab., Tokyo 18: 71-79.
WALLACE, J.B. 1975. Food partitioning in net-spinning Trichoptera larvae: *Hydropsyche venularis, Cheumatopsyche etrona*, and *Macronema zebratum* (Hydropsychidae). Ann. ent. Soc. Am. 68: 463-472.
—. WEBSTER, J.R. & WOODALL, W.R. 1977. The role of filter feeders in flowing waters. Arch. Hydrobiol. 74: 506-532.
WILLIAMS, N.E. & HYNES, H.B.N. 1973. Microdistribution and feeding of the net-spinning caddisflies (Trichoptera) of a Canadian stream. Oikos 24: 73-84.
WILLIAMSON, M. 1972. The analysis of biological populations. London. Edward Arnold.

Discussion

BADCOCK: Did you have any evidence of migration of larger larvae of *Hydropsyche siltalai* from moss to stones? When I was working on populations of *H. siltalai* larvae from moss and stones in the River Allender, Scotland, I interpreted my data as showing that while early instars tended to collect in moss, there was migration of larger larvae from moss to stones, where there was a preponderance of fifth instar larvae, and pupation was mainly under stones.

HILDREW: Fifth instar larvae, at maturity, must search for pupation sites under stones; however, my observations do not relate to this. I feel it is unlikely that many larvae overwinter on moss since it is badly eroded and diminished during winter. Thus new moss must be populated (probably by drift) in the early spring and it is these larvae which are smaller than the population mean.

Proc. of the 2nd Int. Symp. on Trichoptera, 1977, Junk, The Hague

Predation strategy and resource utilisation by *Plectrocnemia conspersa* (CURTIS) (Trichoptera: Polycentropodidae)

C.R. TOWNSEND & A.G. HILDREW

Abstract

Resource utilisation by all five instars of the net-spinning *Plectrocnemia conspersa* was studied in a relatively species-poor, iron-rich stream in southern England during 1974 and 1975. Comparison of the diet of each instar showed that stonefly and chironomid prey were important for all. Microcrustacea formed a significant component of the diet of smaller stages while food of terrestrial origin was more important for the larger instars.

Parallel microdistribution and gut contents analyses enabled us to describe both the aggregative and functional responses in the field of this sit-and-wait predator. All stages showed clear aggregative responses, being more abundant in high prey density patches. There is evidence from the functional responses that predators in these patches gained higher prey returns.

Measures of the half-lives of average prey in the guts of *P. conspersa* were used to estimate total rates of consumption by the predators of stoneflies and chironomids in the field. In most cases these consumption rates accounted for more than the observed seasonal depletion in numbers of prey. Despite methodological inadequacies, the evidence is clear that these invertebrate predators can exert an enormous pressure on their prey.

Introduction

Plectrocnemia conspera (CURTIS) is a caddisfly which has a net-spinning predatory larva. An investigation of its predation strategy and resource utilisation forms part of an intensive study of a species-poor, iron-rich stream in southern England (HILDREW & TOWNSEND, 1976; 1977; TOWNSEND & HILDREW, 1976). The larva can be described as a sit-and-wait predator. It appears not to be simply a filter-feeder of drifting prey, but takes mainly organisms in the benthos which become entangled in its net (HILDREW & TOWNSEND, 1976). Our aim in this paper is to discuss how the predator responds to the seasonal and spatial variations in the availability of its prey.

For the first time in a freshwater ecosystem data is presented on the aggregative and functional responses of a predator in its natural environment. Such data have been provided on a few occasions in terrestrial situations (e.g. HOLLING, 1959; MOOK, 1963), and when available this permits the detailed evaluation in a patchy environment of the total impact of a predator on its prey.

Methods

On each of seven bimonthly sampling occasions during 1974 and 1975, 40 surber samples (1/16 m^2) were taken from random locations within a 500 m stretch of Broadstone stream (Nat. Grid ref. TQ436327), an iron-rich headwater tributary of the river Medway. The samples were immediately fixed in preservative, and later sorted in the laboratory. Numbers and size classes of all species were recorded, and the gut contents of the invertebrate predators were analysed.

Experiments have been performed to determine the half-lives of the typical stonefly and chironomid prey in the guts of fourth and fifth instars of *Plectrocnemia conspersa*. The average for these instars at 13°C is 9.8 hours for stoneflies and 7.3 hours for chironomids. Temperature obviously influences the speed at which items become unrecognisable in the guts. The Q_{10} for these predators was found to be 2.3. With this information it is possible to convert data on gut contents and stream temperature into daily rates of consumption. In the present paper we make the assumption that the half-lives in the gut of the stonefly and chironomid prey typically taken by 1st, 2nd and 3rd instar predators are similar to those recorded for 4th and 5th instars. This approximation is probably not unreasonable, since although prey items of a given size will be recognisable for longer in smaller instars, their characteristic prey is, in fact, much smaller.

Thus the overall impact of the predator on its prey can be estimated.

Results

1. Diet

The seasonal variation in six diet categories of *Plectrocnemia conspersa* is shown in Figure 1. Chironomids and stoneflies remain significant categories throughout the year. However, their relative importance varies seasonally, with more chironomids taken in the summer months. This reflects the greater abundance of chironomids in the environment at this time. It is also a function of a general 'preference' on the part of the predator for chironomid as opposed to stonefly prey. Prey of terrestrial origin are taken at all times of year and are another important component, annually making up 18% of the diet. Oribatid litter mites are the single most important class of terrestrial food. Microcrustacea are the fourth main component of the diet. Their relative importance appears to be greatest in late summer and early winter. Freshwater invertebrates other than chironomids, stoneflies and microcrustaceans are of consistently small significance. This reflects their relative scarcity in the environment.

The composition of the diet of each predator instar is not identical and this is shown in Table 1. Chironomid and stonefly prey are numerically significant for all instars, and in terms of biomass consumed are certainly of even greater importance. Their relative role progressively increases with predator instar number and this pattern is repeated for terrestrial food items. Microcrustaceans, on the other hand, show decreasing numerical importance in the diet with increasing instar number.

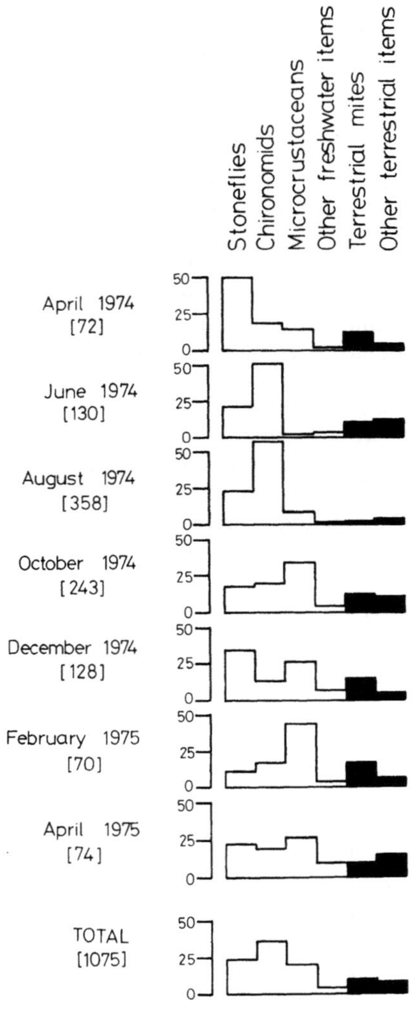

Fig. 1. Seasonal variation in percentage composition of the diet of *Plectrocnemia conspersa* in terms of four categories of freshwater items and two of terrestrial items (solid histogram). Total number of prey items analysed on each sampling occasion is shown in parentheses.

Table 1 also gives a quantitative overall estimate of selectivity. We have used JACOB's (1974) modification of IVLEV's (1961) electivity index,

$$D = \frac{r - p}{r + p - 2rp}$$

where D is the electively index, r is the fraction of a given food category in the diet, and p is the fraction of the same food in the environment. D varies from −1 to 0 for negative selection and from 0 to +1 for positive selection. Jacob's index is an improvement on IVLEV's because it is unaffected by relative abundance of food categories (JACOB, 1974). We only present electivity indices for chironomids and stoneflies as we have no estimates of density in the stream for microcrustacea and terrestrial items. Values for electivity vary from negative selection −.44 for *Leuctra nigra* to positive selection of +.43 for *Brillia modesta*. The contrasts between the two stonefly species and between the two predatory chironomids, Macropelopiini (Tanypodinae) and Pentaneurini (Tanypodinae), are worthy of note. Results from experiments on colonisation of artificial substrates (TOWNSEND & HILDREW, 1976) showed *Nemurella picteti* and Pentaneurini to be especially fast colonisers while *Leuctra nigra* and Macropelopiini were much slower. It may be that the high

Table 1 Percentage composition of diet for each of the five instars of *Plectrocnemia conspersa*

Predator instar	Total prey items	*Leuctra nigra* (Olivier)	*Nemurella picteti* Klapalek	*Heterotrissocladius marcidus* (Walker)	*Micropsectra bidentata* (Goetghebuer)	*Polypedilum albicornis* (Meigen)	Macropelopiini	Pentaneurini	*Prodiamesa olivacea* (Meigen)	*Brillia modesta* (Meigen)	Microcrustacea	Other freshwater items	Terrestrial mites	Other terrestrial items
5	226	15	15	16	0	4	1	4	4	3	3	6	14	15
4	185	10	12	13	2	6	0	2	2	1	19	5	14	16
3	128	5	6	10	1	5	0	3	2	2	39	5	16	6
2	125	10	7	9	1	6	1	3	0	0	42	4	10	9
1	102	5	0	8	2	0	1	3	0	0	66	2	13	1
TOTAL	766	10	10	12	1	4	1	3	2	1	27	5	13	11
D (Electivity)		-.44	+.38	-.06	-.12	+.29	-.10	+.34	+.25	+.43				

positive electivity by the sit-and-wait predators for the former two species reflects their greater mobility in the stream.

2. Aggregative and functional responses in the stream

The parallel microdistribution and gut contents analyses enable us to describe both the aggregative and functional responses of *Plectrocnemia conspersa* in its natural environment (figure 2). The aggregative response is described in terms of biomass of the predator against biomass of stonefly and chironomid prey combined, for individual surber samples. The pattern is clear that *P. conspersa* aggregates in patches of high prey density with much smaller predator biomasses present in low prey density patches (and this is true for all the individual instars). There is a significant correlation ($P < .01$) between predator and prey biomass in all months except August, which provides an interesting exception to the rule. This may be related to the fact that in August numbers and biomass of potential prey are very much higher than on any other sampling occasion. Perhaps the predator's dispersion strategy is adapted to the more constant level of availability of prey during the rest of the year.

The functional responses are expressed in terms of biomass of prey in the guts of unit biomass of predator against biomass of stonefly and chironomid prey combined. Prey biomass was estimated using head capsule width/dry weight regressions. The picture here is not as clear as for the aggregative responses but the results for June, August, October and December suggest strongly that predators in rich patches gain higher returns than predators in poor patches. These months provide the most reliable results because the numbers of prey items involved were greatest. Note that predators in richer patches do better in August even though no aggregative response is apparent.

3. Overall predator impact

The estimates of the half-lives in the gut of typical stonefly and chironomid prey permit us to convert information on numbers of prey items per gut into numbers of prey items taken per 24 hours. There is evidence that *P. conspersa* takes prey at all times of day and night (HILDREW & TOWNSEND, 1976). Daily consumption rates for the two stonefly species separately and for chironomids in toto are expressed in Table 2 as percentage of those present in the benthos. These values are remarkably high for *Nemurella picteti* and the chironomids. The values are smaller on each occasion for the less preferred *Leuctra nigra*. The high values for daily percentage mortality due to predation reflect the unusually high density of *Plectrocnemia conspersa* in the stream, with a maximum average value of 247 individuals per square meter, recorded in August.

It is instructive to compare the estimates of daily percentage predation mortality with the daily percentage mortality needed to account for the seasonal depletion in prey numbers observed. These latter values are indicated in Table 2 for the months where depletion subsequently occurred. In most cases predation accounted for more than the observed depletion. This point is taken up in the discussion.

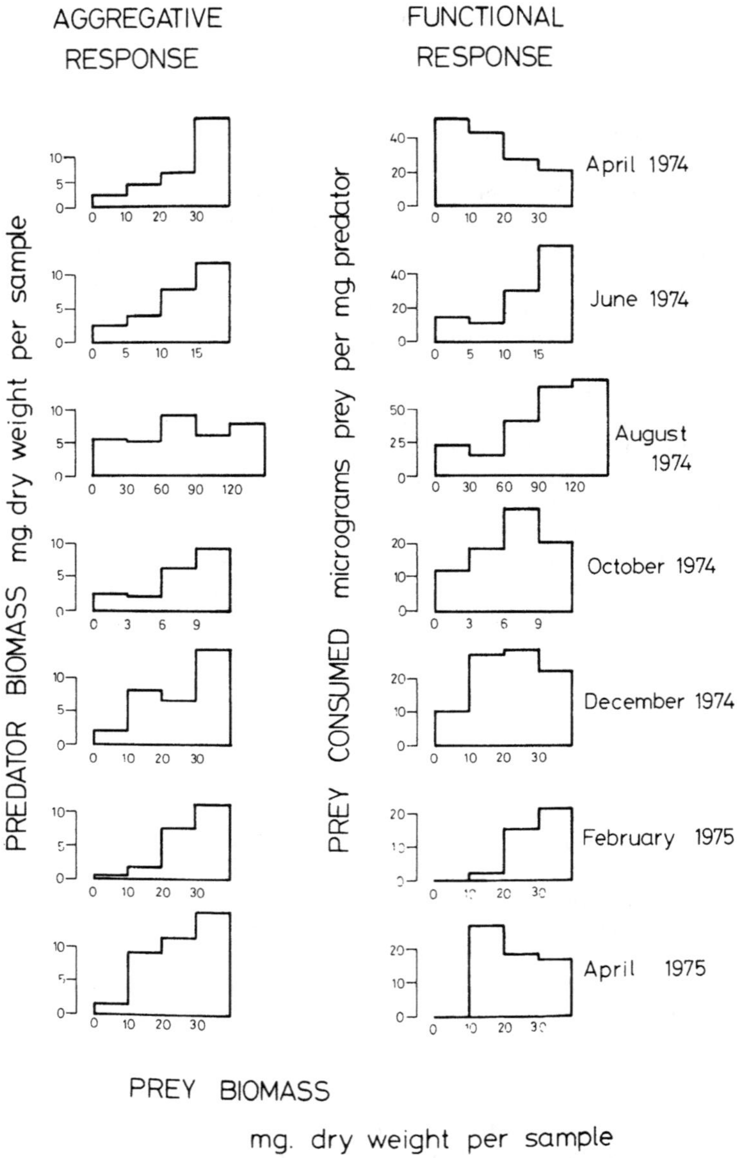

Fig. 2. Aggregative and functional responses of *Plectrocnemia conspersa* for chironomid and stonefly prey combined, on the seven sampling occasions.

Discussion

In a preliminary study, consideration was given to the influence of several environmental factors on the distribution of *Plectrocnemia conspersa* and its prey (HILDREW & TOWNSEND, 1976). We found that while the chironomid and stonefly prey taxa were individually strongly influenced by separate factors related to substrate type and flow rate, distribution of the predators was most closely related to the single factor of prey biomass present, regardless of its taxonomic nature. It is argued, therefore, that the predators are not merely responding to the same physical environmental factors as their prey, but that the pattern of their aggregative response is a result of their reaction to prey density directly. This foraging strategy in a patchy environment is predicted for predators which maximise their prey returns (SCHOENER, 1971; MACARTHUR, 1972) and, indeed, predators in richer patches do consume more prey.

There are no fish in Broadstone stream, but there are important invertebrate predators other than the abundant *P. conspersa.* These include *Sialis fuliginosa* PICT. and the Tanypodinae which both take chironomids and stoneflies in their diet. Our estimates for percentage predation by *Plectrocnemia conspersa* alone generally account for more than the seasonal depletion in prey numbers observed. The reasons for this discrepancy include underestimation of numbers of very small potential prey and the fact that univoltine stream insects go on hatching from eggs for an extended period (MACAN, 1958; HYNES, 1961). A similar discrepancy has been noted for fish and their invertebrate prey (e.g. ALLEN, 1951; see discussion in HYNES, 1970). However, despite the methodological inadequacies, the evidence is clear that invertebrate predators can exert an enormous pressure on their prey.

Recently, BENKE (1976) came to the same conclusion for Odonata larvae in the south-eastern United States. He suggests that the persistence of prey populations in the face of such pressure may be due to the presence of refuges from predators, a possibility also discussed by MACAN (1976) in his long study of the Corixidae of a moorland fishpond. We have shown experimentally (HILDREW & TOWNSEND, 1977) that the preferred substrates of the two common stonefly species in Broad-

Table 2 Estimated daily percentage numerical consumption of three prey taxa. In brackets is daily percentage consumption required to account for observed depletion in prey numbers from one sampling occasion to the next.

	April 74	June 74	August 74	October 74	December 74	February 75	April 75
Leuctra nigra	0.67	1.62	1.07	0.62 (0.90)	0.68 (0.39)	0.09	0.16
Nemurella picteti	1.53	2.25	3.38 (0.99)	2.52 (0.78)	2.94 (0.81)	1.06 (1.37)	2.67
Total chironomids	0.86	3.24	2.24 (3.93)	5.63 (1.65)	3.59 (0.34)	2.71 (0.33)	2.87

stone stream provide refuges from predation by *P. conspersa*. This phenomenon takes on increasing significance for the population dynamics of prey as evidence of high predation pressure in freshwater systems accumulates.

Acknowledgements

We wish to thank the following people who provided us with facilities: Dr. J. PHILLIPSON at Oxford, Professor N.B. MARSHALL at Queen Mary College, Mr. D. STREETER at Sussex University and Professor A.F.G. DIXON at the University of East Anglia.

References

ALLEN, K.R. 1951. The Horokiwi stream, A study of a trout population. Fish. Bull. N.Z. 10: 231-251.

BENKE, A.C. 1976. Dragonfly production and prey turnover. Ecology 57: 915-927.

HILDREW, A.G. & TOWNSEND, C.R. 1976. The distribution of two predators and their prey in an iron rich stream. J. Anim. Ecol. 45: 41-57.

——. & ——. (1977) The influence of substrate on the functional response of *Plectrocnemia conspersa* (CURTIS) larvae (Trichoptera: Polycentropodidae). Oecologia 31: 21-26.

HOLLING, C.S. 1959. The components of predation as revealed by a study of small mammal predation of the European pine sawfly. Can. Ent. 91: 293-320.

HYNES, H.B.N. 1961. The invertebrate fauna of a welsh mountain stream. Arch. Hydrobiol. 57: 344-388.

——. 1970. The ecology of running waters. Liverpool: Liverpool University Press.

IVLEV, V.S. 1961. Experimental ecology of the feeding of fishes. New Haven: Yale University Press.

JACOB, J. 1974. Quantitative measurement of food selection. A modification of the forage ratio and Ivlev's electivity index. Oecologia 14: 413-417.

MACAN, T.T. 1958. Causes and effects of short emergence periods in insects. Verh. int. Verein. theor. angew. Limnol. 13: 845-849.

——. 1976. A twenty-one-year study of the water bugs in a moorland fishpond. J. Anim. Ecol. 45: 913-922.

MACARTHUR R.H. 1972. Geographical ecology. New York: Harper and Row.

MOOK, L.J. 1963. Birds and the spruce budworm. 'The dynamics of epidemic spruce budworm populations'. Mem. Ent. Soc. Can. 31: 268-271.

SCHOENER, T.W. 1971. Theory of feeding strategies. Ann. Rev. Ecol. & Systematics 2: 369-404.

TOWNSEND, C.R. & HILDREW, A.G. 1976. Field experiments on the drifting, colonization and continuous redistribution of stream benthos. J. Anim. Ecol. 45: 759-772.

Discussion

NIELSEN: I found that this species feeds exclusively on animals which are carried into the net by the current. In Norway it has, at least on one occasion, been known to feed on trout fry.

TOWNSEND: We agree that *P. conspersa* takes prey from the drift, but we have found (HILDREW & TOWNSEND, 1976) that it still feeds at a high rate even when there are extremely few organisms drifting.

WALLACE: Is there much cannibalism?

TOWNSEND: Though observed very often in the laboratory, in the field it is unusual.

Proc. of the 2nd Int. Symp. on Trichoptera, 1977, Junk, The Hague

Structure and function of the tracheal gills of *Molanna angustata* CURT.

W. WICHARD

Abstract

The respiratory epithelium of the tracheal gills of *Molanna angustata* CURT. (Molannidae) has numerous tracheoles which are extracellularly located in deep invaginations within the epithelium. The tracheoles are regularly distributed just below the cuticle and run parallel to the longitudinal axis of the gill filaments. The amount of respiratory surface area of the larval tracheal gills depends on both the size and the number of the gills. The size of the tracheal gills is related to the larval body size, whereas the number of the gills is also determined by the environmental oxygen tension. The larvae are regulators above a critical level of oxygen tension, but below it they are conformers.

The larvae of *Molanna angustata* possess abdominal tracheal gills with 1-4 filaments. The gill filaments are tubes consisting of the cuticular and epithelial wall and the central haemocoelic space which is in open communication with the haemocoel of the body. The tracheae of the trachea branchialis (NOVÁK, 1952) run through the central haemolymphic space. They ramify in the distal direction into branches of decreasing diameter, until they approach the base of the epithelium of the wall. The tracheoles arising within the tracheal end cells, or tracheoblasts, represent the terminal parts of this system. They are less than 1 μm, and most often 0.2 μm in diameter and are almost exclusively located within the epithelial wall (Fig. 1). The wall of the tracheal gills consists of a 2-3 μm thick, single-layered epithelium which is covered with a thin cuticle measuring approximately 1 μm across.

The tracheoles are surrounded by a thin cytoplasmic sheath of the tracheoblasts. They run inside the epithelium closely beneath the cuticle, where they are confined in extracellular tubes. These tubes originate from indentations of the basal plasma membrane and remain in connection with the basal face of the epithelium via extracellular clefts. The frequency and distribution of the tracheoles within the respiratory epithelium provide the structural basis for the functional efficiency. As seen in cross-sections and longitudinal-sections through the gill filaments (Fig. 1), the major fraction of the tracheoles are located closely underneath the cuticle and run parallel to the longitudinal axis of the gills. Like the tracheal gills of limnephilid larvae (WICHARD, 1973) the tracheation attains a high structural order in that both the radius of, and the distance between, the tracheoles are of nearly constant size.

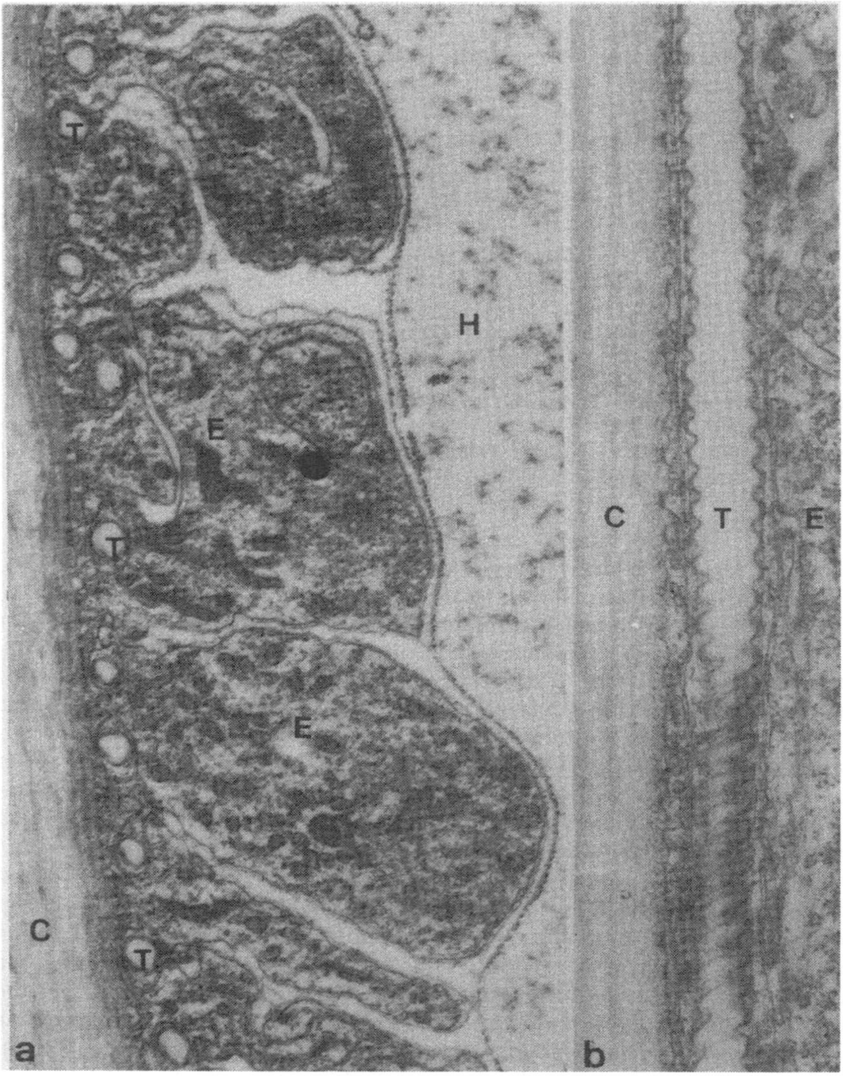

Fig. 1. Cross-section (a) and longitudinal-section (b) through the respiratory epithelium of a gill filament in last instar larva of *Molanna angustata* CURT. (a) 20,000; — (b) 32,000 x; C — cuticle, E — epithelium, H — haemolymphic space, T — tracheole.

This feature of tracheation is based on a minimum of tracheolar material. The subcuticular tracheoles are equidistantly spaced at nearly twice the radius of their catchment area. Under this precondition the total oxygen passing the gill surface is caught by the tracheoles and does not get lost from tracheal respiration by diffusing between the tracheoles. Thus, the amount of respiratory surface area of the tracheal gills is directly proportional to the number and size of the gills. Morphological adaptation under different environmental oxygen conditions does not take place at the cellular level through the rearrangement of tracheoles but at the anatomical level through changes in the number or size of tracheal gills.

Indeed, the number of tracheal gills in limnephilid larvae of different species varies according to the oxygen conditions, but different populations of the same species may have quite different numbers (WICHARD, 1974 a, b). High proportions of oxygen cause a significant decrease in the number of gills, whereas low oxygen tensions result in a significant increase in the number of gills. These morphological adaptations of different populations of the same species occur when the insects moult during larval development. The size of the tracheal gills is related to the larval body size, but the variable numbers of tracheal gills in the last instar larvae of *Molanna angustata* are determined by the environment. (Table 1). In the lake district near Eggstätt in Chiemgau (Bavaria) the populations living in lakes and slow-moving creeks possess different numbers of tracheal gills (Table 2). The observations suggest that the morphological adaptation is an effect of water movement. In still water the respiratory surface is most often surrounded with a low partial

Table 1. Number and arrangement of tracheal gills in last instar larvae of *Molanna angustata* CURT. The scheme describes the right half of the abdominal segments.

segment	dorsal	lateral	ventral
1	3—4		
2	4	1—2	3
3	4	1—2	3
4	3—4	1	3
5	3	0—1	3
6	2—3	0—1	2—3
7	2—3	0—1	2—3
8	2—3		

Table 2. Total number of tracheal gills in last instar larvae of *Molanna angustata* CURT. under different environmental conditions.

	x	S.D.	n	t-test
still water	104.1	1.4	30	t = 10.8***
running water	96.2	3.7	30	

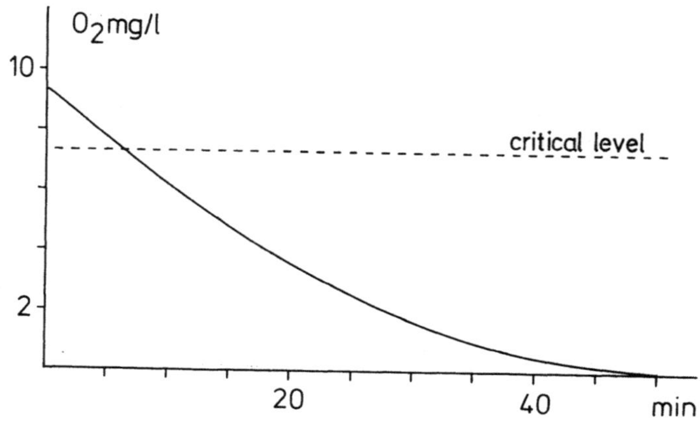

Fig. 2. Experimental loss of oxygen in water by the oxygen consumption of last instar larva of *Molanna angustata* CURT as regulator (above the critical level) and as conformer (below the critical level).

pressure of oxygen, whereas in running water the low oxygen partial pressure is compensated by the water movement.

The morphological adaptation is accompanied by physiological adaptation. The larvae are regulators which are able to maintain their normal oxygen consumption level almost independently of the environmental oxygen tension. When the tension declines, they regulate the gas exchange by movements of the abdomen with permanent loss of energy until a critical level of oxygen tension is reached, below which the larvae become conformers (Fig. 2). Under this condition, the oxygen consumption steadily declines with decrease in environmental oxygen tension.

References

NOVAK, K. 1952. Tracheáln i soustava larev našich Trichopter. Acta Soc. Zool. Bohem. 16: 249-270.

WICHARD, W. 1973. Zur Morphogenese des respiratorischen Epithels der Tracheenkiemen bei Larven der Limnephilini Kol. (Insecta, Trichoptera). Z. Zellforsch. 144: 585-592.

——. 1974a. Zur morphologischen Anpassung von Tracheenkiemen bei Larven der Limnephilini Kol. (Insecta, Trichoptera). I. Autökologische Untersuchungen im Eggstätter Seengebiet im Chiemgau. Oecologia 15: 159-167.

——. 1974b. Zur morphologischen Anpassung von Tracheenkiemen bei Larven der Limnephilini Kol. (Insecta, Trichoptera). II. Adaptationsversuche unter verschiedenen O_2-Bedingungen während der larvalen Entwicklung. Oecologia 15: 169-175.

Proc. of the 2nd Int. Symp. on Trichoptera, 1977, Junk, The Hague

Some aspects of the life histories of Limnephilidae (Trichoptera) related to the distribution of their larvae

P.D. HILEY

Abstract

Twelve species of Limnephilidae were found as larvae in temporary stillwater habitats in Scotland and England. Studies on the life-histories of six of these demonstrated several adaptations associated with colonisation of such habitats. Adults had an ovarian diapause and laid eggs out of water in places where water stood the previous winter. Larvae could remain dormant within the egg-mass for at least 18 weeks, and did not escape until immersed in water. Control experiments suggested that species inhabiting permanent waters do not show these adaptations.

Introduction

Larvae of Limnephilidae were collected from a wide variety of habitats in Scotland and England during 1967-1970. They were identified initially by rearing to the adult stage, but later using the larval key (HILEY, 1976). Information on the larval habitat of each species is given in this key.

Limnephilid larvae were found to colonise two broad habitat categories: permanent waters and temporary waters (i.e. those which dry out for part of the year). NOVAK & SEHNAL (1963) described some adaptations which enabled certain species of Limnephilidae to colonise temporary waters in Czechoslovakia; and the existence of similar adaptations was investigated in British larvae which included several species studied also in Czechoslovakia.

Adult diapause

The ovarian diapause in adults of many species of Limnephilidae has been well documented since its discovery by NOVAK & SEHNAL (1963). They were able to correlate the ovarian diapause in ten species with colonisation of temporary waters. The length of diapause tended to be reduced with increasing altitude, and the reasons for this have been suggested by DENIS (1978). My studies indicated that the following species, found in stillwater habitats in England which dried completely in summer, had an ovarian diapause in England: *Limnephilus stigma* CURT., *L. luridus* CURT., *L. griseus* (L.), *L. affinis* CURT., *L. incisus* CURT., *L. centralis* CURT., *L. sparsus* CURT., *L. auricula* CURT., *L. vittatus* (F.), *L. coenosus* CURT., and *Grammotaulius atomarius* (F.). However, *Rhadicoleptus alpestris* (CURT.)

taken from a small temporary pool at about 350 m above sea level did not show an ovarian diapause. Species like *Limnephilus griseus*, which showed a diapause in Czechoslovakia and in England, do not have a diapause in Iceland (Gislason, pers. comm.). Adults of species inhabiting permanent waters, particularly *Halesus* and the *Potamophylax* group, showed no diapause in England, and pupae frequently contained well-developed ovaries. Some *Stenophylax lateralis* (STEPHENS) pupae had undeveloped ovaries. Some *Stenophylax sequax* (McLACHLAN) pupae had undeveloped ovaries, but the adults were not studied. The larval habitats in England have not been investigated to determine whether or not they are temporary (see BOUVET, 1978).

Oviposition behaviour

NOVAK & SEHNAL (1963) observed *Limnephilus* egg-masses laid under damp vegetation in dried-up pools. This has also been observed for *Stenophylax* in France (BOUVET, 1978), and for some North American species of Limnephilidae (WIGGINS, 1973).

Oviposition was studied at Cragside Estate near Rothbury (Northumberland), which contains a system of man-made lakes and temporary pools in an area of hard rock sparsely clothed with acid moorland vegetation. The area studied in detail was about 200 m above sea level and included two very similar lakes, one with a constant water level, and the other with a water level which fell some 2 m during late summer due to seepage through a cracked dam.

The area was inspected for eggs about once every two weeks from June to October 1969 and less frequently in 1970. Freshly-laid egg-masses were distinguishable from older masses because the jelly was not fully swollen and had a fresh appearance. Masses which subsequently dried were flattened against the substratum in contrast to fresh masses which were globular. Egg-laying started in mid-July and continued into October. The observed times of oviposition by species which laid eggs out of water within this period are given in Table 1 (as determined by rearing). Eggs were laid beneath rocks, logs and other objects, in a humid atmosphere in regions up to 50 m from water. They ceased to be laid when the soil dried suffi-

Table 1. Observed times of oviposition at Cragside

Species	July	Aug.	Sept.	Oct.
Limnephilus marmoratus Curt.	+	+	+	+
L. lunatus Curt.			+	+
L. luridus Curt.			+	
L. griseus Curt.			+	
L. centralis Curt.	+	+	+	+
L. vittatus (F.)		+	+	+

ciently to be friable. Many masses of *L. griseus* eggs were laid in totally dry pools more than 100 m from the nearest water. No eggs were ever found in areas which had not been submerged the previous winter. *Limnephilus lunatus* and *L. marmoratus* were never found in temporary pools though their oviposition on the dried shores of the variable level lake suggests they are capable of colonising habitats which dry completely. *Limnephilus lunatus* eggs were laid within 10 cm of the water's edge, in agreement with GOWER (1967). In Iceland, eggs of *L. griseus* are laid under water (GISLASON, pers. com.).

The adults of these species must have a method of detecting areas which were submerged in the previous winter. They may be attracted by the algal/bacterial film which develops during immersion and remains throughout the dry period.

Field and laboratory observations on *Halesus radiatus* (CURTIS) and *Potamophylax* spp. suggested that these species, which inhabit permanent waters, lay their eggs very close to, or under, water. *Anabolia nervosa* (CURTIS) adults were observed to dive under water in order to lay eggs.

First instar dormancy

NOVAK & SEHNAL (1963), WIGGINS (1973) and BOUVET (1978) have all observed first instar dormancy within the egg-mass of temporary-water limnephilids. Eggs develop to larvae which emerge from the egg-membranes, then become dormant within the jelly. The larvae can remain dormant for several months, and escape from the jelly only when it is immersed in water.

Egg masses collected from Cragside (Table 1) were placed on damp cloths in transparent boxes and kept in a constant temperature of 20°C. Eggs developed to larvae within two to three weeks, and became dormant as previously described. The larvae could be encouraged to move around in the jelly by probing with a needle, but normally remained inactive. Viable larvae were present in egg-masses kept for 18 weeks in this way. They would presumably have survived much longer but for fungal parasites which were favoured by the experimental conditions. The jelly of *L. marmoratus* masses frequently became liquid, presumably through bacterial action. The liberated larvae concealed themselves in the damp cloth and resumed dormancy, surviving 18 weeks at 20°C with few mortalities.

Submerging an egg-mass containing dormant larvae one-quarter in water resulted in some larvae becoming active for an hour or so, but few escaped from the jelly, and once the water had been absorbed by the jelly the larvae became dormant once more. Complete immersion of an egg-mass in water resulted in all the larvae becoming active and the first larvae escaped from the jelly within two minutes. Liberated larvae began to construct cases immediately and shortly afterwards commenced feeding. Immersion of cloths containing *L. marmoratus* larvae had similar results.

There is therefore a mechanism to ensure that dew or raindrops do not cause escape of larvae, and it seems probable that the rate of hydration of the jelly is involved.

Control experiments were conducted with the permanent-water species *Potamophylax cingulatus* (STEPHENS), *Halesus radiatus* and *Athripsodes aterrimus* (STEPHENS), placing egg-masses in the conditions described above. Larvae escaped from the jelly within a few hours of eclosion regardless of the absence of water outside the jelly. These larvae remained active and crawled about on the damp cloths. All such larvae died within seven days if not placed in water and provided with food.

Discussion

It seems likely that latitude affects the adult diapause and oviposition behaviour in a similar fashion to the effect of altitude, and further studies would clearly be most valuable. For example, are Icelandic *L. griseus* capable of behaving in a similar manner to English and Czechoslovakian *L. griseus*? If they do, it suggests that these adaptations allow species of temporary waters to survive the wide variations in climate associated with glacial — interglacial changes.

It would also be interesting to discover whether species from other Trichoptera families which inhabit temporary waters (e.g. Phryganeidae) show the same adaptations.

Increasing agricultural use of marshland, coupled with drainage of temporary pools on pasture is causing a rapid reduction in the number of habitats for temporary-pool Trichoptera. It is important, therefore, to document the distribution of species which depend on such habitats so that appropriate conservation measures may be taken.

Acknowledgements

This forms part of work carried out during the tenure of a Natural Environment Research Council grant to Dr. G.N. PHILIPSON at the University of Newcastle upon Tyne. I am most grateful to Dr. PHILIPSON for his advice during this work.

References

BOUVET, Y. 1978. Adaptations physiologiques et comportementales du cycle biologique des *Stenophylax* (Trichoptera, Limnephilidae) aux eaux temporaires. Proc. 2nd. int. Symp. Trich.

DENIS, C. 1978. Larval and imaginal diapauses in Limnephilidae. Proc. 2nd int. Symp. Trich.

GOWER, A.M. 1967. A study of *Limnephilus lunatus* CURTIS (Trichoptera: Limnephilidae) with reference to its life cycle in watercress beds. Trans. R. ent. Soc. Lond. 119: 283-302.

HILEY, P.D. 1973. The taxonomy of certain caddis-fly larvae, together with an investigation into factors limiting the distribution of selected species. Ph.D. Thesis, University of Newcastle upon Tyne.

—. 1976. The identification of British limnephilid larvae (Trichoptera). Syst. Ent. 1: 147-167.
NOVAK, K. and SEHNAL, F. 1963. The development cycle of some species of the genus *Limnephilus* (Trichoptera). Čas. čsl. Spol. ent. 60: 68-80.
WIGGINS, G.B. 1973. A contribution to the biology of caddisflies (Trichoptera) in temporary pools. Life Sci. Contr. R. Ont. Mus. 88: 1-28.

Discussion

JENKINS: *Crunoecia irrorata* was mentioned as a possible inhabitant of temporary pools. Since this species is often found in hygropetric localities, living in exposed situations on damp moss, might this fact support your theories concerning this sericostomatid in temporary waters?
HILEY: It is certainly interesting and requires further investigation. I am particularly interested in finding out if the temporary-habitat adaptations of Limnephilidae are parallelled in any other family.
CRICHTON: Have you any comment on my own finding that large numbers of limnephilids are caught in light traps on wet nights?
HILEY: It does seem resonable that flight for oviposition should take place on wet nights. Conditions would then be ideal for hydration of the egg-mass jelly.
NIELSEN: In Jutland, *Sericostoma* lives in springs, and larvae can be found throughout the year. On the Danish main island it is characteristic of tiny, temporary streams in beech forests. I do not know how it aestivates. *Lithax obscurus* and *Ironoquia* (?) *dubius* are both rare species in Denmark. They seem to be restricted to forest streams which dry up in some, but not all, summers. Here *Lithax* is protected against competition from *Silo*.

Proc. of the 2nd Int. Symp. on Trichoptera, 1977, Junk, The Hague

A progress report on studies of Irish Trichoptera

J.P. O'CONNOR

Abstract

While Irish caddisflies have been studied intermittently for nearly one hundred and fifty years, our knowledge of the group remains incomplete. Since 1972, seventeen species have been added to the Irish list. During the current investigation, fly-killing aerosol sprays were found to be a particularly valuable aid when trapping adults. Although the Irish fauna is poorer than that of Great Britain, it does nevertheless contain many interesting Trichoptera. Certain of these species appear to be glacial or interglacial relicts.

From the notebook of J. CURTIS, it is evident that Irish caddisflies have been collected since at least the early eighteen-thirties (NEBOISS, 1963). Despite the many years which have subsequently elapsed, our knowledge and understanding of this particular trichopterous fauna remains incomplete and fragmentary.

Most of the earlier records are summarised in a general work on Irish Neuroptera (*sensu lato*) (KING & HALBERT, 1910). These authors list 114 species of Trichoptera and their paper still remains the most authoritative published account of the group in Ireland. In the interim period between their study and the present investigation, new facts were only discovered in an irregular and sporadic manner. Nevertheless several species were added to the Irish list often from research on other groups or from general ecological surveys.

The current study commenced in 1972, and initially collections were concentrated in four selected areas (Fig. 1). The chosen sampling sites include most of the aquatic habitats characteristic of the country. Coverage was later extended to other localities and eventually, it is hoped to publish a detailed account of the Irish Trichoptera. The present paper is therefore only a preliminary review of some of the findings.

The chosen areas were visited at regular intervals. Because of the then unsatisfactory state of larval taxonomy, emphasis was placed on the collection and study of imagines. These were captured by a variety of methods including light-trapping and manual-netting techniques. However pupae and larvae were also collected and identified whenever possible. Several government agencies have since made available caddisfly material accumulated during water-quality monitoring and other investigations, and this is providing a rich source of new records.

During the present survey, it was found that caddisflies are extremely susceptible to certain fly-killing aerosol sprays. It was noted that within seconds of application, the affected insects movements became unco-ordinated and death rapidly ensued. The most effective aerosol contained pybuthrin.

Adults resting on objects situated high above the ground are often difficult to reach with a net. However by directing a short burst of spray slightly upwind of such individuals, they rapidly become paralysed. When they fall to the ground, they may be readily collected.

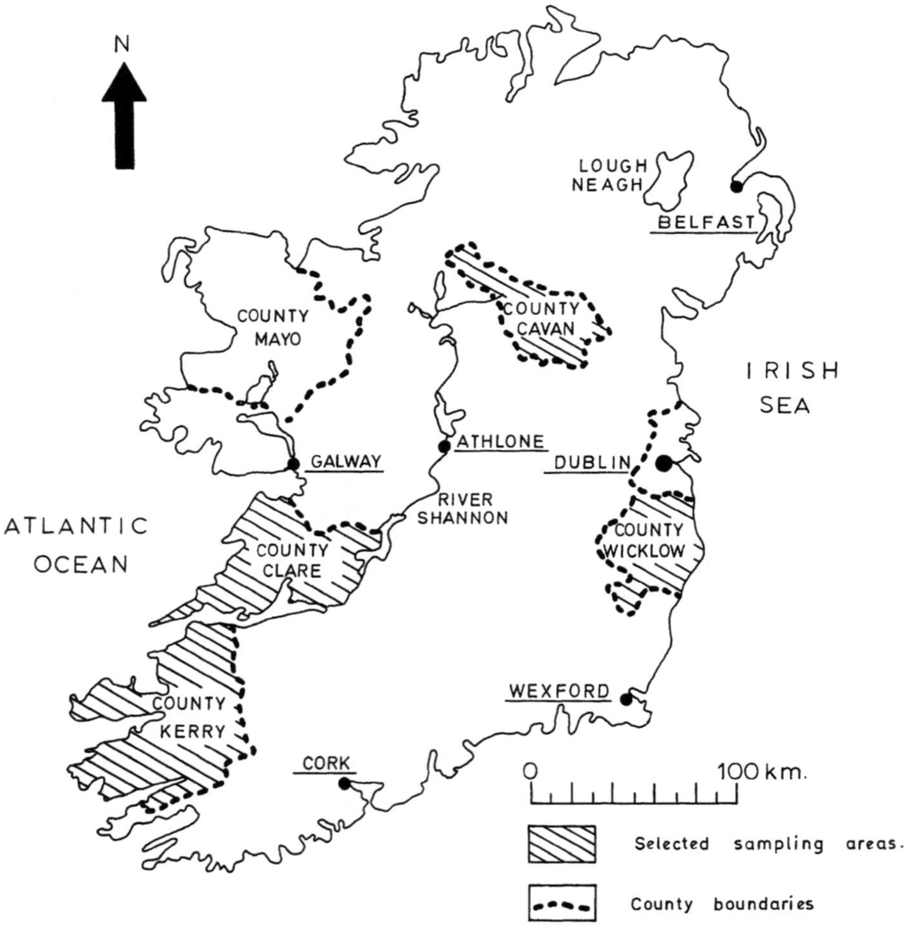

Fig. I. Map of Ireland showing the selected sampling areas, 1972-1974.

When imagines are disturbed, they frequently escape by dropping into dense vegetation where they are inaccessible. These specimens can be either killed in situ or driven from their cover by using a spray. Similarly nocturnal Trichoptera possessing cryptic habits can be easily collected from such hiding places as crevices or vegetation under bridges, culverts etc.

The technique is very effective for dealing with flying caddis including those swarming over water. Indeed, with practice, individuals may be killed without harming nearby specimens. On rivers, the dead insects will be readily carried by the current into a suitably positioned net. Imagines attracted to a light-trap may be immobilised by placing sprayed newspaper in the apparatus prior to the commencement of trapping. The method is also extremely useful for preventing active specimens from escaping from the net after capture. It is very necessary to stress that these aerosols should always be used with restraint, to prevent undue damage to the environment and its associated fauna.

Since 1972, 17 species have been found new to Ireland. While the majority are in the Hydroptilidae and Limnephilidae, other families are represented (Table 1). Among the many interesting species taken, the discovery of *Cyrnus insolutus* McLACHLAN is noteworthy. In the British Isles, it was only previously known from Blelham Tarn in the English Lake District (KIMMINS, 1942). In addition, *Hydroptila martini* MARSHALL has recently been identified from Irish material (MARSHALL, 1977).

Current research is also providing new data on the distribution and taxonomy of previously recorded Trichoptera. *Apatania auricula* (FORSSLUND) for instance is now known to be widespread in the south-west while *Tinodes maculicornis* (PICTET) appears to be common in localities where suitable habitats exist. The larva of *T. maculicornis* is presently been described in co-operation with Dr. E.J. WISE.

Other rediscovered species include *Apatania wallengreni* McLACHLAN, *Limnephilus binotatus* CURTIS, *L. fuscinervis* (ZETTERSTEDT), *L. nigriceps* (ZETTERSTEDT) and *Oecetis notata* (RAMBUR). The last named caddisfly was found by M.A. NORTON. All these insects have been obtained at previously unknown sites suggesting that they are commoner in Ireland than was formerly realised.

While additions have been made to the Irish list, deletions will also be necessary. In the past, misidentifications have resulted in several erroneous records but one example will suffice. An examination of Irish specimens named as *Hydropsyche guttata* PICTET and housed in the National Museum, has revealed that none of these individuals belong to this species.

Many British Trichoptera have not yet been taken in Ireland, a notable example being *Brachycentrus subnubilus* CURTIS. While undoubtedly some species have been overlooked, it is now probable that certain caddisflies are definitely absent. This comparative poverty of species has also been observed for other faunal and floral groups (PRAEGER, 1950). Nevertheless the Irish fauna should not be considered merely an attenuated island one. It does contain interesting and unusual caddisflies. Both *A. auricula* and *T. maculicornis* have not been found in Great Britain (CRICHTON, 1971; EDINGTON & ALDERSON, 1973) while other species including *L. fuscinervis* appear to be commoner here than in our nearest neighbour.

The events of the Ice Age must have had a considerable effect on the composition and distribution of the Irish trichopterous fauna. It is to be expected that many established species were annihilated during this period and that much of the present fauna results from post-glacial immigrations.

However *A. auricula*, *T. maculicornis* and certain other species appear to be either glacial or interglacial relicts. The first named insect teems in an area already noted for similar survivors (CORBET et al., 1960; MURRAY, 1976; O'CONNOR, unpublished work). Elsewhere the species has been reported from Latvia, Sweden, Norway and Eastern Fennoscandia (HICKIN, 1967; SVENSSON & TJEDER, 1975), a geographical distribution which has been described as curious (KIMMINS, 1951; SCHMID, 1954).

Table 1. Additions to the Irish trichopterous fauna (1972-1977).

1. *Glossosoma conformis* Neboiss: Recorded in Co. Wicklow.
2. *Cyrnus insolutus* McLachlan: Species taken in Co. Kerry (O'Connor and Wise, in press) and Co. Clare.
3. *Lype reducta* (Hagen): Larva identified from Co. Kerry (Edington and Alderson, 1973). Adult also trapped at larval site of capture.
4. *Hydropsyche siltalai* Döhler: A common species (O'Connor and Bracken, in prep.).
5. *Allotrichia pallicornis* (Eaton): Specimens collected in Co. Dublin and Co. Wicklow.
6. *Hydroptila cornuta* Mosely: The species is known from Co. Wicklow (O'Connor and Bracken, in press), Co. Cavan, Co. Clare and Co. Kerry.
7. *H. martini* Marshall: Irish specimens have been identified as belonging to this species (Marshall, 1977). This material was collected in Co. Cavan.
8. *H. pulchricornis* Pictet: Occurs in Co. Kerry (O'Connor and Wise, in press), Co. Cavan and Co. Clare.
9. *Orthotrichia angustella* (McLachlan): Found in Co. Kerry (O'Connor and Wise, in press) and Co. Clare.
10. *O. costalis* (Curtis): Locally abundant in Co. Clare.
11. *Apatania muliebris* McLachlan: Species collected in a spring in the west of Ireland (Fahy, 1972). Adult netted in Co. Kerry.
12. *Ecclisopteryx guttulata* (Pictet): Frequents lotic habitats in Co. Wicklow.
13. *Limnephilus elegans* Curtis: Imagines collected in Co. Kerry (O'Connor and Wise, in press) and Co. Mayo.
14. *Mesophylax impunctatus* McLachlan: Known from Co. Kerry (Hiley, 1976; Wise and O'Connor, in press) and Co. Clare.
15. *Hydatophylax infumatus* (McLachlan): An adult light-trapped in Co. Cavan.
16. *Beraeodes minutus* (L.): A larva taken in Co. Kerry (O'Connor and Wise, in press).
17. *Lasiocephala basalis* (Kolenati): Found in Co. Kerry (Wise and O'Connor, in press).

During the Pleistocene, Ireland was influenced by three separate glaciations but little is known of how the first of these affected the country (O'RIORDAN, in press). The Munsterian (Old Drift) and Midlandian (New Drift) cold stages followed and these periods may correspond to the British Wolstonian and Devensian eras (MITCHELL, 1976).

During most of these glacial episodes, parts of the present range of *A. auricula* remained icefree. Indeed even at the period of maximum glaciation, nunataks evidently existed and coastal regions, marginal to the ice-front, have also been postulated (PRAEGER, 1950; ORME, 1970; MITCHELL, 1976). Any of these areas could have provided a refuge.

By contrast, *T. maculicornis* has been reported abroad from Switzerland, France and Portugal (STEINMANN, 1972). This caddisfly may have entered Ireland during one of the interglacial epochs perhaps subsequent to the Munsterian glaciation. During the Midlandian era, large areas in the south and south-east remained uncovered (MITCHELL, 1976). Similar but smaller pockets existed in the north, east and west. *T. maculicornis* along with other Trichoptera could certainly have survived this last major glacial event in such localities while they disappeared elsewhere.

Current research is revealing that many Irish caddisflies have interesting geographical ranges including the localised occurrence of certain species. At present it is uncertain whether these observed distributions result from environmental or historical factors. Fluctuating changes in population densities could also be a contributory factor. It is hoped that future work will elucidate these problems and other mysteries still surrounding the Irish Trichoptera.

Acknowledgments

I wish to thank all my friends and colleagues who by their encouragement and kindness have made this work possible. I am also grateful to the Department of Fisheries, Dublin, for providing me with a Fisheries Science Studentship which I held at University College, Dublin.

References

CORBET, P.S., LONGFIELD, C. & MOORE, N.W. 1960. Dragonflies. London: Collins.

CRICHTON, M.I. 1971. A study of caddis flies (Trichoptera) of the family Limnephilidae, based on the Rothamsted Insect Survey, 1964-68. J. Zool., Lond. 163: 533-563.

EDINGTON, J.M. & ALDERSON, R. 1973. The taxonomy of British psychomyiid larvae (Trichoptera). Freshwat. Biol. 3: 463-478.

FAHY, E. 1972. Some records of Trichoptera from Ireland. Ir. Nat. J. 17: 199-203.

HICKIN, N.E. 1967. Caddis larvae. Larvae of the British Trichoptera. London: Hutchinson.

HILEY, P.D. 1976. The identification of British limnephilid larvae (Trichoptera). Syst. Ent. 1: 147-167.

KIMMINS, D.E. 1942, *Cyrnus insolutus* McL. (Trichoptera), new to Britain, Entomologist 75: 66-68.
KIMMINS, D.E. 1951. *Apatidae inornata* (WALLENGREN) and *A. auricula* (FORSSLUND) (Trichoptera). Ann Mag. nat. Hist. 4: 410-416.
KING, J.J.F.X. & HALBERT, J.N. 1910. A list of the Neuroptera of Ireland. Proc. R. Ir. Acad B 28: 29-112.
MARSHALL, J.E. 1977. *Hydroptila martini* sp.n. and *Hydroptila valesiaca* SCHMID (Trichoptera: Hydroptilidae) new to the British Isles. Entomologist's Gaz. 28: 115-122.
MITCHELL, F. 1976. The Irish landscape. London: Collins.
MURRAY, D.A. 1976. *Buchonomyia thienemanni* FITTKAU (Diptera, Chironomidae), a rare and unusual species recorded from Killarney, Ireland. Entomologist's Gaz. 27: 179-180.
NEBOISS, A. 1963. The Trichoptera types of species described by J. CURTIS. Beitr. Ent. 13: 582-635.
O'CONNOR, J.P. & BRACKEN, J.J. (in press) A comparative limnological study of two Irish lakes (Lough Sillan, Co. Cavan and Lough Dan, Co. Wicklow). Ir. Fish. Invest. Ser. A.
——. & ——. (in preparation) Notes on *Hydropsyche instabilis* (CURTIS) and *H. siltalai* DOHLER (Trichoptera) in Ireland.
——. & WISE, E.J. (in press) The Trichoptera of the Killarney lakes. Proc. R. Ir. Acad.
O'RIORDAN, C.E. (in press) Guide to the extinct terrestrial mammals of Ireland in the National Museum. Dublin: Govt. Stationery Off.
ORME, A.R. 1970. The world's landscapes. 4. Ireland. London: Longmans.
PRAEGER, R.L. 1950. Natural history of Ireland. London: E.P. (Republished edition, 1972).
SCHMID, F. 1954. Contribution a l'etude de la sous — famille des Apataniinae (Trichoptera, Limnophilidae) II. Tijdskr v. Ent. 97: 1-74.
STEINMANN, H. 1972. Keys to the families and genera of European Annulipalpia (Trichoptera). Fol. Ent. Hung. 25: 445-486.
SVENSSON, B.W. & TJEDER, B. 1975. Check-list of the Trichoptera of North-Western Europe. Ent. scand. 6: 261-274.
WISE, E.J. & O'CONNOR, J.P. (in press) The Trichoptera of Rivers Flesk and Laune. Proc. R. Ir. Acad.

Proc. of the 2nd Int. Symp. on Trichoptera, 1977, Junk, The Hague

Observations on caddis larvae in *Stratiotes* vegetation

L.W.G. HIGLER

Abstract

Studies on the aquatic macrofauna on *Stratiotes* plants, both living and artificial, provide information on the distribution patterns of caddis larvae in littoral vegetation. The life cycles of two related species demonstrate a temporal separation of the same larval sizes. The life history of a species of the family Polycentropodidae with its food and predators is elucidated.

The qualitative and quantitative distribution of aquatic invertebrates on *Stratiotes* plants in autochtonous swamp successions has been investigated (HIGLER, 1977). Caddis larvae are the second most numerous insect order. In 73 samples 24 species were found totalling nearly 4,000 larvae, pupae or empty cases (Table 1). The distribution patterns of the most abundant species can be related to structural characteristics of the investigated vegetation. According to these characteristics five sections (A to E) can be distinguished from the shore to the open water (Fig. 1). In the first section near the shore (A) a water layer of 30 to 40 cm stands over a thick sapropel layer of about 2 meters. In summer oxygen conditions are often bad with oversaturation in the afternoon and deficiency or lack in the night. The presence of toxic agents such as NH_3, NO_2^- and H_2S can make living conditions still worse. The *Stratiotes* plants in this section, growing emersed in summer and autumn, are generally smaller than in the B-section. The plants in the B-section, emersed as well, grow near each other in water 40 to 60 cm deep over a sapropel layer of $1\frac{1}{2}$ meters or more. This section can reach extensive areas of hundreds of square meters with no other aquatic plants, except for *Lemna trisulca* and sometimes *Utricularia* species. The C-section is formed by a zone of a few meters with a very high plant density of emersed plants. The next section (D) consists of plants that stay submerged through the year. The plant density is much lower than in B, the water depth is 60 to 80 cm and the sapropel layer is thinner. The submerged plants of the E-section grow wide apart from each other in water with a depth of about one meter.

The width of the lines of the distribution patterns in Fig. 1 indicates the preference for the sections; the dotted parts indicate incidental presence of one or more species. The niche breadth (column B in Table 1) is largest for those groups of species that can stand different circumstances. In this respect the boundary submerged/emersed is an important threshold. The groups 5 and 6, confined to the submerged part of the vegetation and group 3, confined to the B-section proper have a much lower niche breadth than the other groups. Group 1 which only occurs in the emersed sections has to face the more difficult circumstances in the A-section and consequently a higher niche breadth than for group 4 could be expected. The larvae of the family Polycentropodidae with comparable food demands and predators show a partial or complete spatial separation. The same phenomenon can be observed in the representatives of the family Hydroptilidae. The species that can be found together, like the two *Holocentropus* species (Fig. 2), or the *Cyrnus* species, have different life cycles, pointing to a temporal separation. The larvae of different sizes have different food demands, as is expressed in Fig. 3 where the life history of a polycentropodid species has been represented.

The cycle starts with an egg mass from one female. The highest possible number

Table 1. Caddis species collected in 73 samples from *Stratiotes* vegetation in The Netherlands, 1972.
F = percentage of samples where the species was observed
N = total number of specimens collected
$\bar{n}/m^2$ = number of specimens per square meter (arithmetic mean)
B = niche breadth according to the formula $B = 1/\Sigma\ p_i^2$
D = distribution as indicated in Fig. 1.

	F	N	$\bar{n}/m^2$	B	D
Hydroptila pulchricornis PICT.	1	1			6
Agraylea multipunctata CURT.	15	139	35.6	2.27	3
Agraylea sexmaculata CURT.	23	308	38.3	8.70	3
Orthotrichia costalis (CURT.)	7	12			6
Oxyethira flavicornis (PICT.)	44	564	58.2	12.37	4
Tricholeiochiton fagesii (GUIN.)	75	865	42.8	17.11	2
Cyrnus flavidus McL.	23	64	10.8	7.20	5
Cyrnus insolutus McL.	13	75	28.4	2.70	6
Cyrnus crenaticornis (KOL.)	10	28	12.1	5.06	6
Holocentropus picicornis (STEPH.)	77	1311	61	14.25	2
Holocentropus dubius (RAMB.)	40	196	17.4	11.58	1
Ecnomus tenellus (RAMB.)	12	26	8.1	4.77	5
Phryganea grandis L.	4	3			
Phryganea bipunctata RETZ.	1	1			
Agrypnia pagetana CURT.	3	4			
Limnephilus flavicornis (FAB.)	12	13	3.6	3.58	2?
Oecetis furva (RAMB.)	47	102	10	13.75	2
Oecetis ?lacustris (PICT.)	1	1			
Athripsodes aterrimus (STEPH.)	1	1			
Athripsodes senilis (BURM.)	3	2			
Leptocerus tineiformis CURT.	5	5			
Triaenodes bicolor (CURT.)	23	35	5.2	9.12	3
Mystacides longicornis (L.)	12	16	2.9	6.15	2?
Mystacides ?nigra (L.)	1	1			

(N) within this cycle can be reduced by water mites, snails or by an infection with fungi (P1). The fresh hatched larvulae are eaten by *Hydra*, *Chaoborus* larvae, water mites and *Cymatia coleoptrata* (P2). Their prey consists of microfauna like rotifers or small crustaceans (Fa). If the remaining larvae have become bigger, they eat Cladocera, Copepoda and Ostracoda (Fb). At that stage, however, they are eaten by water spiders, *Erpobdella* and *Ilyocoris* (P3). In the next stage predators like Odonata nymphs, beetle larvae and *Notonecta* (P4) are concerned, while fishes (P5) can be the main predators if their presence is allowed by the environment. In addition to small crustaceans more Oligochaeta are eaten and smaller chironomid larvae as well (Fc and Fd). The biggest larvae eat cladocerans, *Stylaria lacustris* and chironomid larvae. Together with the larger aquatic carnivores, birds (*Chlidonias niger*) and amphibians (P6) can be predators. The pupal stage is fairly safe, where probably only some parasites (P7) can be harmful, but the free-swimming pupa (f) is a defenseless prey to invertebrate carnivores (P4), amphibians and birds (P6) and

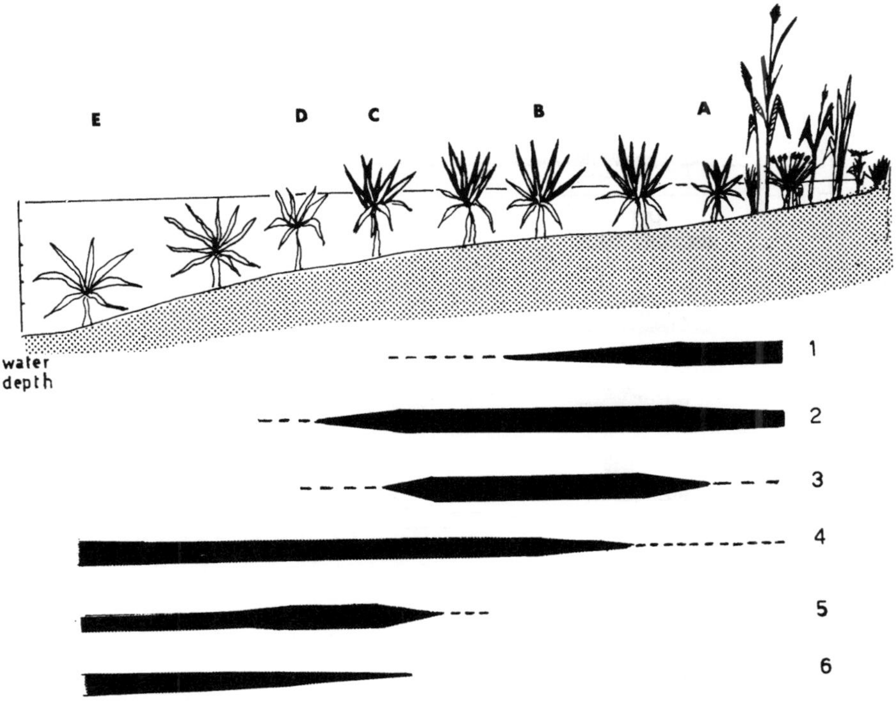

Fig. 1. Distribution patterns of caddis larvae in an autochtonous swamp succession with *Stratiotes aloides*.

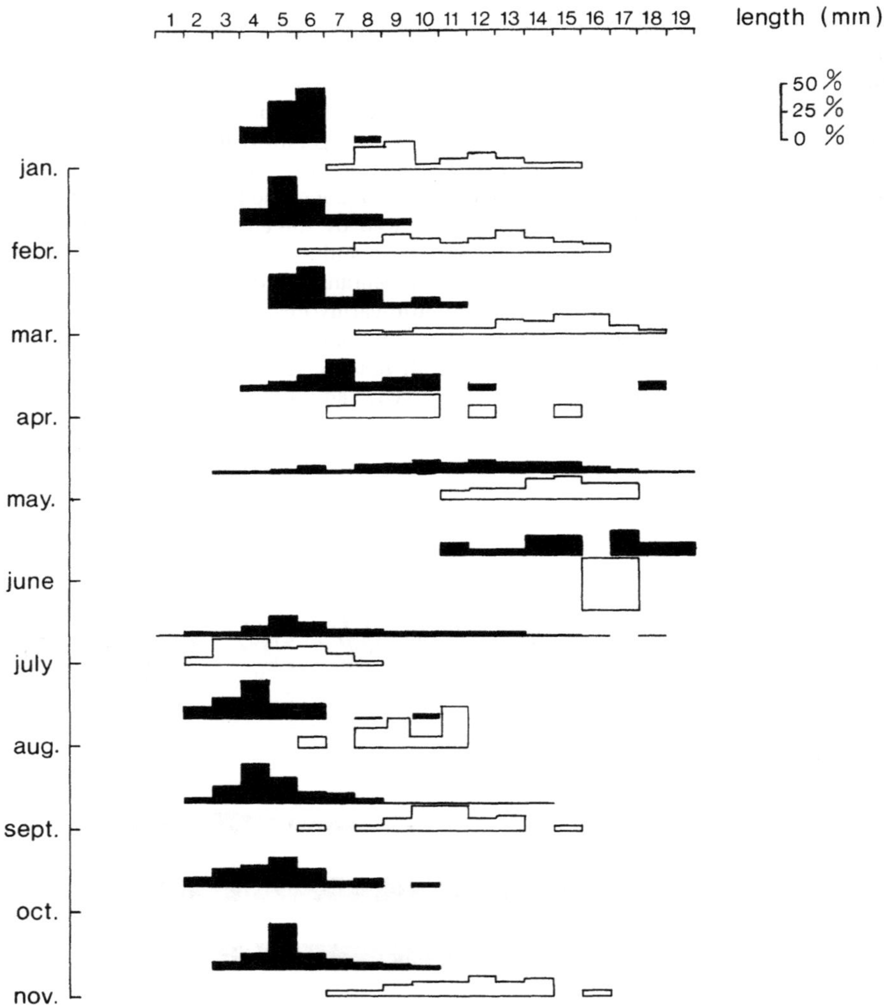

Fig. 2. Life cycles of *Holocentropus picicornis* (■) and *Holocentropus dubius* (□).

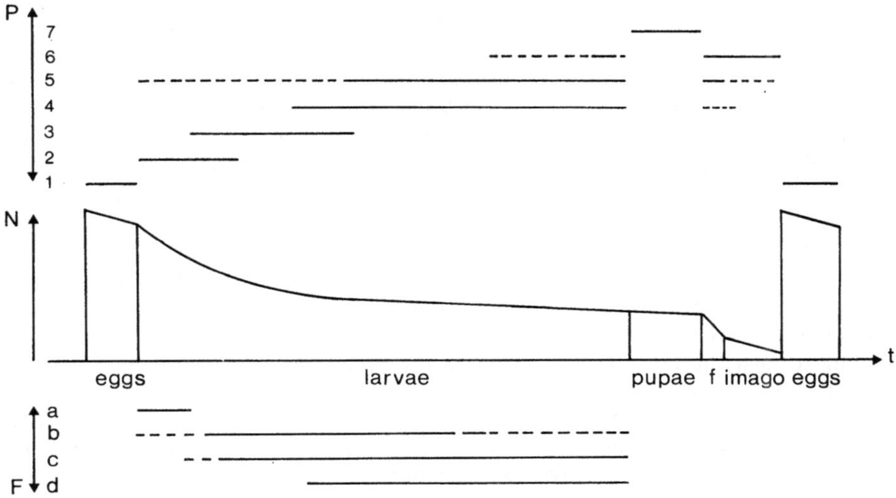

Fig. 3. The life history of a polycentropodid species with its predators (P) and food (F). Explanation in the text.

Table 2. Average number per m² of plant surface of different sampling dates. Artificial plants only.

	Venematen							Hol	
Orthotrichia costal.	162	107	43	14	5.6	1.8	0.2		
Oxyethira flavicorn.	19	11	8	4.5	3	2.8	1.9	1	0.2
Hydroptila pulchric.	2.6	2	0.3			1.7	0.3		
Tricholeiochiton fag.	0.1	0.1	0.1			1	0.1	0.1	
Holocentropus picic.	1.9	1.1	0.5			3	0.5	0.4	0.2
Holocentropus dubius	0.3	0.1				9.2	2.7	2.2	0.4
Cyrnus flavidus	14	4.6	4.1	2.5	2.2	48	27	22	1.2
Cyrnus crenaticorn.	26	25	11	10	3.3	4.7	4.2	1.9	0.3
Cyrnus insolutus	3.3	1.6				6	5.3	2.3	0.6
Ecnomus tenellus	229	58	55	49	5	30	13	1.8	
Triaenodes bicolor	0.1					2.2	0.3	0.3	

especially fishes (P5). Finally, the adult caddisfly is eaten by birds and amphibians and sometimes by fishes (*Scardinius erythrophthalmus*). If one adult pair is left, the same cycle can start again.

In four broads we have put plastic *Stratiotes* plants in and outside the vegetation. The artificial plants between living *Stratiotes* and those in open water near the shore- and marsh-vegetation were colonized rapidly by the same species as have been found on the living plants. On the plants in the open water we found the species of the groups 4, 5 and 6, but generally in much higher numbers than on the living plants of the E-section. The distance to the littoral vegetation is a limiting factor for the colonization by *Holocentropus* species, Leptoceridae and *Tricholeiochiton fagesii*. The colonization of artificial plants in the midwater of larger waterbodies (distance to littoral vegetation tens of meters) is dominated by the *Cyrnus* species (in particular *C. crenaticornis*), *Ecnomus tenellus* and the Hydroptilidae *Orthotrichia costalis*, *Hydroptila pulchricornis* and *Oxyethira flavicornis*. In Table 2 we show some provisional results of experiments with artificial plants in two broads, where *Stratiotes* had disappeared. The plants in the broad Venematen were situated from midwater to shore over a distance of 50 meters with 2 to 4 meters of free water between the plants. Their depth varied from 0 to 80 cm, that is from floating to near the bottom. The plants in the broad Hol were put in two parallel rows at a distance of 2 to 4 meters from the reeds. The numbers in table 2 represent the arithmetic means per square meter of plant surface of five (Venematen) or four (Hol) sampling dates through the year. We only used the species that were caught regularly and for the ease of survey we have put the numbers in sequence from high to low.

The broad Venematen is a larger waterbody (about 40 ha), the other broad is smaller ($\frac{1}{2}$ ha) and it has much aquatic vegetation, submerged as well as emerging. The difference in dimensions seems to play an important role, while the distance to the shore is of particular importance in the larger waters without much vegetation in midwater. *Orthotrichia costalis* and *Ecnomus tenellus* were collected in surprisingly high numbers. Once we found 266 *Orthotrichia* and 268 *Ecnomus* on one plant (plant surface 1 m^2). *Cyrnus crenaticornis* seems to prefer the open water in the larger broad; *C. flavidus* is found in high numbers in the smaller one. It is noteworthy that *Holocentropus dubius* was caught in higher numbers in the broad Hol than *H. picicornis*. In the light of the results of Table 1, one would expect the reverse. This observation, however, is confirmed by the results of MACAN & KITCHING (1976). In their study on the colonization of squares of plastic suspended in midwater they found both *Holocentropus dubius* and *Cyrnus flavidus* outside the vegetation on the artificial substrata. These experiments provide valuable information on the activities of a number of invertebrates (at night?) outside the littoral vegetation.

In addition to the species lists of the two tables some species must be mentioned that were found only on the artificial plants, but near the shore (or living *Stratiotes*).

Agraylea cognatella (McL.)	first record for The Netherlands
Cyrnus trimaculatus (CURT.)	tens on a few plants
Tinodes assimilis (McL.)	a single observation
Lype phaeopa (STEPH.)	,, ,, ,,
Anabolia nervosa (CURT.)	,, ,, ,,
Ceraclea senilis (BURM.)	a few times.

References

HIGLER, L.W.G. 1977. Macrofauna-cenoses on *Stratiotes* plants in Dutch broads. RIN-verhandeling nr. 11.

MACAN, T.T. & KITCHING, A. 1976. The colonization of squares of plastic suspended in midwater. Freshwat. Biol. 6: 33-40.

Discussion

EDINGTON: Do you think that the polycentropodid species, which live at different distances from the shore, are ecologically separated by having access to different copepod and cladoceran species as a food supply?

HIGLER: Cladocerans occur in very high numbers in the emersed part of the vegetation, while copepods are found predominantly in the submerged part. However, the life conditions with respect to oxygen regime, accessibility for fish and plant density in relation to water depth, are of much greater influence for distribution than the composition of the micro-fauna. I do not believe that any preference for cladocerans or copepods exists in polycentropodid larvae.

JONES: What was the time-scale of the invasion of your artificial plants? Have you any detail on the succession of invading species? What was the timing of your introduction of these plants?

HIGLER: The colonization was investigated over two years, and at intervals of one to several months. The first examination was one month after introduction.

JENKINS: The larval biology of *Ecnomus tenellus* is imperfectly known in Great Britain. I would be most interested to know whether you have observed larvae of this species constructing galleries, or whether they are free-living? I have taken them abundantly on *Chara* sp. and on submerged wood from a Welsh lake, where the larvae appear to be free-living.

HIGLER: I have not observed *Ecnomus* larvae in galleries on the plants. The sampling method does not guarantee that nets, if present, are discovered, but I do not expect these larvae to have nets like *Holocentropus.*

HILDREW: If the polycentropodid larvae are not limited by food supply then either, the life history differences are trivial as far as coexistence is concerned, or the species share limiting resources other than food according to size class. Do you think population density is resource-limited?

HIGLER: In those situations, where the species occur together in the *Stratiotes* vegetation, I consider population density as a limiting factor.

Proc. of the 2nd Int. Symp. on Trichoptera, 1977, Junk, The Hague

Continuous rearing of the limnephilid caddisfly, *Clistoronia magnifica* (BANKS)

N.H. ANDERSON

Abstract

Seven successive generations of *Clistoronia magnifica* (BANKS) were reared in laboratory cultures in 46 months at 15.6°C. Rearing techniques are described and the biology of the species, as observed in the laboratory, is discussed and compared with field data.

No obvious deterioration of vigour was noted; pupal and adult weights have remained constant or increased slightly up to the fifth generation. Laboratory-reared specimens were heavier than field material, but the laboratory colony produced fewer eggs per female.

Introduction

In my study of the distribution and biology of Oregon Trichoptera (ANDERSON, 1976a), considerable effort was expended on laboratory rearing to associate immature and adult stages. The rearing programme was gradually expanded to screen for species that were amenable to laboratory culture and could then be used as models for basic studies of growth, development and feeding responses. The desirability of growth and trophic relations studies that are based on the entire larval interval, rather than on short-term feeding trials or gut analysis, is obvious. *Pseudostenophylax edwardsi* (BANKS) was initially considered a candidate species because it had been reared for more than a generation (ANDERSON, 1974). However, later data indicate that reared adults were undersized, and developmental times in the laboratory were extremely variable. Other limnephilid genera that I have reared with mixed success are *Psychoglypha*, *Hesperophylax*, *Hydatophylax* and *Limnephilus*. Compared with these genera, *Clistoronia magnifica* (BANKS) has been remarkably easy to culture; seven generations have now been reared in 46 months. A brief description of rearing methods and the use of enchytraeid worms as a dietary supplement is given in ANDERSON (1976b). The present paper summarizes the results of the laboratory rearing and describes the biology of the species as observed under these conditions.

Field biology

Clistoronia is a genus of five species, confined to North America and largely occurring in mountainous regions of the west (WIGGINS, 1977). *C. magnifica* occurs from British Columbia to Oregon. Most of the Oregon records are from the Cascade Mountains, from ponds or lakes above 1000 m; there is one exceptional record from Astoria, at sea level on the Pacific Coast (ANDERSON, 1976a).

WINTERBOURN (1971) conducted life history studies at Marion Lake, British Columbia (elevation approximately 300 m). He described *C. magnifica* as an early-season univoltine species whose larvae inhabit submerged marginal vegetation and open sediments. Adults emerge from early May to late June but females are not reproductively mature at that time; adults live through the summer and return to the lake during August and September to mate and lay eggs. Oviposition occurs on submerged plants or logs. Larval development through the five instars occurs from August to January, with all larvae being in the final instar from January to April.

No complete life cycle data are available for the Oregon population. In Lost Lake, Linn Co. (elevation 1200 m), the population consisted of 34% mature larvae, 60% prepupae and 6% pupae in early May, when water temperature was 11°C and about two weeks after the ice had melted. The laboratory culture was initiated from egg masses collected from nearby lakes in late July and August 1973. The globular, gelatinous masses (25-30 mm dia.) were collected by a scuba diver on the lake bottom at 3 to 5 m depth.

Rearing methods

The routine rearing of larvae was in aerated shallow pans of tap water with sand substrate at 15.6°C and long days (16 h light: 8 h dark). Pans were washed and water replaced once or twice a week to remove faeces and excess decomposing food. Overcrowding resulted in retarded development and cannibalism, so large groups were subcultured, especially for the final instar. About 50 larvae could be reared to maturity in a 25 x 40 x 5 cm pan.

Material provided for food and case-building included *Alnus* leaves, conifer needles, wheat grains and, occasionally, green grass. The leaves, conditioned in water for 1-2 weeks to allow leaching and microbial colonization, provided food and substrate for the larvae. Water-soaked needles were used primarily for case-construction. The dry wheat grains were a necessary supplement to the detrital food for normal growth. They were provided in small amounts (20-40 kernels at one time) to avoid excessive decomposition and consequent deoxygenation.

A supplement of enchytraeid worms increased the growth rate (ANDERSON, 1976b), so most rearings in the third to sixth laboratory generations included the worm supplement. The most common cause of mortality was due to toxic products and/or deoxygenation from decomposing foods, especially worms. This could be reduced by avoiding excess feeding, frequent cleaning of the pans, or rearing in a 'drippery', a series of trays with a variable water exchange (ANDERSON, 1973).

When mature larvae had sealed off their cases, they were placed in small wire cages partially submerged in the drippery. At emergence the adults were transferred to a 0.16 m^3 screen oviposition cage. The cage was streaked with dilute honey for food and contained a pan of water with an exposed rock or stick to enable the adults to reach the water.

Adult behaviour

At 15.6 °C, the adults are long-lived, with several individuals surviving from 4-6 weeks. Though they are strong fliers, much of the daytime is spent resting on the sides or corners of the cage. They do not seem to select concealed sites, such as folds in paper towels. A trait that facilitates handling of *C. magnifica* is that escaped adults tend to reappear on or near the cage by the next day, presumably in response to pheromone attraction. Resting adults show little response to jarring of the cage, or to moving objects; they can be easily captured with the fingers or forceps. From the lack of activity during daylight hours, I infer that adults are nocturnal. However, the absence of a twilight period in the laboratory may result in atypical behaviour.

Males and females are similar in size, weight, and emergence time. Generally more females than males were obtained. Mating occurred within 1-2 days after emergence. Pairs remained in copulation for several hours during the day. Both males and females were observed to mate more than once.

Females emerge with undeveloped ovaries and require a preoviposition period of about two weeks. WINTERBOURN's (1971) observations suggest that females have a reproductive diapause over the summer. However under long-day conditions there was no evidence of diapause during the seven laboratory generations.

Egg masses occur on substrates at or below the water line or sometimes loose in the pan. The only instance of oviposition that I have observed was at 0700 hrs. The greenish egg mass, 6 mm dia., was extruded from the body before the female crawled completely under the water enveloped in an air bubble. She was submerged for 5 min. while attaching the mass to a stick; she then walked rapidly up the stick, shook her wings without attempting to fly and assumed a resting postion on the cage. Females also can rest on the water without being trapped in the surface film. This behaviour suggests that egg masses could be deposited directly into the water and then sink to the bottom. In the field, egg masses also occur both attached and loose. WINTERBOURN (1971) indicates that they occur on plants or other submerged substrates, whereas the material used to start my cultures was collected by a scuba diver from the bottom of the lake.

In laboratory cages, some small masses were deposited on damp paper and on the floor of the cage, well away from the water. These masses would expand and develop normally if placed in water but the eggs desiccated if left where deposited.

After contact with water, the gelatinous mass expands to a large sphere with the eggs arranged in lines. The mass that I observed being deposited required 8 hours to achieve full size. Though field-collected masses may contain over 300 eggs, those deposited in the laboratory contained only 50-200 eggs. Females deposit more than one egg mass but the later ones are usually quite small and may contain 50-100% non-viable eggs.

Eggs and larval development

Eggs, incubated at 15.6°C, hatched within 2-2½ weeks. The larvae remained within the matrix for a few days, moving around slowly within the dense medium. Case-making begins within a few hours after the larvae have left the egg mass. Larvae continued to emerge from a single egg mass for over a week. Whether this irregular development is typical of field situations is not known.

The initial case, made of bits of debris, is a crude tube that is gradually made stronger and tidier in successive days. After completion of the initial case, the young larvae frequently floated at the surface. This planktonic drifting could perhaps be a dispersal mechanism.

Conifer needles were the predominant material for cases of early and mid-instar larvae, though some incorporated a component of sand, perhaps because of a shortage of the preferred material. In instar IV there was frequently a significant component of sand, but after the moult to instar V the cases were rebuilt using conifer needles. In the latter part of the final instar all larvae were in sand-grain cases. WIGGINS (1977) describes the case of mature *C. magnifica* larvae as composed of small pieces of wood arranged irregularly to form a cylinder with little curvature or taper. He indicated that a new case of fine rock fragments was constructed before pupation.

Five larval instars can readily be distinguished by head capsule measurements (Table I). Head colour darkens with successive instars; the change is particularly apparent between instar IV and V. In the latter, the head is dark brown to black with some lighter areas, especially in the fronto-clypeal area (WIGGINS, 1977).

Laboratory rates of development at 15.6°C for instars I-IV were comparable with field rates. WINTERBOURN (1971) stated that the first four instars could be completed in 10 weeks and most of the second laboratory generation were in instar IV by the ninth week (Fig. 1). Rate of development is dependent on both temperature and food quality, as will be discussed further in later sections. Even under

Table 1. Head-capsule measurements of larval instars of *Clistoronia magnifica* (Second laboratory generation).

Instar	N	Mean (mm)	Range (mm)
I	6	0.38	0.37—0.40
II	5	0.52	0.48—0.55
III	10	0.86	0.82—0.89
IV	7	1.42	1.34—1.47
V	8	2.06	1.94—2.20

constant rearing conditions, the duration of the final instar is greater than that of the other instars combined. In the field, the final instar is further prolonged because it is the overwintering stage.

The data for Fig. 1 are derived from a cohort of larvae emerging within 3-5 days. The first three instars occurred as relatively discrete units but thereafter there was considerable overlap of instars. The duration of the fourth and fifth stadia was quite variable between individuals, which results in an extended period of pupation and of adult emergence. Environmental cues in the field, such as changing photoperiod and temperature, may result in more synchronous emergence than in the laboratory. Data for the total emergence are not available because, due to space limitations, the rearing of late individuals of one generation was terminated when the next generation was beginning to be reared.

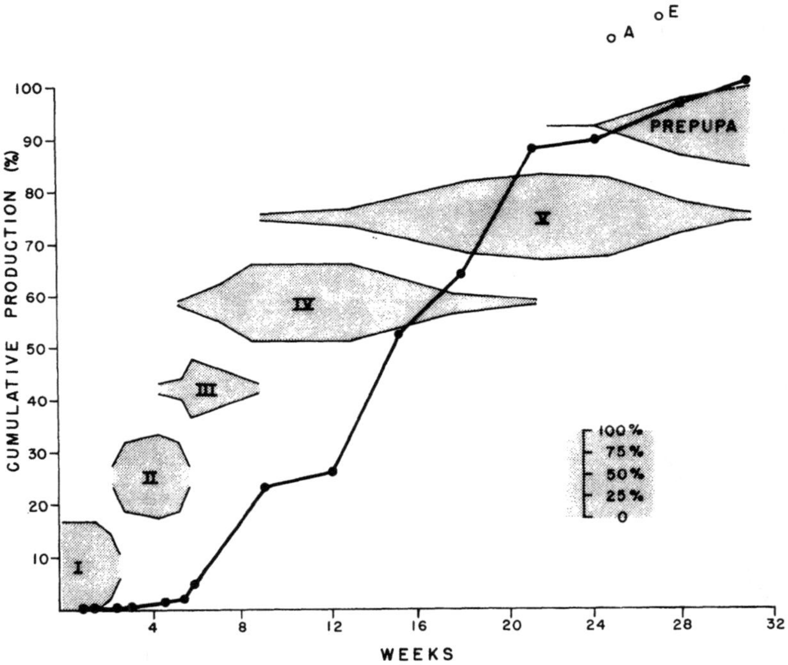

Fig. 1. Cumulative production of *Clistoronia magnifica* second laboratory generation compared with population structure (expressed as percent in each instar). Reared at 15.6°C. Occurrence of first adults and eggs is indicated by A and E.

Prepupal-pupal stages

At the end of the feeding period the final-instar larva attaches its sand case to a substrate and seals off both ends with a silk grating. The prepupal stage was completed in about one week at 15.6°C. The larval exuviae are packed in the back of the case but frequently several of the sclerites were pushed out of the case. This behaviour is atypical of most limnephilids which have a grating sufficiently fine to retain the sclerites.

As is pointed out by WIGGINS (1977), the term prepupa as used above includes both the interval of the resting larva and the period of larval-pupal apolysis. Thus this one week includes both the prepupal and pharate pupal stages. However, for comparing the performance of various treatments (e.g. effects of temperature), I do not distinguish between these two events. The most convenient markers are the closing of the case, and the casting of the exuviae when the individual is recognizable as a typical pupa. The timing of the latter can be observed without injury to the individual either by noting when sclerites are pushed out of the case, or by removing a small 'window' to observe the body form. The sexes can be distinguished by the maxillary palpi: 3 segments in males and 5 in females.

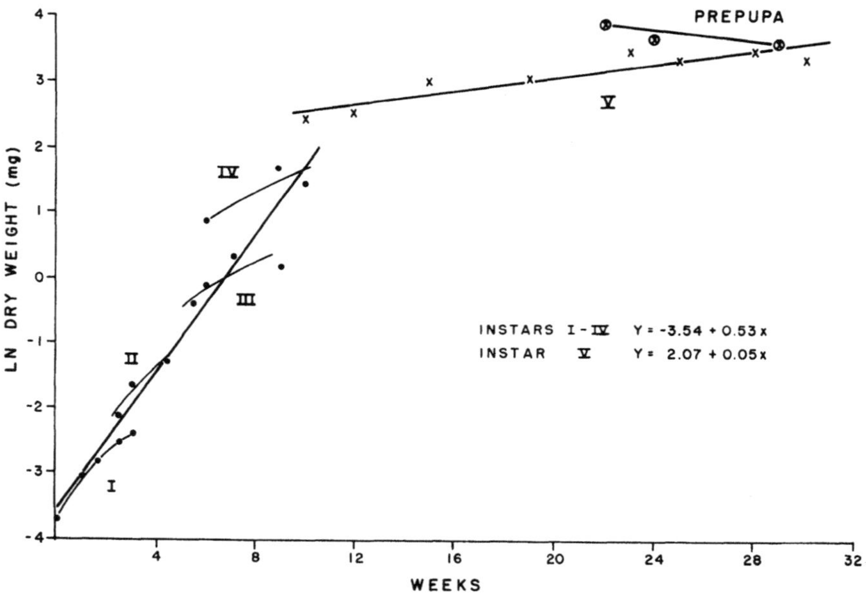

Fig. 2. Growth rates of larval instars of second laboratory generation of *Clistoronia magnifica* reared at 15.6° C.

The major advantages of using newly-moulted pupae for dry-weight comparisons are: (1) timing is predictable as one week after the case is sealed; (2) as a non-feeding stage, variability due to amount of gut content is not a factor; (3) the sexes are recognizable; (4) compared with adults, the pupae are non-mobile and the potential mortality of the three week pupal period is avoided.

Adult emergence in the laboratory occurred during the night. The pharate adult cut through the end of the case using the pupal mandibles, swam to the surface and crawled up the wire cage, where it then cast the pupal exuvia. Though most of the adults obtained under standard rearing procedures were good specimens, there were always some inferior individuals. Malformations ranged from slightly crumpled wings, through complete failure to cast the pupal exuviae, or death as pharate adults. The factors responsible for these abnormalities require further investigation, though low dissolved oxygen reduces the number of perfect adults, and dietary deficiencies may also be a factor (ANDERSON, 1976b). However, good adult emergence is not just related to larval growth. Within groups reared under uniform conditions there was no significant difference in weights of pharate adults, those with crumpled wings, or perfect specimens.

Larval growth

Larval weight of individuals of the second laboratory generation is plotted against time to illustrate within-instar and between-instar growth patterns (Fig. 2). On a log scale, growth rates can be fitted to two distinct linear regressions: instars I-IV with an instantaneous growth rate of 7.6% per day, and instar V of 0.7% per day. R^2 values are 0.92 and 0.85, respectively. For all instars there is a tendency for rapid growth early in the stage and slower growth before moulting. The values given are somewhat lower than for later generations because experience showed that this culture was somewhat crowded. Also in later generations the cultures were terminated before the slower larvae matured, whereas this series was followed for 31 weeks.

Though the instantaneous growth rate of early instars is an order of magnitude greater than that of the final instar, the latter stage is when the majority of growth and feeding occurs. This is illustrated by calculating production for the second laboratory generation, using RICKER's (1958) method:

Production = Growth rate X Mean biomass

Cumulative production is compared with the population structure in Fig. 1. This culture was sampled frequently for instar composition, but the first count of total numbers was at week 6. To arrive at an initial population, mortality for the first six weeks was assumed to be 40%. On this basis, the cumulative production was less than 5% of the total production when 80% of the population had completed instar II. If the actual mortality for the first six weeks was doubled (undoubtedly an overestimate in this culture) the contribution of the first two instars is still less than 10%. Thus it is evident that even though the early instars are abundant and have a

Table 2. Duration of Larval stage and pupal weights of *Clistoronia magnifica* reared at various temperature regimes and 16 Hours light.

Temperature (°C)	Lab generation	No.	Time to prepupa (weeks) Mean	Range	Pupal dry wt. (mg) (Mean ± 95% C.I.)
15	4	26	18	16—20	37.90 ± 2.38
15	5	23	18	15—23	39.14 ± 1.23
15 (to 10 at 15 wks)	5	6	18	17—19	40.16 ± 2.65
15 (to 10 at 15 wks)	5	11	33	24—51	39.67 ± 2.43
10	4	8	46	43—52	42.39 ± 5.14
10 (to 15 at 40 wks)	4	5	42	41—44	42.95 ± 3.65
20	5	14	22	20—24	35.09 ± 2.61
"Accelerated Field"*	5	21	24	23—26	41.38 ± 1.82

*Temperature and photoperiod: 10 wks -- 20°, 14,5-13.5 hrs; 1 wk -- from 20° to 5°, 12 hrs; 7 wks -- 5°, 10 hrs; 3 wks -- 10°, 14.5 hrs; remainder -- 15°, 15 hrs.

high instantaneous growth rate, their biomass is so small that they are unimportant in production estimates. This calculation is relevant to field situations for many invertebrates where there is a problem in accurately sampling early-instar larvae. Over 60% of the production of this population of *C. magnifica* occurred between weeks 12 and 21, when the age structure was dominated by instar IV and early-to mid-instar V.

Effects of temperature

During the fourth and fifth laboratory generations, a series of experiments were conducted to determine the temperature range tolerated by *C. magnifica* larvae, and the effects of temperature on rate of development and mature weight (Table 2). Experiments were initiated with first-instar larvae at 10° and 20°C in addition to the standard 15.6° (given as 15°). They were reared to pupae, oven-dried at 60°C, cooled in a desiccator and weighed.

As expected, larvae developed very slowly at 10°, but mortality was low and the weight of each instar was similar to that of larvae reared at 15°. Some final-instar larvae weighed 70 mg, the heaviest encountered in any of my cultures. At week 40, when larvae had not pupated in twice the time required at 15°, it seemed possible that a temperature increase was necessary to induce pupation. Five larva were transferred to 15° and the first of these sealed off its case the following week (Table 2). Thus the higher temperature accelerated development, but shortly thereafter the larvae at 10° also began to seal their cases. Though pupae from the 10° series had the highest mean weights, the 95% confidence intervals overlap with those of the 15° series.

Final-instar larvae transferred from 15° to 10° at week 15 (when the first sealed cases occurred in the stock culture) exhibited a split pupation period (Table 2). Six of these were prepupae by week 19, in synchrony with the 15° series, whereas another 11 were delayed to 24-51 weeks. Apparently those that were physiologically near maturity pupated on schedule, whereas the others were delayed almost as much as continuous rearing at 10°. Rearing at 20° throughout the life cycle was much poorer than at lower temperatures. Early-instar larvae developed rapidly and were of average weight. Mortality was high in instar V and the duration of this stadium was increased. Individuals that survived to pupation were significantly smaller than those reared at lower temperatures. The difference in response between the early instars and the final instar can be explained in terms of the field life cycle. The early instars are present during late summer, when surface water temperature may exceed 25°, whereas the mature larvae are the overwintering stage.

The 'accelerated field' series were reared with a temperature cycle similar to that experienced under field conditions, but compressed by one third to 24 weeks from the approximate nine months that larvae require for a univoltine cycle. The early instar larvae were reared at 20°, and then temperature was reduced through steps to

5°, and increased again up to 15°. This temperature regime resulted in pupae that were comparable in weight to those reared at either continuous 10° or 15°.

Comparison of generations

Weights of post-larval stages of the successive generations are compared with those of field-collected *C. magnifica* in Fig. 3. Using weight as a criterion, it is apparent that the laboratory cultures are not inferior to a field population. In most comparisons the former are significantly heavier. Though there is a tendency for highest weights in the fourth and fifth generation, the within-generation differences are not

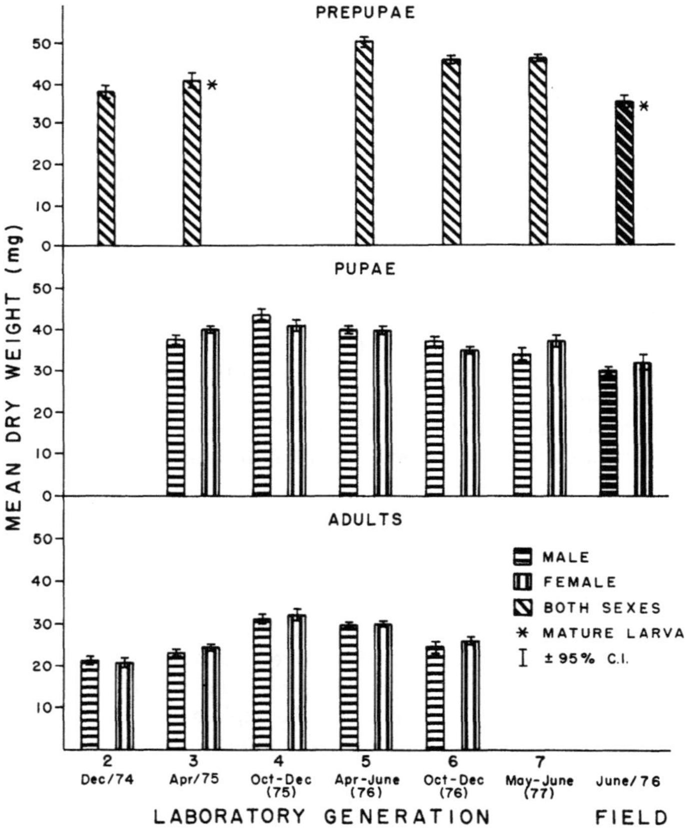

Fig. 3. Comparison of weights of post-larval stages of six generations of reared *Clistoronia magnifica* with field-collected material.

significant. I conclude from the data in Fig. 3. that current procedures are suitable to rear *C. magnifica* of a predictable size.

Though it has required 46 months to produce seven generations in the laboratory, it seems likely that two generations per year could be achieved. The trial and error involved in developing procedures and diets has resulted in some delays in the cycle. Table 3 is a generalized scheme of the timing and expected growth patterns that would result in a generation within six months. The range in weights are the extremes observed for cultures fed on suitable diets. Larvae at the low end of the range would either die or require a longer time than the durations listed for each stadium. Small prepupae or pupae would also have low viability and, if they matured, would produce adults of low vigour and fecundity. As is apparent from the range of developmental times in Fig. 1, there should be no problem in developing cultures with overlapping generations in order to have all stages available on a continuous basis.

A laboratory culture is exposed to intense selection pressures. The *C. magnifica* culture was initiated from a few egg masses, and succeeding generations have been started from less than 10 pairs. The current stock is highly inbred as no wild stock have been added in 7 generations. Despite the inbreeding, there has been no apparent decline in the vigour, or increase in abnormal adults. Though there is always some unexplained mortality in each generation, I have seen no obvious manifestations of diseases.

As *C. magnifica* has proven to be amenable to laboratory culture, it opens the

Table 3. Generalized growth and development pattern of *Clistoronia magnifica* to produce two generations per year. Based on seven generations reared at 15.6°C.

Stage	Dry Weight (mg)		Duration (weeks)
	Mean	Range	
Egg	0.02	—	2.5
Instar I, non-fed	0.02	—	
I, late	0.06	0.04—0.10	2
II, early	0.1	0.07—0.22	
II, late	0.3	0.15—0.55	2
III, early	0.7	0.45—1.2	
III, late	1.5	0.6—2.7	2
IV, early	3.5	2—6	
IV, late	7.5	4—13	2.5
V, early	11	9—12	
V, late	48	30—65	9
Prepupa, early	47	28—55	1
Pupa, early	38	27—47	3
Adult, early	26	17—40	—
Adult (preoviposition)	—	—	2

door to several exciting directions of research in ecology, behaviour and physiology. However, the extrapolation of laboratory data to field situations must be approached with caution. The merits and limitations of research based on laboratory colonies are aptly phrased by COLE (1966) in discussing mass culture techniques for body lice: '... it must be realized that an individual from a laboratory colony is no longer the same ecologically or perhaps even physically. Its reactions and responses to stimuli, (e.g. insecticides), may be more or less different than they would be in the wild state. This principle applies to all species. Still, dependable studies are possible with laboratory colonies; indeed, they are vitally necessary to modern research.'

Acknowledgements

I am indebted to Dr. T.J. TAYLOR for providing the egg masses to start my rearing programme. I thank Dr. ED GRAFIUS for help and discussions on several phases of the study. Miss STACIE KRUER provided much needed assistance in culturing the caddisflies and saved the colony from extinction on more than one occasion. This work was supported by NSF Grant GB36810X.

This is Technical Paper No. 4625, Oregon Agricultural Experiment Station. Contribution No. 284 from the Coniferous Forest Biome.

References

ANDERSON, N.H. 1973. The eggs and oviposition behaviour of *Agapetus fuscipes* CURTIS (Trich., Glossosomatidae). Entomologist's mon. Mag. 109: 129-131.

——. 1974. Observations on the biology and laboratory rearing of *Pseudostenophylax edwardsi* (Trichoptera: Limnephilidae). Can. J. Zool. 52: 7-13.

——. 1976a. The distribution and biology of the Oregon Trichoptera. Oregon Agric. exp. Sta., Techn. Bull. 134, 152 pp.

——. 1976b. Carnivory by an aquatic detritivore, *Clistornia magnifica* (Trichoptera: Limnephilidae). Ecology 57: 1081-1085.

COLE, M.M. 1966. Body lice. in C.N. SMITH (ed.) Insect colonization and mass production. New York and London: Academic Press. pp. 15-24.

RICKER, W. 1958. Handbook of computations for biological statistics of fish populations. Bull. Fish. Res. Bd. Canada 119.

WIGGINS, G.B. 1977. Larvae of the North American caddisfly genera. Univ. Toronto Press.

WINTERBOURN, M.J. 1971. The life histories and trophic relationships of the Trichoptera of Marion Lake, British Columbia. Can. J. Zool. 49: 623-635.

Discussion

MACKAY: What was the photoperiod used during rearing? Could the absence of ovarian diapause be caused by the long photoperiod?
ANDERSON: I did no experimental work on effects of photoperiods on induction of diapause. For continuous rearing I selected a photoperiod of 16 hours light and 8

hours dark. This photoperiod was selected because it is well known from work with terrestrial insects (aphids, leafhoppers, etc) that long days are conducive to continuous culturing.

MORSE: You have eliminated about 4 months preoviposition time ordinarily experienced in field populations. What do you suppose is the function of this period in natural populations?

ANDERSON: I do not know why a species from a permanent water habitat should apparently have an ovarial diapause. It could be a mechanism to avoid predation on the egg stage during the summer months. Notice that the limited field data suggest that the preoviposition period is quite variable; eggs were collected as early as July in Oregon, but not until August or September in British Columbia.

CRICHTON: You report on the small number of deformed adults in the cultured specimens, but have you any information on the number occurring in the field?

ANDERSON: No.

WINTERBOURN: In addition to differences in larval growth and adult life span, several other life history features differed in Anderson's and my studies (WINTERBOURN 1971). In Marion Lake, British Columbia, most egg masses appeared to be attached to the undersurfaces of lily pads (*Nuphar*), whereas in Oregon eggs were found on the lake bed. Marion Lake larvae also possessed cases built from plant fragements throughout the final instar; inorganic materials were not available. This provides further evidence of a high degree of ecological flexibility in this species.

Proc. of the 2nd Int. Symp. on Trichoptera, 1977, Junk, The Hague

Preliminary observations on spatial distribution patterns of stream caddisfly populations

V.H. RESH

Abstract

Spatial heterogeneity may influence sampling variability of stream caddisfly populations. The mean number of *Cheumatopsyche pettiti* (BANKS) larvae/Surber square foot sample in a riffle of uniform depth and substrate size (Rock Creek, Carroll County, Indiana, USA) was calculated for sample sizes ranging from 2 to 52 with 30 replicates for each sample size. With a sample size of 2, means ranged from 1.5 to 14.5 larvae/square foot, a departure from the population mean of 6.2. Hyporheic distributions, resource orientation, and population age structure may influence the negative binomial distribution pattern of *C. pettiti.* Larvae of *Dicosmoecus gilvipes* (HAGEN) in the McCloud River, Shasta County, California, USA, exhibited non-aggregated distribution patterns in areas of uniform substrate size and aggregated patterns in areas of mixed substrate size. A reduction in sampling variability may reflect the differences in microenvironmental variation between uniform and mixed substrate areas. Spatial distribution patterns may change temporally, *e.g.* the value of *k* for a population of *Ceraclea ancylus* (VORHIES) in Brashears Creek, Spencer County, Kentucky, USA ranged from 0.12 to 0.23 during larval development but increased to 0.39 during pupation. Both taxonomic and biometric considerations are necessary in designing ecological studies.

Populations of caddisflies and other aquatic insects exhibit distinct patterns in both time and space. Temporal patterns may reflect the phenology of individual species, mortality rates, or population recruitment. Spatial patterns may be influenced by abiotic (e.g. substrate, current) and biotic factors (e.g. territoriality, location of food sources). Furthermore, these patterns may also be interrelated, e.g. when pre-pupation movements result in a change in spatial arrangement. The purpose of this report is to provide information on spatial distribution patterns of stream caddisfly populations and to relate these patterns to problems of sampling variability.

The spatial pattern of a multiple cohort population of the hydropsychid caddisfly *Cheumatopsyche pettiti* (BANKS) was analyzed in a riffle of uniform depth (10-12 cm) and substrate particle sizes (ϕ-5, ϕ-6), located in Rock Creek, Carroll County, Indiana, USA. Fifty-two Surber square foot samples were collected from 26 randomly chosen locations in the riffle. The frequency distribution of the number of larvae for each of the 52 samples and the calculation of statistics *k*, *U*, and *T* (ELLIOTT, 1971) indicate a non-random distribution that best agrees with the

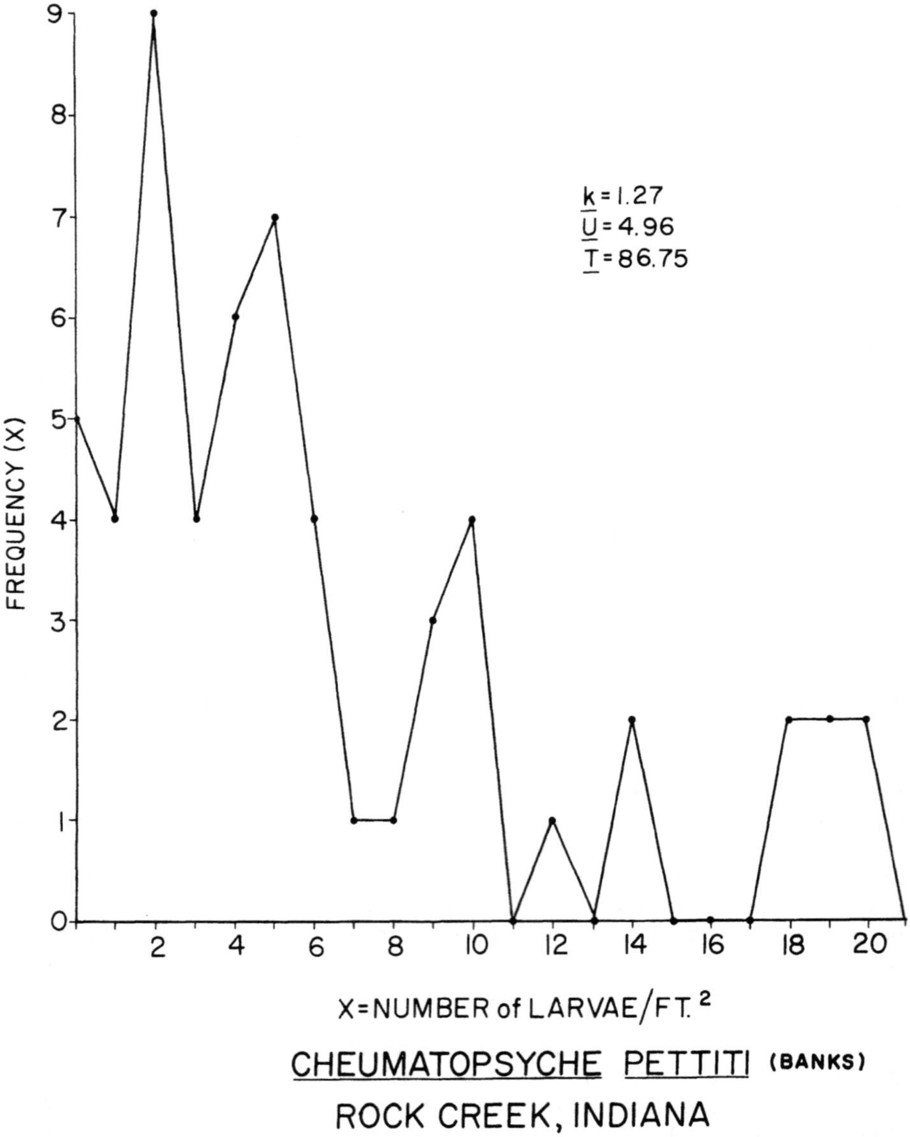

Fig. 1. Frequency distribution of *C. pettiti* larvae/Surber sample (ft^2) in Rock Creek, Carroll Co., Indiana, USA. Formulae for clumping statistics k, U, and T are in ELLIOTT (1971).

spatial pattern predicted for a negative binomial distribution (Fig. 1). Analysis of the predominant caddisfly species from the 100 Surber samples collected by NEEDHAM & USINGER (1956, Table 3) indicates similar non-random distribution patterns: *Sericostoma* (k = 6.31), *Glossosoma* (k = 0.42), *Hydropsyche* (k = 0.65), *Brachycentrus* (k = 0.67), *Lepidostoma* (k = 0.62), and *Rhyacophila* (k = 1.19).

Spatial heterogeneity of benthic populations may greatly influence sampling variability. In order to illustrate this interaction, a data matrix was constructed using the number of *C. pettiti* larvae in each of the 52 quantitative samples (Fig. 1) and a hypothetical sampling regime was developed in which a mean population estimate was calculated for sample sizes ranging from 2 to 52 and arranged in increments of 2. Numbers were replaced in the matrix and could be drawn and included in the calculations more than once for a given value of n. The procedure was repeated 30 times for each sample size, resulting in mean estimates calculated for 780 analyses.

The results of these manipulations indicate that with a small sample size the variability of mean estimates is very large (Fig. 2). For example, means range from 1.5 to 14.5 larvae of *C. pettiti*/sample with a sample size of 2, a significant departure from the sample mean of 6.2. There are several factors that could be involved in producing the spatial patterns of *C. pettiti* (Fig. 1) and the resulting sampling variability (Fig. 2), including: 1) inconsistent underestimations of population size because of hyporheic distributions; 2) microhabitat preference of the net-spinning larvae; and 3) instar specific patterns which may produce clumped distributions when all larvae of *C. pettiti* are considered together as a single population.

A reduction in sampling variability may result from more narrowly defining the sampling site to areas with similar physical characteristics (ALLEN, 1959). LAMBERTI & RESH (unpublished data) examined the spatial distribution patterns of a univoltine single cohort population of the limnephilid caddisfly *Dicosmoecus gilvipes* (HAGEN) in the McCloud River, Shasta County, California, USA. In areas of uniform substrate size, *D. gilvipes* had a non-aggregated distribution, whereas in areas of mixed substrate sizes, aggregated, negative binomial patterns occurred. The reduction in microenvironmental variation within these uniform substrate ribbons may influence these distribution patterns. The corresponding reduction in sampling variability indicates the potential value of these considerations in the development of future sampling regimes.

The spatial pattern of *C. pettiti* presented above (Fig. 1) represents a measurement of instantaneous population distribution. However, these patterns may change over time. A univoltine single cohort population of the leptocerid caddisfly *Ceraclea ancylus* (VORHIES) in Brashears Creek, Kentucky, USA, exhibited a different spatial pattern (which can be identified by a change in k) when examined during pupation in May than had been observed during the previous larval period (Fig. 3, see RESH, 1975, for sampling methods). Problems of sampling variability that are present in examining instantaneous population distributions become compounded in an analysis of population dynamics over time. This is especially true in calcula-

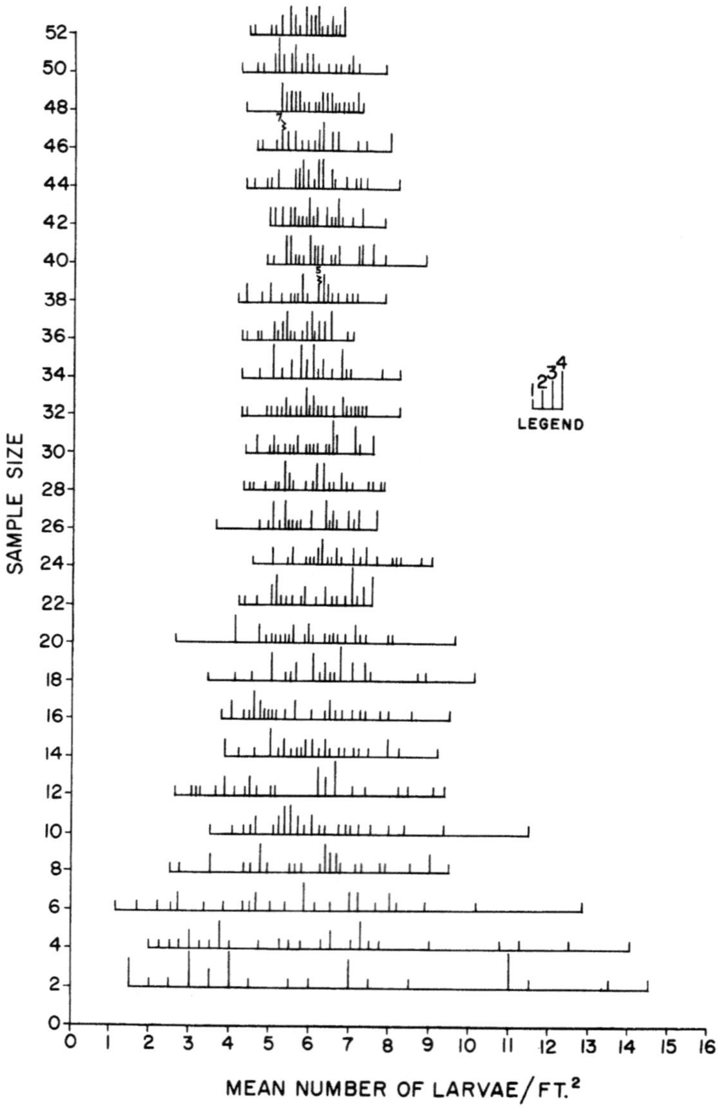

Fig. 2. Sample size influence on mean number of *C. pettiti* larvae/Surber sample (ft^2) in Rock Creek, Carroll Co., Indiana, USA. For each sample size the horizontal line refers to the range of means calculated, with the vertical lines referring to the number of times an individual mean (to the nearest tenth) was calculated.

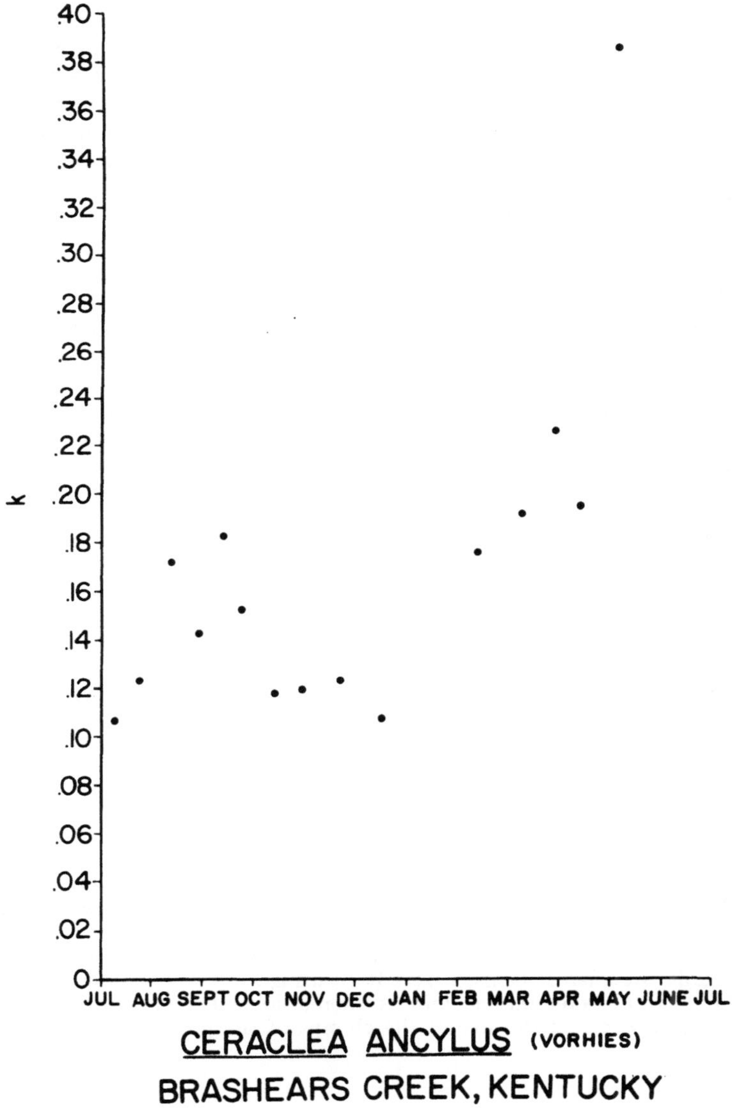

Fig. 3. Changes in *k*, 1971-1972, for *C. ancylus* in Brashears Creek, Spencer Co., Kentucky, USA.

tions of secondary production of aquatic insects since both population and standing stock dynamics must be accurately measured over the duration of the life cycle.

Many of the papers presented in these proceedings have dealt with taxonomic problems of Trichoptera. Taxonomy must be considered as an integral part of any ecological study. However, the biometric components of such a study must be taken into account. Without either part of this matched pair, taxonomy and biometrics, the quantitative interpretation of the ecological interactions of caddisflies and other aquatic insects are subject to a wide range of error. With these points in mind, quantitative sampling regimes must be devised that consider in detail the spatial patterns of the population under examination.

Acknowledgements

I thank GARY LAMBERTI for providing unpublished data. The research leading to this report was funded by the Office of Water Research and Technology, USDI, under the Allotment Program of Public Law 88-379, as amended, and by the University of California Water Resources Center, as part of Office of Water Research and Technology Project No. A-063-CAL and Water Resources Center Project UCAL-WRC-W-519.'

References

ALLEN, K.R. 1959. The distribution of stream bottom faunas. Proc. N.Z. ecol. Soc. 6: 5-8.

ELLIOT, J.M. 1971. Some methods for the statistical analysis of samples of benthic invertebrates. Scient. Pubs. Freshwat. biol. Ass. 25: 1-144.

NEEDHAM, P.R. & USINGER, R.L. 1956. Variability in the macrofauna of a single riffle in Prosser Creek, California, as indicated by the Surber sampler, Hilgardia 24: 383-409.

RESH, V.H. 1975. The use of transect sampling techniques in estimating single species production of aquatic insects. Verh. int. Verein. theor. angew. Limnol. 19: 3089-3094.

Discussion

MACKAY: Small streams are often highly heterogeneous in substrate. How can we obtain the large number of samples necessary for precise estimates of production etc., without damaging the stream?

RESH: Quantitative sampling designs should also consider the dimensions of the sampling unit. In small streams, I have been using a 15 cm^2 sampling device. The critical factor to be taken into account is the relationship between the dimension of the sampling device and the size of the clumps of the population under examination.

Proc. of the 2nd Int. Symp. on Trichoptera, 1977, Junk, The Hague

Problems concerning some previous descriptions of larvae of *Ceraclea fulva* (Rambur) and *C. senilis* (Burmeister) (Trichoptera: Leptoceridae)

I.D. WALLACE

Abstract

Larvae of four species of leptocerid are regularly found feeding upon freshwater sponge in Britain: *Ceraclea albimacula* (RAMBUR), *C. fulva* (RAMBUR), *C. nigronervosa* (RETZIUS) and *C. senilis* (BURMEISTER). *C. albimacula, C. fulva* and *C. senilis* larvae have been previously described from European and Russian material but *C. nigronervosa* has not been adequately described.

British material of *C. albimacula* agrees with the previous description. However, definitely proven British material of *C. fulva* and *C. senilis* differs considerably from previous descriptions.

British larvae of *C. fulva* resemble descriptions of *C. senilis*;
British larvae of *C. senilis* resemble descriptions of *C. fulva*;
British larvae of *C. nigronervosa* resemble a description of *C. fulva*.
Possible reasons for these observations are discussed.

Introduction

In Britain four species of *Ceraclea* STEPHENS feed upon freshwater sponge as larvae. They are *C. albimacula* (RAMBUR), *C. fulva* (RAMBUR), *C. nigronervosa* (RETZIUS) and *C. senilis* (BURMEISTER).

Larvae of these four species collected in Britain have been reared to the adult stage to confirm their identities. Individual larvae were reared separately, enabling direct association between the resultant adult and the final instar larval and the pupal exuviae which were carefully retained.

This definitely identified British material was compared with previous descriptions in the literature. British larvae and pupae of *C. albimacula* (previously referred to *alboguttatus* HAGEN, according to MORSE 1975) were in agreement with the single previous description, by KRAWANY (1937). There are no previous detailed descriptions under the name of *C. nigronervosa*, although NIELSEN (1948) described the case which resembles that of British larvae. Larvae of *C. fulva* and *C. senilis* collected in Britain were markedly different from previous descriptions of those species. Several existing descriptions of *C. fulva* and *C. senilis* are only detailed enough to allow comparisons with British material on a few characters; however the descriptions of SILFVENIUS (1905), SILTALA* (1907) and LEP-

* SILFVENIUS changed his name to SILTALA

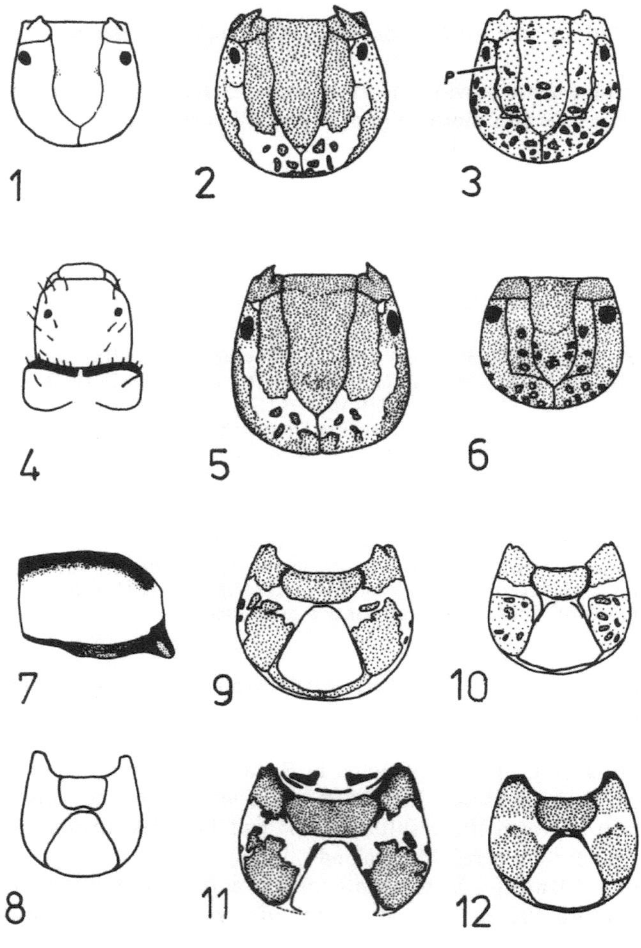

Figs. 1-3. British material; head, dorsal view, setae omitted.
Fig. 1. *C. fulva*; Fig. 2. *C. nigronervosa*; Fig. 3. *C. senilis*. ('p' = parafrontal ecdysial line)
Figs. 4-6. Non-British material.
Fig. 4. *C. senilis* (by LESTAGE), head and pronotum, dorsal view;
Fig. 5. *C. fulva* (by LEPNEVA), head, dorsal view, setae omitted;
Fig. 6. *C. fulva* (by SILFVENIUS), head, dorsal view, setae omitted.
Fig. 7. British material; *C. fulva*, pronotum, right half, setae omitted.
Figs. 8-10. British material; head, ventral view, setae omitted.
Fig. 8. *C. fulva*; Fig. 9. *C. nigronervosa*; Fig. 10. *C. senilis*.
Figs. 11-12. Non-British material; head, ventral view, setae omitted.
Fig. 11. *C. fulva* (by LEPNEVA); Fig. 12. *C. fulva* (by SILFVENIUS).

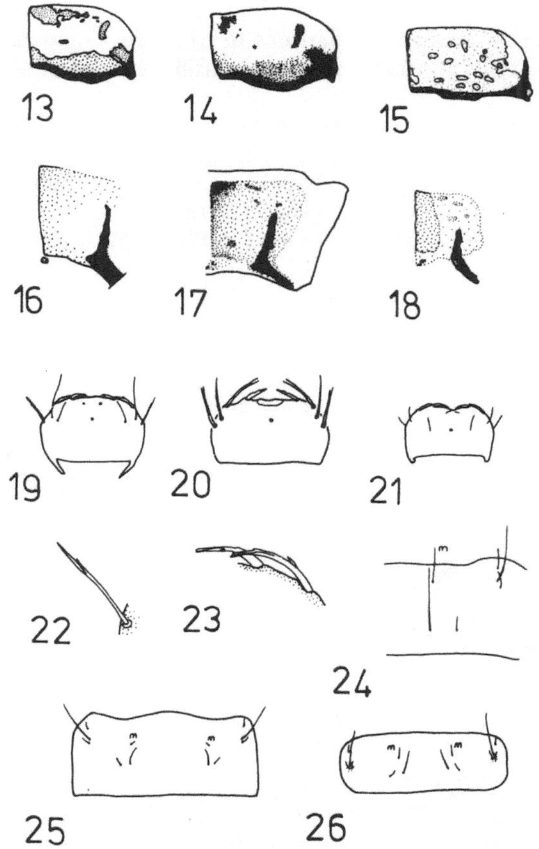

Figs. 13 & 15. British material; pronotum, right half, setae omitted.
Fig. 13.*C. nigronervosa*; Fig. 15. *C. senilis.*
Fig. 14. Non-British material; *C. fulva* (by LEPNEVA), pronotum, right half, setea omitted.
Figs. 16 & 18. British material; mesonotum, right half, setae omitted.
Fig. 16.. *C. nigronervosa*; Fig. 18. *C. senilis.*
Fig. 17. Non-British material; *C. fulva* (by LEPNEVA), mesothorax, right half, setae omitted.
Figs. 19 & 21-23. British material; labrum, any colouration omitted.
Fig. 19. *C. nigronervosa*; Fig. 21. *C. fulva*; Figs. 22 & 23. *C. nigronervosa* anterior edge setae in detail.
Fig. 20. Non-British material; *C. fulva* (by LEPNEVA), labrum, any colouration omitted.
Figs. 24-26 British material; metanotum, any colouration omitted.
Fig. 24. *C. fulva*, right half; Fig. 25. *C. nigronervosa*;
Fig. 26.*C. senilis.* ('m' = medioanal seta)

NEVA (1966) allow detailed comparisons to be made. Comparisons are set out in the table (pages 343-345), which includes a consideration of *C. nigronervosa.*

Conclusions drawn from the table of comparisons

From the comparisons, the following facts emerge: British larvae of *C. fulva* resemble *C. senilis* of KLAPÁLEK (1888), SILFVENIUS (1905), ULMER (1909), LESTAGE (1921) and LESTAGE (1926); British larvae of *C. senilis* resemble *C. fulva* of SILFVENIUS (1905), SILTALA (1907), ULMER (1903), ULMER (1909)

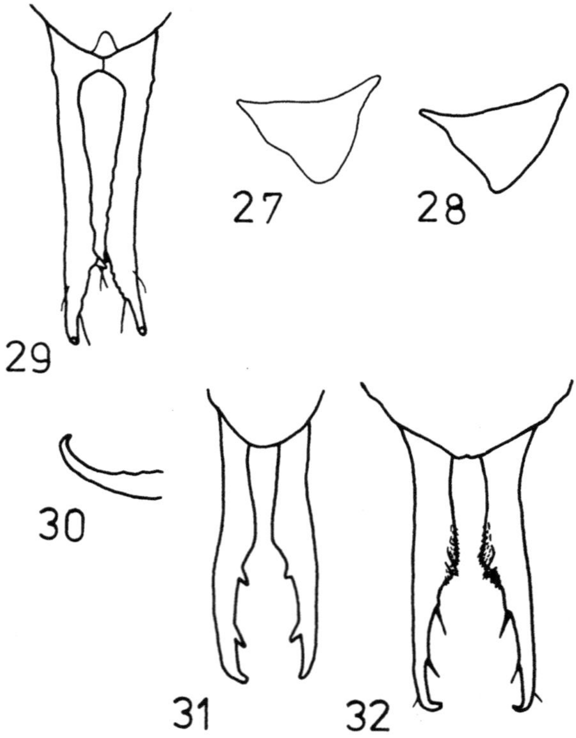

Figs. 27 & 28. British material; protrochantin, outline only.
Fig. 27. *C. senilis*; Fig. 28. *C. nigronervosa.*
Figs. 29 & 30 & 32. British material; pupa, posterior rods.
Fig. 29. *C. senilis*, dorsal view; Fig. 30. *C. senilis*, posterior tip of rod, lateral view; Fig. 32. *C. fulva.*
Fig. 31. Non-British material; *C. senilis* (by LESTAGE), pupa, posterior rods, outline only.

and LESTAGE (1921); British larvae of *C. nigronervosa* resemble *C. fulva* of LEPNEVA (1966). LEPNEVA's 1966 description of *C. senilis* cannot be matched with any British species and is not included in the table. Superficially, the account resembles British material of *C. fulva* and thus *C. senilis* of other authors. The white dorsal head colour and the dark anterior part of the pronotum are the same. However, LEPNEVA's figure shows parafrontal ecdysial lines. (One such line is indicated as 'p' in Figure 3 of this paper.) Parafrontal ecdysial lines are absent in British material of *C. fulva* and are stated as being absent in *C. senilis* by SILTALA (1907). Studies by RESH et al. (1976) indicate that the absence of parafrontal ecdysial lines is characteristic for *C. fulva* and closely related species of *Ceraclea*. The ventral apotome colour differs between British *C. fulva*, other accounts of *C. senilis* and LEPNEVA's *C. senilis*. The ventral apotome is also a different shape in British *C. fulva* and LEPNEVA's *C. senilis*.

In contrast to the larvae, most existing descriptions of the pupae are in agreement with British material. The exceptions are: LEPNEVA (1966) who states that the antennal retaining lappet has three or four setae in *C. fulva* (British material has three only); LEPNEVA who states that *C. senilis* has hook-bearing plates on the seventh abdominal segment (British material has not); LESTAGE (1926) who figures the pupal rods of *C. senilis* being much thinner in the posterior part than the anterior part. His figures (Fig. 31 of this work) resemble e.g. British *C. fulva* (Fig. 32). In British material of *C. senilis* the pupal rods taper uniformly almost to the tip (Fig. 29) and have the posterior tips curved towards the dorsal side (Fig. 30) rather than inwards as in other species.

Discussion

The conclusions drawn from the comparisons between the previous foreign descriptions of *C. fulva* and *C. senilis* and British material of those two species and *C. nigronervosa* could have a number of explanations.

Two or more species could be masquerading under each of the present names *fulva* and *senilis*. A similar possibility is that both species could have two or more very distinct races or subspecies. In both these cases it is curious that there is general similarity between the British pupae and previous descriptions. It could however be argued that the relative inactivity of the pupa and the short time spent at that stage would encourage a lesser degree of morphological divergence.

Another possibility, which this author thinks is the most likely, is that during rearing larvae to the adult stage an incorrect association has taken place. In this work, collection of the larval and pupal exuviae in the course of rearing an individual adult ensures correct association. Methods such as preserving some of a group of apparently similar larvae and rearing the remainder through so the adult may be identified, can lead to larvae of one species being ascribed to a second species. This is a particularly acute problem in the Leptoceridae where the larval exuvia is ejected from the case, with the result that pupal cases containing pupae are of little use in larval taxonomy. Mature pupae can be identified by the genitalia of the enclosed adult. This could explain why, unlike the larvae, British pupae are in agreement with most previous descriptions. However, the figures of the whole pupa of *C. senilis* by LESTAGE (1926) which are obviously of mature pupae have posterior rods unlike those of British pupae.

Another likely possibility is that the initial describers of the larvae of *C. fulva* and *C. senilis*, KLAPÁLEK (1888) and SILFVENIUS (1905) respectively, could have made an incorrect association which was perpetuated by later authors who did not bother to rear out material and confirm the identities. In this respect it is interesting that LEPNEVA (1966), although describing some pupae from her own material, e.g. *Ceraclea annulicornis* (STEPHENS), uses figures by other authors for the pupae of *C. fulva* (except for the case) and *C. senilis*. Using previous keys, larvae of *C. nigronervosa* would be identified as *C. fulva*. This could explain why LEPNEVA's 1966 description of *C. fulva* resembles British *C. nigronervosa* whereas the other descriptions of *C. fulva* resemble British *C. senilis*.

Acknowledgements
The work was financed through a special research grant awarded to Dr. G.N. PHILIPSON of the Department of Zoology, the University of Newcastle upon Tyne by the Natural Environment Research Council of Great Britain. I would like to thank Dr. PHILIPSON for his help and for criticising the manuscript. I would also like to thank Dr. M.I. CRICHTON for confirming the identification of some voucher specimens of reared adults of *C. fulva* and *C. senilis*.

References
KLAPÁLEK, F. 1888. Untersuchungen über die Fauna der Gewässer Böhmens. I. Metamorphose der Trichopteren 1 Serie. Arch. naturw. LandDurchforsch. Böhm. 6: 1-64.

KRAWANY, H. 1937. Trichopterenstudien. Die Metamorphose von *Leptocerus alboguttatus* HAG., *Synagapetus armatus* McLACH und *Rhyacophila hirticornis* McLACH. Int. Revue ges. Hydrobiol. Hydrogr. 34: 1-14.

LEPNEVA, S.G. 1966. Larvae & pupae of Integripalpia Trichoptera. Fauna of the U.S.S.R. Zool. Inst. Akad. Nauk. S.S.S.R. 95: 1-700 (trans. Israel Program Sci. Trans. Inc. 1971).

LESTAGE, J.A. 1921. Trichoptera in ROUSSEAU, E. Les larves et nymphes aquatiques des insectes d'Europe. Bruxelles: 343-959.

——. 1926. Notes trichoptérologiques (10). Une larve de Trichoptère spongillicole. Annls. Biol. lacustre 14: 241-248.

MORSE, J.C. 1975. A phylogeny and revision of the caddisfly genus *Ceraclea* (Trichoptera, Leptoceridae). Contrib. Am. ent. Inst. 11: 1-197.

NIELSEN, A. 1948. Trichoptera, Caddis-Flies with a description of a new species of *Hydroptila*. Biological studies on the River Susaa by BERG, K. Folia limnol. scand. 4: 133.

RESH, V.H., MORSE, J.C. & WALLACE, I.D. 1976. The evolution of the sponge feeding habit in the caddisfly genus *Ceraclea* (Trichoptera: Leptoceridae). Ann. ent. Soc. Am. 69: 937-941.

SILFVENIUS, A.J. 1905. Beiträge zur Metamorphose der Trichopteren. Acta Soc. Fauna Flora fenn. 27: 1-168.

SILTALA, A.J. 1907. Trichopterologische Untersuchungen. I. Über die postembryonale Entwicklung der Trichopteren-larven. Zool. Jb. Neapel. suppl. 9: 309-626.
ULMER, G. 1903. Über die Metamorphose der Trichopteren. Abh. naturw. Ver. Bremen 18: 1-154.
——. 1909. Trichoptera, in BRAUER, A. Die Süsswasserfauna Deutschlands, parts 5-6 A., Jena: 1-326.

Discussion

MALICKY: Your results illustrate once more that association between field-collected larvae and field-collected adults is always problematic. The only correct way to get true associations is by rearing larvae from eggs obtained from a known female.

HILEY: Your talk emphasises a problem I have found in other groups, notably Limnephilidae. Even where larvae have plenty of distinguishing characters, the only reliable rearing technique is that of direct association of larvae with adults. Batch rearing is bound to cause wrong associations in some cases.

EDINGTON: It should be emphasized that various descriptions of European material are not independent. Later descriptions often copy earlier ones, and mistakes have been perpetuated in this way.

A comparison of characteristic features of British larvae of *Ceraclea fulva*, *C. nigronervosa* and *C. senilis* with previous foreign descriptions of *C. fulva* and *C. senilis*. (Previous accounts are:— KLAPÁLEK 1888, designated in Table as 'K'; SILFVENIUS 1905 & SILTALA 1907, 'S'; ULMER 1903 & 1909, 'U'; LESTAGE 1921 & 1926, 'L'; LEPNEVA 1966, 'Lp'.)

BRITISH MATERIAL	NON-BRITISH MATERIAL
Head dorsal view	
C. fulva (Fig. 1) Pale, without spots, parafrontals absent	*C. senilis* (K,S,U,Lt) (Fig. 4) Pale, without spots, parafrontals absent (S) not apparent in figure (Lt)
C. nigronervosa (Fig. 2) Distinct orange and brown areas, parafrontals present	*C. fulva* (Lp) (Fig. 5) Brown and reddish-orange areas, parafrontals present
C. senilis (Fig. 3) Pale brown with darker spots, parafrontals present	*C. fulva* (S,U,Lt) (Fig. 6) Brownish with spots, parafrontals present
Head ventral view	
C. fulva (Fig. 8) Pale without spots	*C. senilis* (K,S,U,Lt) Pale, or not mentioned so therefore presumed pale, as dorsal
C. nigronervosa (Fig. 9) Ventral apotome and much of surface brown. Ventral apotome width more than 2.5 x height	*C. fulva* (Lp) (Fig. 11) Ventral apotome dark brown, other coloured areas brown or reddish-brown. Ventral apotome width more than 2.5 x height
C. senilis (Fig. 10) Ventral apotome and much of surface brown. Ventral apotome width only 2 x height	*C. fulva* (S,U,Lt) (Fig.12) Brown colour present. Ventral apotome width only 2 x height

BRITISH MATERIAL	NON-BRITISH MATERIAL

Pronotum

C. fulva (Fig. 7) Pale with dark anterior margin	*C. senilis* (K,S,U,Lt) (Fig. 4) Pale without spots, dark anterior part
C. nigronervosa (Fig. 13) Pale with reddish-brown areas and spots (no dark anterior margin)	*C. fulva* (Lp) (Fig. 14) Whitish with reddish areas and indistinct spots. Narrow reddish-brown anterior margin
C. senilis (Fig. 15) Brown with colourless areas (no dark anterior edge)	*C. fulva* (S,U,Lt) Brownish with spots. Anterior margin brown, not black

Mesonotum

C. fulva Pale with a few indistinct spots and dark postero-lateral bars	*C. senilis* (K,S,U,Lt) Pale, with black bars and faint spots
C. nigronervosa (Fig. 16) Faintly brownish, mainly alongside central suture	*C. fulva* (Lp) (Fig. 17) Brownish along middle suture
C. senilis (Fig. 18) Generally pale brown with darker patch in centre	*C. fulva* (S,U,Lt) Brownish, middle darker than rest

Metanotum

C. nigronervosa (Fig. 25) Medioanal seta ('m' in figure) short. Compare with *C. fulva* (Fig. 24) and *C. senilis* (Fig. 26)	*C. fulva* (Lp) Medioanal seta short

Metasternum

C. fulva Approximately 10 setae on each side	*C. senilis* (S) Numerous setae, over nine, on each side
C. senilis From 3 — 7 setae on each side	*C. fulva* (S,1907) 4 setae on each side. However 'S', 1905 states there is only 1 seta on each side
(*C. nigronervosa* Usually 2 setae on each side)	

Labrum

C. nigronervosa (Figs. 19, 22,23) Edge setae with branched tips. (Compare with *C. fulva*, Fig. 21 which has normal setae	*C. fulva* (Lp) (Fig. 20) Anterior edge setae with branched tips

Anal proleg claw accessory hooks

C. fulva 2 books	*C. senilis* (S,Lt) 2 hooks
C. nigronervosa 1 or 2 additional hooks	*C. fulva* (Lp) 1 hook
C. senilis 1 hook	*C. fulva* (S,Lt) 1 hook

BRITISH MATERIAL	NON-BRITISH MATERIAL
Protrochantin	
C. senilis (Fig. 27) Noticeably pointed distal end. (Compare with *C. nigronervosa*, Fig. 28)	*C. fulva* (S) Pointed. Presumably this was regarded as noteworthy as the protrochantin is not mentioned for related species
'Purs' on legs mid tibia	
C. fulva 1 *C. nigronervosa* 2 — 6 (average 4) *C. senilis* 3 — 6 (average 4)	*C. senilis* (S) 0 — 1 *C. fulva* (Lp) 3 *C. fulva* (S) 3 — 6
Mid tarsus	
C. fulva 1 *C. nigronervosa* 1 *C. senilis* 2— 3	*C. senilis* (S) 1 *C. fulva* (Lp) 2 *C. fulva* (S) 1 — 4
Hind tibia	
C. fulva 1 — 3 *C. nigronervosa* 5 — 9 (average 7) *C. senilis* 4 — 7 (average 5)	*C. senilis* (S) 4 *C. fulva* (Lp) 4 *C. fulva* (S) 4 — 7

General notes on the figures

1. All figures are of final instar larvae.

2. Figures of non-British material are re-drawn and adapted from SILFVENIUS (1905), LESTAGE (1926) and LEPNEVA (1966). To enable easy comparison in this paper, most figures are re-drawn to approximately the same size as the corresponding figures of British material.

Author index

Subject index

MIX
Papier aus verantwortungsvollen Quellen
Paper from responsible sources
FSC® C105338

If you have any concerns about our products,
you can contact us on
ProductSafety@springernature.com

In case Publisher is established outside the EU,
the EU authorized representative is:
Springer Nature Customer Service Center GmbH
Europaplatz 3, 69115 Heidelberg, Germany

Printed by Libri Plureos GmbH
in Hamburg, Germany